Carbon Allotropes

The study of corrosion-resistant coatings using nanostructured materials is developing and growing at a very rapid rate. *Carbon Allotropes: Advanced Anticorrosive Coatings Materials* summarizes current research and technology on the significance of carbon allotropes in anticorrosive coatings. The book describes the synthesis, characterization, inhibitory mechanism, and industrial applications of carbon allotropes as a corrosion-resistant coating. The usage of carbon allotropes as nanostructure materials based on corrosion-resistant coatings used in contemporary industrial platforms is discussed in this book. Additionally, this book assesses the potential of carbon allotropes as the foundation for corrosion-resistant nanostructure materials and coatings. For scientists and engineers looking for cutting-edge ideas for nanostructures based on carbon allotropes as corrosion-resistant coating materials, this book is an invaluable resource.

Salient Features

1. This book covers nanostructured carbon allotropes as materials for corrosion-resistant coatings.
2. The use of carbon allotropes as anticorrosives has revolutionized modern industrial practices.
3. This book offers academics and entrepreneurs an influential foundational, interdisciplinary, and primary literature reading.
4. The most cutting-edge coatings based on carbon allotropes are also highlighted, along with current manufacturing and operating challenges.
5. This book maintains ongoing research on corrosion-resistant coatings using nanostructured materials while incorporating fresh advancements across the entire area.

Carbon Allotropes
Advanced Materials for Anticorrosive Coatings

Edited by
Jeenat Aslam,
Chandrabhan Verma,
Chaudhery Mustansar Hussain,
and Ruby Aslam

CRC Press
Taylor & Francis Group
Boca Raton London New York

CRC Press is an imprint of the
Taylor & Francis Group, an **informa** business

First edition published in 2024
by CRC Press
6000 Broken Sound Parkway NW, Suite 300, Boca Raton, FL 33487-2742

and by CRC Press
4 Park Square, Milton Park, Abingdon, Oxon, OX14 4RN

CRC Press is an imprint of Taylor & Francis Group, LLC

Library of Congress Cataloging-in-Publication Data
Names: Aslam, Jeenat, editor. | Verma, Chandrabhan, editor. | Hussain, Chaudhery Mustansar, editor. | Aslam, Ruby, editor.
Title: Carbon allotropes : advanced materials for anticorrosive coatings / edited by Jeenat Aslam, Chandrabhan Verma, Chaudhery Mustansar Hussain, Ruby Aslam.
Description: Boca Raton : CRC Press, 2024. | Includes bibliographical references and index.
Identifiers: LCCN 2023013365 (print) | LCCN 2023013366 (ebook) | ISBN 9781032348162 (hardback) | ISBN 9781032348179 (paperback) | ISBN 9781003323976 (ebook)
Subjects: LCSH: Corrosion and anti-corrosives. | Protective coatings. | Metals--Anodic oxidation. | Carbon composites. | Carbon compounds.
Classification: LCC TA462 .C37 2024 (print) | LCC TA462 (ebook) | DDC 620.1/1223--dc23/eng/20230816
LC record available at https://lccn.loc.gov/2023013365
LC ebook record available at https://lccn.loc.gov/2023013366

ISBN: 978-1-032-34816-2 (hbk)
ISBN: 978-1-032-34817-9 (pbk)
ISBN: 978-1-003-32397-6 (ebk)

DOI: 10.1201/9781003323976

Typeset in Times
by MPS Limited, Dehradun

Contents

Preface

Carbon allotropes are of tremendous interest for various applications due to their superior thermo-physical properties, high thermal stability, high electrical conductivity and variety in morphology. The use of carbon allotropes, such as the graphene family, carbon nanotubes, carbon quantum dots (CQDs) and carbon dots (CDs), carbon nanorods (CNRs), and micro- and nanocontainers in various corrosion protection coating systems has been shown to improve not only the coating systems' anticorrosion resistance but also their mechanical and tribological properties. This book reviews the most recent research on the use of carbon allotropes in anticorrosion materials. The book is structured into three parts, each of which contains several chapters, to condense the detailed explanation of anticorrosive applications of carbon allotropes and to offer the reader a sensible and expressive design of the issue.

Part 1 investigates corrosion and how carbon allotropes are nanostructured. Topics covered in Chapters 1–5 are the fundamental and economic impact of corrosion, present and past advancements and future directions of corrosion prevention, carbon allotropes in corrosion prevention, fundamental, synthesis, characterization and properties functionalization of carbon allotropes, covalent and noncovalent functionalization, and characterization of carbon allotropes.

Part 2 discusses the recent advancements of carbon allotropes in anticorrosive coatings. Topics included in Chapters 6–10 are graphene and graphene oxides, carbon nanotubes (CNTs), quantum dots (QDs) and carbon nanorods (CNRs), advanced micro- and nanocontainers, and computational-based advancements in anticorrosive coatings.

Part 3 talks about anticorrosive self-healing coating materials and industrialization. Topics covered in Chapters 11–13 are the concept of self-healing coatings, graphene-reinforced composites as protective coatings for oil and gas pipelines, and the industrial and economic impact of carbon allotropes as anticorrosive coatings.

The objective of this book is to provide the most recent developments in carbon allotropes for the field of anticorrosive applications. This book is written for a highly diverse audience who work in chemical engineering, advanced materials research, and other related subjects. This book will be a priceless resource for libraries in academic and professional institutions, government and non-profit organizations, solitary research groups, and scientists. This book is intended to be a resource for advanced undergraduate and graduate students, researchers, and scientists looking for carbon allotropes for anticorrosive applications to meet current research demands. All chapters were edited by renowned academic and professional researchers, scientists, and subject matter specialists.

We thank all authors on behalf of CRC Press (Taylor and Francis) for their excellent and sincere work. An enormous amount of gratitude is extended to Dr. Renu Upadhyay (Commissioning Editor), Miss Jyotsna Jangra (Editorial Assistant), and the Editorial Team of CRC Press (Taylor and Francis) for their unwavering support and assistance during this endeavor. Finally, thanks to CRC Press (Taylor and Francis) for releasing the book.

(Editors)
Jeenat Aslam, Chandrabhan Verma,
Chaudhery Mustansar Hussain & Ruby Aslam

Editor's Biographies

Jeenat Aslam is currently working as an associate professor at the Department of Chemistry, College of Science, Taibah University, Yanbu, Al-Madina, Saudi Arabia. She earned her Ph.D. degree in surface science/chemistry from the Aligarh Muslim University, Aligarh, India. Materials & corrosion, nanotechnology, and surface chemistry are the primary areas of her research. Dr. Jeenat has published a number of research and review articles in peer-reviewed international journals like ACS, Wiley, Elsevier, Springer, Taylor & Francis, Bentham Science, and others. She has authored over thirty book chapters and edited more than twenty books for the American Chemical Society, Elsevier, Springer, Wiley, De-Gruyter, and Taylor & Francis.

Chandrabhan Verma works at the Department of Chemical Engineering, Khalifa University of Science and Technology, Abu Dhabi, United Arab Emirates. He obtained his Ph.D. in material science/chemistry at the Indian Institute of Technology (Banaras Hindu University) Varanasi, India. He is an American Chemical Society (ACS) member and serves as a reviewer and editorial board member for various internationally recognized American Chemical Society (ACS) and ACS, RSC, Elsevier, Wiley and Springer platforms. Dr. Verma is the associate editor-in-chief of the *Organic Chemistry Plus Journal*. He is the author of several research and review articles published in ACS, Elsevier, RSC, Wiley and Springer, etc. He has a total of more than 9744 citations with an H-index of 55 and an i-10 index of 140. Dr. Verma has edited many books for ACS, Elsevier, RSC, and Wiley. Dr. Verma received several awards for his academic achievements.

Chaudhery Mustansar Hussain is an adjunct professor and director of laboratories in the Department of Chemistry and Environmental Science at New Jersey Institute of Technology (NJIT), Newark, New Jersey, United States. His research is focused on the applications of nanotechnology and advanced materials, environmental management, analytical chemistry, smart materials and technologies, and other various industries. Dr. Hussain is the author of numerous papers in peer-reviewed journals as well as a prolific author and editor of around 150 books, including scientific monographs and handbooks in his research areas. He has published with Elsevier, American Chemical Society, Royal Society of Chemistry, Springer, John Wiley & Sons, and CRC Press.

Ruby Aslam is a research associate fellow under CSIR-HRDG, New Delhi, in the Department of Applied Chemistry, Aligarh Muslim University, Aligarh, India. She received her M.Sc., M. Phil., and Ph.D. degrees from the same university. She has authored/co-authored several research papers in international peer-reviewed journals of wide readership, including critical reviews and book chapters. Dr. Ruby has edited more than five books for American Chemical Society, Springer, Wiley, De-Gruyter, and CRC Press.

Contributors

M.I. Abdou
Production Department
Egyptian Petroleum Research Institute
Nasr City, Cairo, Egypt

Ekemini D. Akpan
Centre for Materials Science, College of Science, Engineering and Technology
University of South Africa
Johannesburg, South Africa

Ruby Aslam
Corrosion Research Laboratory
Department of Applied Chemistry
Faculty of Engineering and Technology
Aligarh Muslim University
Aligarh, UP, India

Elyor Berdimurodov
Faculty of Chemistry
National University of Uzbekistan
Tashkent, Uzbekistan

Avni Berisha
Department of Chemistry
Faculty of Natural and Mathematics Science
University of Prishtina
Prishtina, Kosovo

Khalid Bouiti
Laboratory of Molecular Spectroscopy Modelling, Materials, Nanomaterials, Water and Environment, CERNE2D, ENSAM
Mohammed V University in Rabat
Morocco

Omar Dagdag
Department of Mechanical Engineering
Gachon University
Seongnam, Republic of Korea

Abdelouahed Dahrouch
Laboratory of Molecular Spectroscopy Modelling, Materials, Nanomaterials, Water and Environment, CERNE2D, Faculty of Science
Mohammed V University in Rabat
Morocco

Eno E. Ebenso
Centre for Materials Science
College of Science, Engineering and Technology
University of South Africa
Johannesburg, South Africa

A.M. Fadl
Production Department
Egyptian Petroleum Research Institute
Nasr City, Cairo, Egypt

Richika Ganjoo
Department of Chemistry
Faculty of Technology and Science
Lovely Professional University
Phagwara, Punjab, India

Maryam Sirati Gohari
Research Department of Ceramic, Materials, and Energy Research Center
Alborz, Iran
And
Department of Materials Science and Engineering
School of Engineering Shiraz University
Shiraz, Iran

S.A. Haddadi
Surface Coating and Corrosion Department
Institute for Color Science and Technology
Tehran, Iran

Souad El Hajjaji
Laboratory of Molecular Spectroscopy Modelling, Materials, Nanomaterials, Water and Environment, CERNE2D, Faculty of Science
Mohammed V University in Rabat
Morocco

Rajesh Haldhar
School of Chemical Engineering
Yeungnam University
South Korea

M. Izadi
Department of Materials Engineering
Hamedan University of Technology
Hamedan, Iran

Sirus Javadpour
Department of Materials Science and Engineering
School of Engineering
Shiraz University
Shiraz, Iran

Seong-Cheol Kim
School of Chemical Engineering
Yeungnam University
South Korea

Ashish Kumar
Department of Chemistry
NCE Department of Science and Technology
NCE Bihar Engineering University
Government of Bihar
Bihar, India

Sumit Kumar
PG Department of Chemistry
Magadh University
Bodh Gaya, Bihar, India

Najoua Labjar
Laboratory of Molecular Spectroscopy Modelling, Materials, Nanomaterials, Water and Environment, CERNE2D, ENSAM
Mohammed V University in Rabat
Morocco

Nabil Lahrache
Laboratory of Molecular Spectroscopy Modelling, Materials, Nanomaterials, Water and Environment, CERNE2D, ENSAM
Mohammed V University in Rabat
Morocco

M. Mahdavian
Surface Coating and Corrosion Department
Institute for Color Science and Technology
Tehran, Iran

Sheerin Masroor
Department of Chemistry
A. N. College
Patliputra University
Patna, Bihar, India

Mohammad Mobin
Corrosion Research Laboratory
Department of Applied Chemistry
Faculty of Engineering and Technology
Aligarh Muslim University
Aligarh, UP, India

I. Mohammadi
Department of Materials Engineering
Faculty of Engineering
Tarbiat Modares University
Tehran, Iran

Fetah Podvorica
Department of Chemistry
Faculty of Natural and Mathematics Science
University of Prishtina
Prishtina, Kosovo
And
Materials Science – Nanochemistry Research Group
NanoAlb – Unit of Albanian Nanoscience and Nanotechnology
Tirana, Albania
And
Kosova Academy of Sciences and Arts
Prishtina, Kosovo

Taiwo W. Quadri
Centre for Materials Science
College of Science, Engineering and Technology
University of South Africa
Johannesburg, South Africa

Alimorad Rashidi
Nanotechnology Research Center
Research Institute of Petroleum Industry (RIPI)
Tehran, Iran

Seyed Ali Rezaei
Research Department of Ceramic, Materials, and Energy Research Center
Alborz, Iran

Swati Sahu
Institute of Renewable Energy Technology & Management
School of Studies in Electronics & Photonics
Pt. Ravishankar Shukla University
Raipur, India

Shveta Sharma
Department of Chemistry
Faculty of Technology and Science
Lovely Professional University
Phagwara, Punjab, India

Madhulata Shukla
Department of Chemistry
G.B. College
Ramgarh, Kaimur
Veer Kunwar Singh University
Bihar, India

Abhinay Thakur
Department of Chemistry
School of Chemical Engineering and Physical Sciences
Lovely Professional University
Phagwara, Punjab, India

Sanjay Tiwari
Institute of Renewable Energy Technology & Management
School of Studies in Electronics & Photonics
Pt. Ravishankar Shukla University
Raipur, India

Saman Zehra
Corrosion Research Laboratory
Department of Applied Chemistry
Faculty of Engineering and Technology
Aligarh Muslim University
Aligarh, UP, India

Part 1

Corrosion and Carbon Allotropes Nanostructured

Design and Developments (State-of-the-Art)

1 Corrosion

Fundamental and Economic Impact

Sheerin Masroor
Department of Chemistry, A. N. College, Patliputra University, Patna, Bihar, India

CONTENTS

1.1 INTRODUCTION

The biggest problem that society, researchers, and employers of industries are facing today is corrosion. The intensity of the problem can be understood by the fact that around $15 billion per year goes to waste in the United States due to this phenomenon. In general, about 40% of metal must be replaced to ensure functioning of instruments or machinery. By examining the chemistry, nature, and existence of metals found in the environment, one can conclude that metals, except for noble metals, don't exist in native states. They are present as complex compounds such as oxides, carbonates, and sulphides. This shows their presence is energetically more favourable. So, for this reason, general metals regress in the extreme combined form via the process of corrosion. It is inescapable for the proper functioning of metals and their alloys [1].

The term *corrosion* can be understood as the deterioration of a material, mainly made of metal, due to the reactions going on around it in the nearby environment [2]. The corrosion cycle is given in Figure 1.1.

The destruction caused by this phenomenon can be best separated into two categories:

i. Dry corrosion happens in the absence of a liquid or when the liquid is in the form of a dew point, where the metal is allowed to react with gases or vapours, mainly at elevated temperatures.
ii. Wet corrosion occurs when the metal reacts with an electrolyte solution. It causes severe deterioration [3].

DOI: 10.1201/9781003323976-2

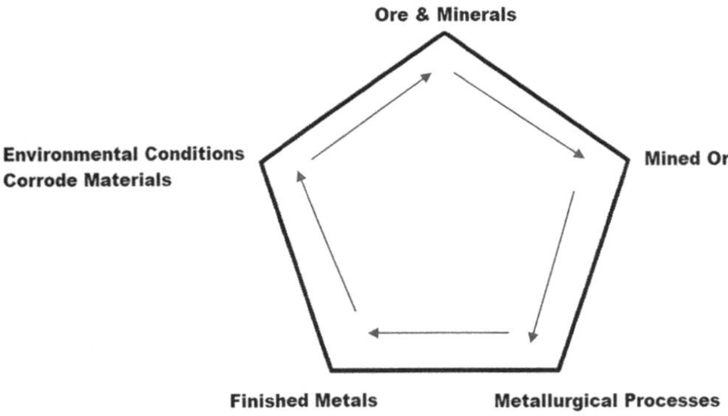

FIGURE 1.1 Corrosion cycle.

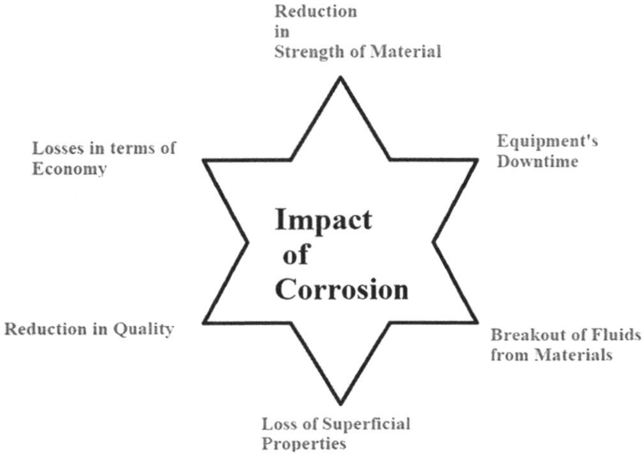

FIGURE 1.2 Impact of corrosion.

The impact of corrosion is depicted in Figure 1.2. This chapter emphasises the fundamentals of corrosion, with its mechanism, forms, and other influences.

1.2 CHEMISTRY OF CORROSION

This phenomenon occurs through electrochemical or chemical processes. The best example is the dissolution of metal in acid. Corrosion occurs by chemical or electrochemical means. Electrochemical processes require the involvement of metals, different chemicals, and aqueous media, which all must be present at a single time, making a cell and alternatively producing electricity. Through these cells, specific metals are changed back to their original compounds by applying energy in the form of electricity. Figure 1.3 depicts a corrosion cell, which contains the main four (4) components: anode, cathode, conductor, and electrolyte (the current medium, such as soil or water).

The electrolyte is the current carrying medium, such as water or soil. The positive electrode (anode) and the negative electrode (cathode) are in contact with the electrolyte. At the anode, oxidation occurs, which is called *corrosion*. This leads to the discharge of electrons and the making of ions [4]. Although at the cathodic site no corrosion occurs due to the reduction reaction, that is the reception of electrons that were previously generated by the anode [5]. The purpose of a

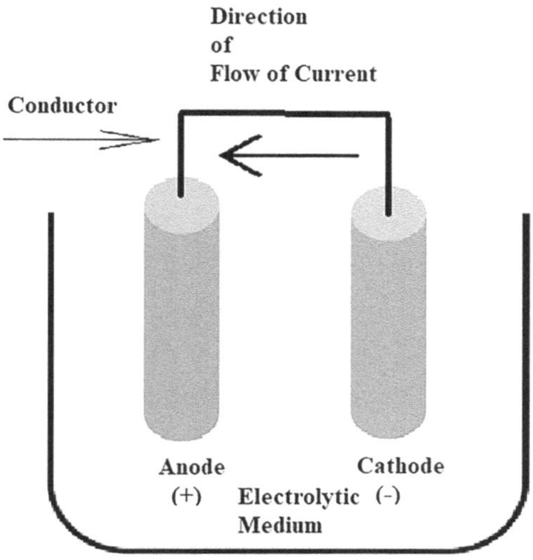

FIGURE 1.3 Basic corrosion cell.

conductor is to connect the anode and cathode so that the current starts flowing along the circuit. The nature of the corrosion cell is electrical, which follows Ohm's law. This states, "The amount of flowing current (corrosion loss) decreases as the resistance of the circuit increases and current flow increases as the potential difference between the anode and cathode increases. So, the amount of corrosion experienced is directly proportional to the electrolyte's ability to react, the potential difference between the anode and cathode, and the amount of resistance in the external metal circuit" [6]. Multiple factors control the functioning of the electrolyte and corrosion cell:

i. Temperature: It has a great effect on the nature of electrolytes. The resistivity of the electrolytes becomes low as the temperature increases. Alternatively, as the temperature lowers, the resistivity becomes high.
ii. pH: pH shows the degree of acidity or alkalinity of the liquid solution. It can also be explained as the relative amounts of hydrogen ions (H^+) and hydroxyl ions (OH^-).
iii. Ionization: The minute electrically charged particles known as ions are responsible for the flow of electricity. This causes the conversion of metals from their specific metallic form to their corroded form. The formation of these ions happens by the addition or removal of an electron from the basic atom. The intactness of an electron gives a positively charged ion, whereas inclusion provides a negatively charged ion. Compounds split or break by dissociation into their elemental form.
iv. Resistivity: It shows the measure of electrolytic power to obstruct the flow of electric current. Depressed resistivity can explain the flow of electricity in a free manner, whereas elevated resistivity gives poor conductivity of electrolytes [7].
v. Potential Difference: It is an important property that affects the corrosion cell. It tells about the potential difference between the metals forming corrosion cells. The potential of different metals is placed in a specific series known as the Electromotive Series given in Table 1.1. These differences in the potentials of metals provide indications of the effect on the corrosion cell.

The resistance in the external circuit, which connects the anode and cathode, has a great effect on the corrosion of the cell. If the resistance in the given circuit becomes elevated, the current flowing rate develops a low value; hence, corrosion is reduced.

TABLE 1.1
Electromotive Series of Elements

Type of Reaction	Element	Potential of Electrode
Anodic	Lithium (Li^+/Li)	−3.04
	Magnesium (Mg^{2+}/Mg)	−2.36
	Aluminium (Al^{3+}/Al)	−1.66
	Zinc ($Zn^{2+}Zn$)	−0.76
	Chromium (Cr^{2+}/Cr)	−0.74
	Iron (Fe^{3+}/Fe)	−0.44
	Cadmium (Cd^{2+}/Cd)	−0.40
	Nickel (Ni^{2+}/Ni)	−0.21
	Tin (Sn^{2+}/Sn)	−0.14
	Lead (Pb^{2+}/Pb)	−0.14
	Hydrogen ($2H^+/H_2$)	−0.00
Cathodic	Copper (Cu^{2+}/Cu)	+0.34
	Silver (Ag^+/Ag)	+0.79
	Platinum (Pt^{2+}/Pt)	+1.20
	Gold (Au^{3+}/Au)	+1.49
	Fluorine ($F_2/2F^-$)	+2.87

1.3 MECHANISM OF CORROSION

1. Mechanism of Dry Corrosion

   It mainly happens due to the simple chemical action of atmospheric gases such as carbon dioxide (CO_2) and sulphur dioxide (SO_2).

   It mainly occurs via three routes:

   a. Oxidation Corrosion: The mode of action is direct by oxygen on the metal surface, which is in the absence of moisture. It mainly occurs at standard temperature and pressure (STP).
   b. Liquid Metal Corrosion: The liquid metal flows over solid metal at elevated temperatures, causing the solid metal to become weak. For example, sodium metal acts as a coolant in a nuclear reactor, leading to corrosion of cadmium.
   c. Corrosion by Some Other Gases: Some specific gases present in the vicinity of metal certainly react with metal and make a shielding or non-shielding layer on the surface of metals. The metal and gases chemically react with each other and undergo corrosion.

2. Mechanism of Wet Corrosion

   It is also called electrochemical corrosion and happens when there is a difference in electrode potential. For these metals, some part becomes the anode (+ve charge), and the other becomes the cathode (−ve charge), present in the electrolyte. It is the main type of corrosion that leads to maximum damage. This occurs via the formation of three types of cells:

   a. Galvanic Cell: The formation of this type of cell depends upon the connection of two dissimilar metals, leading to a difference in potential existing in a single electrolyte. Figure 1.4 depicts a galvanic cell.

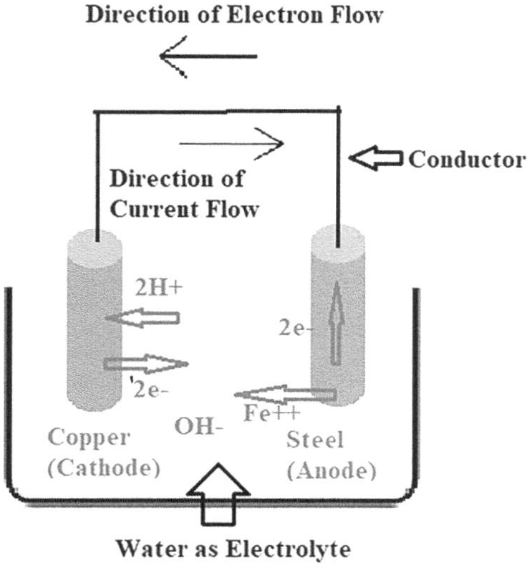

FIGURE 1.4 Assembly of a galvanic cell.

b. Concentration Cell: In this cell, the anode and cathode are made from the same material, but the electrolyte is different in nature. The flow of current is maintained due to multiple levels of oxygen concentration, various types of chemicals present, or varied values of the same chemical in the electrolytic medium. The difference in electrolytes is due to variations found in the soil of buried pipelines from wet to swampy to dry to sandy. Figure 1.5 depicts the concentration cell.

c. Electrolytic Cell: In addition to the four prime components of the corrosion cell, an external power supply is needed to generate electrical energy. This energy source is fixed in the conductor between the anode and cathode, leading to the flow of current. These connections create electrolytic cells, as depicted in Figure 1.6.

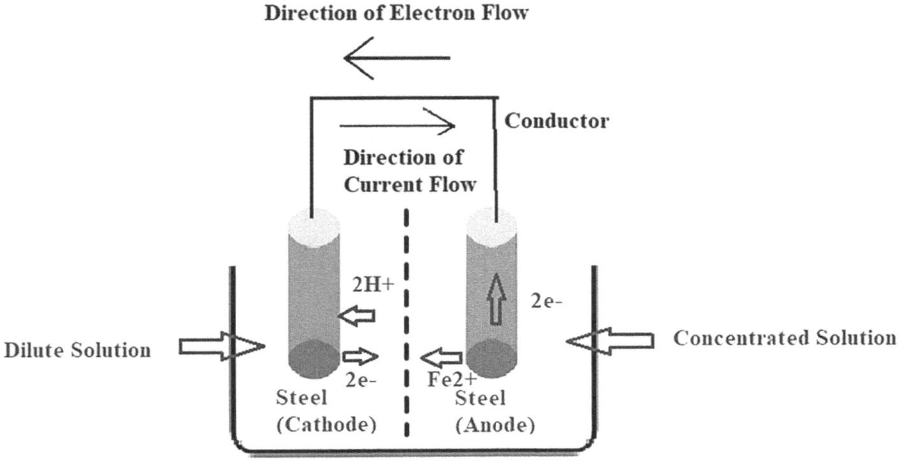

FIGURE 1.5 Assembly of a concentration cell.

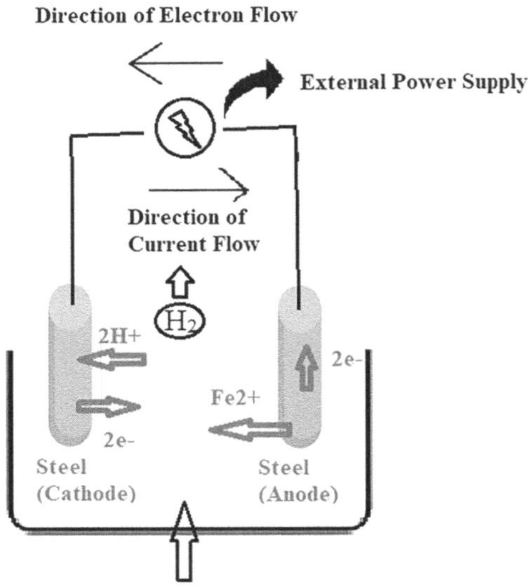

FIGURE 1.6 Assembly of an electrolytic cell.

1.4 CALCULATION OF CORROSION RATE

The calculation of the corrosion rate can be defined in terms of the thickness formed on the metal or the weight loss of the metal, which corrodes uniformly under the area that has been exposed.

The numerical data obtained by the fitting corrosion model is mainly designated as corrosion current. A practical way is to think of corrosion rates in millimetres per year. The mechanism of the generation of a corrosion rate from corrosion current is given as [8].

Let's assume the presence of an electrolytic dissolution reaction that involves a chemical species, *M,* is undergoing the reaction

$$\mathrm{M} \rightarrow \mathrm{M}^{n+} + ne^-$$

Now, relate the current flow to mass by applying Faraday's Law.

$$Q_S = n\, F\, M_S$$

where Q_S = Charge in coulombs obtained from the reaction of species *M*, n = Number of electrons transferred per molecule or atom of *M*, F = Faraday's constant = 96 485.34 coulombs/mole, and M_S = Number of moles of species *M* reacting.

The important form of Faraday's Law needs an idea of equivalent weight. So, the equivalent weight (EW_M) is the mass of the species *M* that will combine with one Faraday of charge.

$$\mathrm{EW} = \mathrm{AW}/n$$

where AW = atomic weight of the species.

Mole fraction is applied as the weighting factor. If the dissolution is not uniform, you may have to measure the corrosion products to calculate EW.

Putting it into Faraday's Law we get:

$$W_M = EW_M Q_M / F$$

TABLE 1.2
The Value of Corrosion Constant (K) With the Units

Units for Corrosion Rate	K
mm/year	3272
μm/year	3.272×10^6
mils/year	1.288×10^5

where W_M is the mass of species M that has reacted.

In special cases where the corrosion may occur uniformly along the metal's surface, we can calculate the corrosion rate in terms of distance per year.

The conversion of weight loss to a corrosion rate (CR) is given by the equation:

$$\text{CR} = I_{\text{corr}} K \text{EW}/D\, A$$

Here, CR= Corrosion rate, I_{corr} = Corrosion current in Amperes, K = Constant that defines the units for the corrosion rate, EW= Equivalent weight in grams/equivalent, A = Area of the sample in cm², and D = Density, $Q = I\,t$ = Charge (t is the time in seconds and I is current).

Substitute the value of Faraday's constant and apply it to the previous equation.

1.5 CORROSION RATE CONSTANTS

Table 1.2 provides the value of K, which is used in the corrosion rate equation to calculate corrosion rates.

1.6 FORMS OF CORROSION

The corrosion may happen in multiple ways in differing forms. The differences mainly depend upon three factors [9]:

a. Corrodent's Nature: It may be wet (liquid or moisture) or dry (high-temperature gases).
b. Corrosion's Mechanism: It may be direct reactions of chemicals with metal or electrochemically.
c. Nature of Corroded Metals: It may be uniform if the metal corrodes at the same rate over the entire exposed surface, or it is localised if only specific areas are affected.

Various forms of corrosion have been summarised in Table 1.2.

S.No	Name of Corrosion	Pictorial Presentation of Metal with Corrosion Types
1.	No Corrosion	
2.	Uniform Corrosion	

(Continued)

S.No	Name of Corrosion	Pictorial Presentation of Metal with Corrosion Types
3.	Galvanic Corrosion	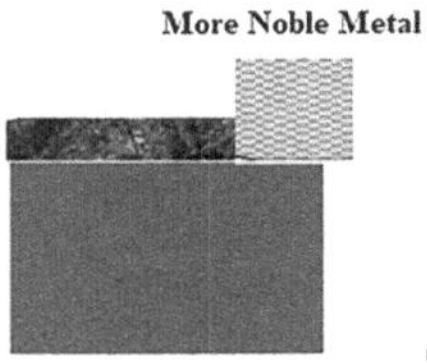
4.	Erosion Corrosion	
5.	Fretting Corrosion	
6.	Crevice Corrosion	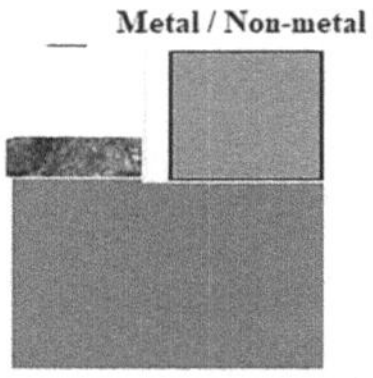
7.	Pitting Corrosion	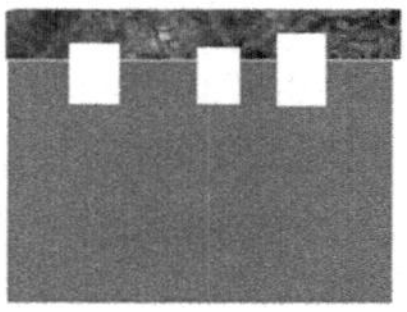
8.	Exfoliation Corrosion	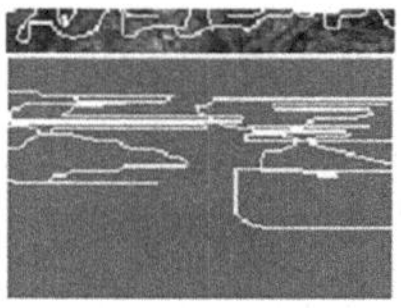
9.	Dealloying	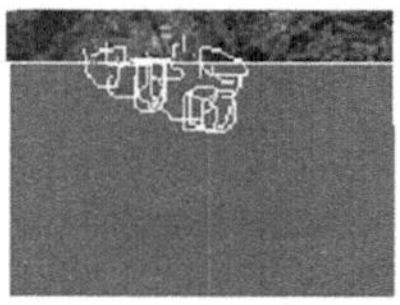
10.	Intergranular	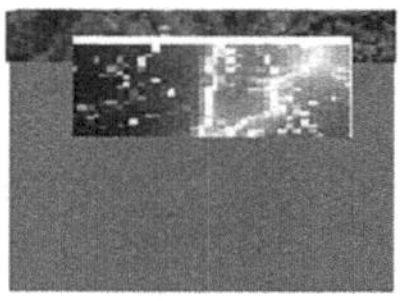

S.No	Name of Corrosion	Pictorial Presentation of Metal with Corrosion Types
11.	Stress Corrosion	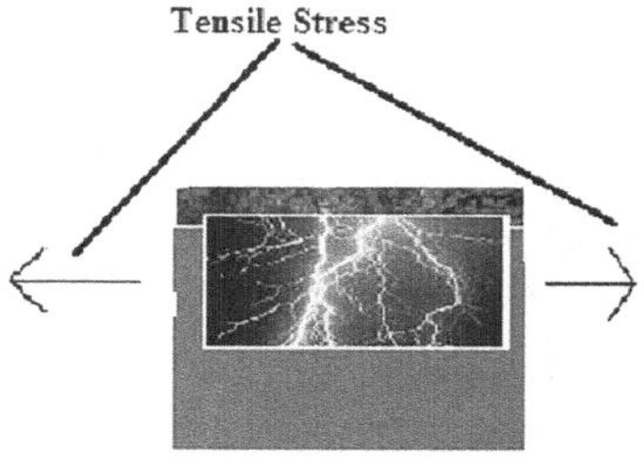
12.	Corrosion Fatigue	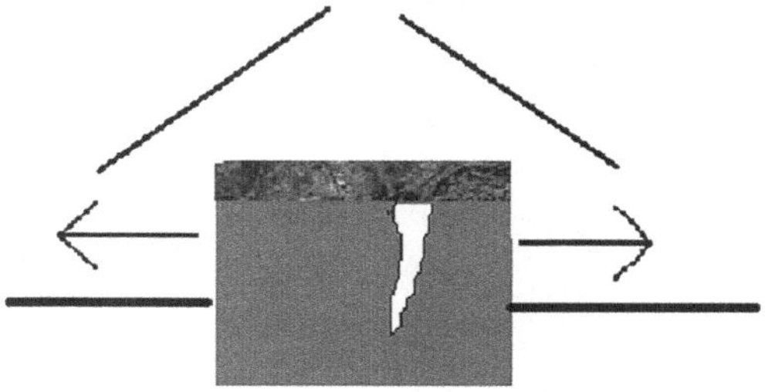

1.7 CORROSION CONTROLLING METHODS

The most general way to control corrosion is to understand the corrosion mechanism involved and to remove one or more of the elements of the corrosion cell. The corrosion controlling methods are based on changing the nature of metals:

a. **Use of Protective Coatings**
- The most general option is to protect the metal from an external corrosive environment.
- Coatings provide long-term protection under a broad range of corrosive conditions, extending from atmospheric exposure to full immersion in a strongly corrosive solution.
- The strength and integrity of a structure can be maintained.
- They provide protection by providing a physical barrier and preventing the aggressive medium from attacking the metal surface.
- The negative impact of using coatings are the mechanical and thermal damages, high cost, etc.
- There are two types of coating: ***anodic coatings*** (coating material is anodic to the metal on which the coating is to be produced, e.g., iron or steel is protected by coating of zinc and the process is called galvanisation) and ***cathodic coatings*** (coating material is cathodic to the metal on which the coating is to be produced, e.g., iron or steel is protected by a coating of tin, called *electroplating*).

b. **Use of Anodic Protection**
- This is an electrochemical method of controlling corrosion and is dependent on the phenomenon of passivity (loss of chemical reactivity experienced by certain metals and alloys under particular environmental conditions).

- It is most often used in highly corrosive environments to protect metal immersed in solution with uncommonly acidic or basic qualities.
- In this technique, the corrosion of a metal surface is controlled by making it an anode of an electrochemical cell and controlling the electrode potential in a zone where the metal is passive.
- Anodic protection is used to protect metals that exhibit passivation in environments where the current density in the freely corroding state is significantly higher than the current density in the passive state over a wide range of potentials.

c. **Use of Inhibitors**

- It is the most prevalent, effective and economic method to protect metallic surfaces against corrosion in aggressive media in closed systems.
- Corrosion inhibitors are chemical substances that, when added in small concentration to an environment, effectively reduce the corrosion rate of a metal exposed to that environment.
- They are considered as a retarding catalysts by (i) increasing or decreasing the anodic and/or cathodic reaction, (ii) decreasing the diffusion rate for reactants to the surface of the metal, and (iii) decreasing the electrical resistance of the metal surface. Inhibitors are often easy to apply and offer the advantage of in-situ application without causing any significant disruption to the process.
- Inhibitors are classified according to their composition or mechanism of protection such as: ***Adsorption type corrosion inhibitors*** (largest class of corrosion-inhibiting substances, organic compounds, for example, organic amines), ***Vapor phase corrosion inhibitors*** (volatile corrosion inhibitors [VCIs], act in pressure of 0.1 and 1.0 mm Hg, for example, dicyclohexyl ammonium nitrite and benzothiazole), ***Inorganic inhibitors*** (for example, Pb^{2+}, Mn^{2+}, Cd^{2+} helps to inhibit corrosion of iron in acidic medium), ***Organic inhibitors*** (organic substances containing N, S or O atoms in the molecule), ***Mixed inhibitors*** (formulation can be composed of two or more inhibitors, interfere with both the anodic and cathodic reactions), ***Anodic inhibitors*** (compounds that reduce the anodic area by acting on the anodic sites and polarizing the anodic reaction), ***Cathodic inhibitors*** (act on the cathodic sites and polarize the cathodic reactions), and ***Green corrosion inhibitors*** (efficient, economically viable, and environmentally compliant methods to prevent corrosion have been one of the major challenges to corrosion researchers; for example, extracts of *Azaricta indica*, *Fenugreek* leaves, *Zenthoxylum alatum*, *Opuntia*, *Nypa fruticans*, *Ocimum viridis*, *Phyllanthus amarus*, *Chamomile*, *Halfabar*, and *Murraya koenigii*).
- Carbon Allotropes
    - They are considered as green inhibitors because they are made from minimal, renewable resources with less, minimal, or negligible impact on the environment. Important examples include graphene and its oxides, carbon nanotubes (CNT), carbon dots and fullerenes.

1.8 ECONOMIC IMPACT OF CORROSION

It is said that corrosion can be best described as a necessary evil and hence inevitable in society.

This is because the resultant product formed has a more complex nature and leads to failures of metallic or non-metallic structures. This destruction will cause a heavy burden on the economy of any country [10]. In addition, on average around 10% of total metal found in the world is currently deteriorated due to corrosion, which affects the economy of the nation and other important treasures, such as transportation, big infrastructure, military values, manufacturing plants, and production of metals. It is estimated that the cost of corrosion for any industrialized country or nation is around 3%–4% of the Gross Domestic Product (GDP) [11]. Some countries, such as developing countries of Africa, Asia, Latin America, and Australia/Oceania, are spending around ten times the GDP to fight corrosion.

To accept this contention, a study was done in the year 2003 in the most advanced country, the United States, to gather the cost of corrosion. The obtained result showed that the United States spent around 13 times the GDP than did Nigeria and got around $41.9 billion. The cost of corrosion is mainly related to the challenges that show up in the engineering materials, designs, and structures.

Direct replacement and maintenance costs, along with additional costs, relate to the use of expensive materials. The economic cost of corrosion is divided into two groups [12].

a. Direct Losses/Cost: This includes the cost of protection and prevention along with research and development. This can be quantitatively analyzed.

b. Indirect Losses/Cost: Losses such as those due to spills or fire cannot be quantitatively evaluated. This corrosion loss may range between 10% and 40%.

The cost of corrosion of metals in the United States is around $300 billion per year at current prices. Around one-third of these costs could be prevented by the use of corrosion-obstructed materials and the application of corrosion-related best technical practices. Different studies calculated that in the year 1975, metallic corrosion cost around $82 billion, or 4.9% of the GNP. Later, it was reported that approximately 60% of that cost was unavoidable. The remaining $33 billion, or 40%, occurred due to poor practices and is an avoidable cost. The last two decades have shown economic growth and inflation in the price and overall enhancement of GNP to many folds.

1.9 CONCLUSION

This chapter highlights the fundamentals of corrosion along with its impact on the economy. It has been said that the phenomenon is causing direct and indirect impacts on national economies and also is affecting the world's economy. It was also estimated that the total avoidable cost ranges from 10% to 40%. Different studies are ongoing to gather information on the direct costs of corrosion, although the results of corrosion loss are more often due to indirect losses.

REFERENCES

1. R. W. Staehle. Editorial. *Corrosion*, Vol. 31, No. 1, Jan. 1975, pp. 1–4.
2. A. deS. Brasunas. Scope and Language of Corrosion. *Basic Corrosion Course*, National Association of Corrosion Engineers, Houston, 1970.
3. M. G. Fontana and N. D. Greene. *Corrosion Engineering*, McGraw-Hill, New York, 1967, pp. 1–115, 282.
4. O. C. Mudd. Control of Pipe-Line Corrosion. *Corrosion*, 1945, pp. 2–4.
5. F. L. Que. Introduction to Corrosion. *Basic Corrosion Course*, National Association of Corrosion Engineers, Houston, 1969, pp. 2–4.
6. W. T. Bryan, ed. *Designing Impressed Current Cathodic Protection Systems with Durco Anodes*, Durian, Dayton, 1970.
7. Corrosion Control. *In Engineering and Design Manuals*, U.S. Army Corps of Engineers, Aug. 1, 1962.
8. ASTM Standard G102-89. *Standard Practice for Calculation of Corrosion Rates and Related Information from Electrochemical Measurements*, Annual Book of ASTM Standards, ASTM International, West Conshohocken, Vol. 3.02, 2006.
9. The Effects and Economic Impact of Corrosion. *Corrosion: Understanding the Basics*, Edited by J. R. Davis, ASM International (ASM TECHNICAL BOOKS), 2000. ISBN electronic: 978-1-62708-250-1. DOI: 10.31399/asm.tb.cub.9781627082501

10. A. O. Oniwon. Keynote Address. *Journal of Corrosion Science & Technology*, Vol. 1, No. 1, 2004, pp. iii–iv.
11. E. Bardal. *Corrosion and Protection: Engineering Materials and Processes*, Verlag London, Spinger, 2003
12. Okoroafor, C. Cathodic Protection as a Means of Saving National Asset. *Journal of Corrosion Science & Technology*, Vol. 1, No. 1, 2004, pp. 1–6.

2 Corrosion Prevention

Present and Past Advances and Future Direction

Nabil Lahrache
Laboratory of Molecular Spectroscopy Modelling, Materials, Nanomaterials, Water and Environment, CERNE2D, ENSAM, Mohammed V University in Rabat, Morocco

Abdelouahed Dahrouch
Laboratory of Molecular Spectroscopy Modelling, Materials, Nanomaterials, Water and Environment, CERNE2D, Faculty of Science, Mohammed V University in Rabat, Morocco

Khalid Bouiti and Najoua Labjar
Laboratory of Molecular Spectroscopy Modelling, Materials, Nanomaterials, Water and Environment, CERNE2D, ENSAM, Mohammed V University in Rabat, Morocco

Souad El Hajjaji
Laboratory of Molecular Spectroscopy Modelling, Materials, Nanomaterials, Water and Environment, CERNE2D, Faculty of Science, Mohammed V University in Rabat, Morocco

CONTENTS

DOI: 10.1201/9781003323976-3

INTRODUCTION

Carbon can form covalent bonds with other carbon atoms or with metal or non-metal atoms of a different nature. These bonds can be hybridized: sp, sp^2or sp^3. Carbon can be found in different types of compounds and structures. There are molecules composed of carbon, such as diamond and graphite, which have different properties from carbon atoms because the carbon atoms are linked differently in the two molecules. In recent years, other naturally occurring carbonaceous materials have been discovered and identified as carbon nanomaterials [1].

Carbon nanoallotropes can be classified according to their morphological characteristics (presence of voids in the internal structures of compounds such as fullerenes and carbon nano-corns. These products can also be classified according to the type of covalent bond linking the carbon atoms or according to the dimensions of the nanostructures. If the three dimensions of the molecule are of the nanometric order it is said to be of zero dimension.

Examples of 0D products: fullerenes.

One-dimensional (1D) nanomaterials have dimensions outside the nanoscale like the molecules of nanoallotropes.

Two-dimensional (2D) nanomaterials are molecules that have sheet structures with a thickness in the nanometer range (glass fibres, glass fibres and glass fibres), these products have nanometre thicknesses (graphene, graphene nanoribbons and multilayer graphene) [1].

CORROSION EFFECT

Corrosion of metals is a redox reaction of metals, the half oxidation equation of the metal is:

$$M \rightarrow M^{n+} + ne^-$$

And the half equation for the reduction of dissolved oxygen in an acid solution is:

$$O_2 + 2\ H_2O + 4e^- \rightarrow 4\ OH^-$$

Then the reaction that represents corrosion is [2].:

$$2M + O_2 + 2H_2O \rightarrow 2MOOH + 2H^+$$

Material surfaces corrode if they are in contact with an aggressive environment:

- Spike corrosion [2].
- Galvanic corrosion [3].
- Corrosion by concentration pile [4].
- Corrosion by differential aeration [5].
- Intergranular corrosion [6].
- Cavernous Corrosion [7].
- Corrosion erosion cavitation [8].
- Selective Corrosion [9].
- Stress corrosion [10].
- Fatigue-corrosion [11].
- Hydrogen embrittlement [12].
- Fretting corrosion [13].
- Bio-corrosion [14], [15], [16] and [17].
- Hot corrosion (fluxing) [18].
- corrosion by molten metals [19].

Corrosion Inhibition Effect

An inhibitor is a species that slows down or stops a chemical reaction. Thus, a corrosion inhibitor is a substance that, together with a corrosive medium, reduces the attack by corrosive media.

Inhibitors are species that are used to fight against corrosion. To find effective inhibitors to stop corrosion three points must be considered [19]:

1. Decomposition complications.
2. Estimate the damage caused by corrosion and check that it exceeds the cost of the inhibitors.

A corrosion inhibitor is an element that at low concentrations reduces or prevents corrosion [20].

Over time, considerable effort has been made to find corrosion inhibitors useful in many corrosive environments.

Examples: nitrogen-based compounds, sulphur complexes, aldehydes, thio-aldehydes, acetylenic compounds and several alkaloids such as palavering, strychnine, quinine and nicotine are good corrosion inhibitors [21].

Synthetic Inhibitors

Artificial inhibitors are compounds that are synthesised by chemical reactions and are not natural.

Examples:

- Benzyl Chloride Quaternary compounds
- Diethylsulfate Quaternary compounds
- Amine Ethoxylates Compounds
- Imidazoline compounds
- Organic boron compounds [22].

Green Inhibitors

Green corrosion inhibitors are adsorbable compounds that are not harmful to society and the environment. Thus a "green inhibitor" is an "ecological inhibitor".

Examples: Tea, pomegranate, Emblica officinalis, eucalyptus oil, rosemary, 2-hydroxy-1,4-naphthoquinone resin … .. [23]

Corrosion inhibitors adsorb onto the surface of the metal, forming a protective layer that prevents aggressive and corrosive products from interacting with the metal surface, thus reducing the rate of corrosion of the metal [24].

Organic coatings can protect metal surfaces from corrosion:

1. Organic inhibitors adsorb and form a layer that prevents the diffusion of corrosive agents (such as water and oxygen);
2. Organic inhibitors form electrostatic bonds with electrons on the metal surface;
3. Organic inhibitors block the parts of the anodes and cathodes located at the metal/inhibitor layer interface [25] and [26].

The defects of the layers formed after adsorption of organic inhibitors are:

- Creation of cracks, air bubbles etc. on these layers;
- Coatings formed by organic inhibitors have a low resistance which allows the formation of cracks and reduces their life span [26] and [27].

0D Carbon Nanostructures Inhibition Effect

Studies have shown that some 0D carbon nanostructures can corrode and have proposed some corrosion inhibitors [28–32]. Organic corrosion inhibiting compounds usually contain electronegative heteroatoms (such as oxygen, sulphur, nitrogen or phosphorus), aromatic rings or π-bonds that can adsorb onto the metal surface by the formation of coordinated covalent bonds or by electrostatic interaction with the metal surface

Organic corrosion inhibiting compounds are molecules containing:

- electronegative heteroatoms (such as oxygen, sulphur, nitrogen or phosphorus),
- aromatic rings or
- π-bonds

So these molecules can form electrostatic interactions with free electrons on the metal surface [31].

Xiao et al. used water-soluble carbon quantum dots as lubricants. They investigated the action of these quantum dots on the tribological properties of water by studying the contact of the two products Si_3N_4-steel and Si_3N_4-Si_3N_4. they found that the friction coefficient decreased in both types of tribopairs after the addition of water-soluble carbon. Thus, this study showed that water-soluble carbon quantum dots enhance tribofilm formation on the Si_3N_4 disc surface for Si_3N_4-Si_3N_4 contacts, which confirms that nanoparticles increase friction and thus inhibit corrosion [28].

Hu, Y. & al. showed that the use of water-soluble carbon dots resulted in low coefficients of friction and wear resistance, so they showed that these products have a high inhibiting effect [30].

Mingjun Cu & al showed the inhibitory effect of nitrogen-doped carbon dots of carbon steel by hydrochloric acid solution and proved that this inhibition depended on the concentration of nitrogen-doped carbon dots [21] and [32].

1D Carbon Nanostructures Inhibition Effect

The material transfer films created by carbon nanotubes can be protective barriers for surfaces. No corrosion inhibitors for carbon nanotubes have been presented to date [33].

2D Carbon Nanostructures Inhibition Effect

Studies have shown that 2D carbon nanostructured products contain a single layer of atoms and have demonstrated that graphene forms a protective layer against gas and liquid infiltration [34] and [35].

The adsorption layers and graphene-based tribofilms have the ability to act as barriers which will serve to minimise the occurrence of corrosion on the tribosurfaces [36–43].

Researchers have shown that the use of 0.07% (mass) graphene oxide reduces the corrosion of stainless steel. They proved that this corrosion inhibition is due to the formation of a solid protective layer on the steel surface [38].

A study has shown that graphene is a good metal corrosion inhibitor that can also reduce the oxidation of other products such as aggressive compounds and showed that the addition of graphene decreased the oxidation of MoS_2 [44].

Other Corrosion Inhibition Effect

Carbon-based nanomaterials increase the tortuosity of the path of corrosive agents through the coating. This leads to highly corrosion-resistant nanocomposite coatings. The corrosion resistance of nanocomposite coatings can be explained as follows: Carbon-based nanomaterials act on the polymer by reducing its porosity and thus prevent the passage of aggressive agents such as oxygen and water [25].

Carbon-based nanomaterials agglomerate in organic inhibitors because they have a surface that makes van der Waals interactions with the organic inhibitors [45].

The loading of the nanofiller layer, and the use of carbons and surfactants improve the dispersion of the nanofillers in the polymer. This modification of the surface composed of carbon-based nanofillers avoids the agglomeration of the nanofillers in the polymer and also allows the elimination of the porosity of the coating and thus increasing the inhibition [46].

Modification of (carbon-based) nanomaterials is a method to improve the loading efficiency of the polymer. This can be achieved by using a carbon-based functional group [47–49], and thus improve inhibition. The carbon-based functional groups grew on the surface of the metals participate in the polymerisation of the polymer [50]. The addition of the nanofillers to the polymer decreases the porosity of the coating formed on the metal surface and thus increases the inhibition [50].

Chemically modified nanofillers decrease the contact with water and improve the adhesion strength of the coatings to the metal. These molecules decrease the sorption of water on the coatings and thus decrease the contact of water with the coating formed on the metal surface. If the amount of corrosive agents in contact with the coating formed on the metal surface is reduced, corrosion is reduced and the time of detachment of the coating from the metal surface is increased [32].

The negative charges of the carbon-based nanofillers inhibit the permeability of Cl^- and OH^- anions into the inhibitors [51].

The molecules of carbon-based nanomaterials can be divided into three groups: conductors (such as graphene), semiconductors (carbon dots) and insulators (such as graphene oxide). Conductive carbon-based nanomaterials are used in conductive polymers to increase the inhibition of conductive polymers (so-called cathodic inhibition), such as zinc-rich epoxies, which ensure electrical contact between the zinc atoms and the inhibiting polymer, [52–54].

Recent Advancements in Corrosion Inhibitor Systems Through Carbon Allotropes: Past, Present, and Future

Role of Nanotechnology as Next Generation of Corrosion Inhibitor Systems: Theoretical Aspects

Nanotechnology is defined as the design, production and application of nanoscale molecules. (scale: 1–100 nm) [55] and [56].

Nanomaterials are used in the industrial, biomedical, agricultural, food and environmental fields because they have large surface areas, high chemical and optical reactivity and high mechanical strength [57] and [58].

These products can also be used as corrosion inhibitors for metals and alloys [59] and [60].

Researchers have shown that the use of nanomaterials as inhibitors is more suitable than organic or inorganic inhibitors [61] and [62].

Nanomaterials are widely used in biology and industry because they have great physical, chemical and mechanical properties [63] and [64].

At low concentrations, nanomaterials provide great protection against corrosion but these molecules are not widely used because their solubility in electrolytes is limited and their ability to agglomerate (the power of solute molecules to combine and form larger molecules) [65] and [66].

Nanomaterials are used as inhibitors of covalent and/or non-coving polar functions because these products have hydroxyl, epoxy and carboxyl functions, etc. These polar functions increase the dispersibility of nanomaterials by increasing the inter-molecular interactions of nanomaterials with the electrolytes contained in aqueous solutions. Polar functional groups of nanomaterials are chemically modified to improve the attachment of heterocyclics to the surface of nanomaterials [67] and [68].

Nanomaterials can use the free electrons of heteroatoms (N, O, S and P) or those of the bond π, or those of the aromatic or heteroaromatic rings to form a layer inhibiting corrosion.

- In the first stage the nanomaterials are transferred from the solution to the metallic surface and
- In the second stage there is adsorption of molecules on the surface of the metal due to the covalent bonds formed [69], [70] and [71].

Real-world Corrosion Inhibitor Systems Applications and Nanotechnology

Carbon Nanotubes CNTs as Corrosion Inhibitors

High mechanical strength, high thermal and chemical resistance, dispersal capacity and high interaction with the metal surface give carbon nanotubes and graphene oxides and their derivatives great inhibitory power [72].

Carbon nanotubes and their derivatives can be used as corrosion inhibitors, and these molecules are also used as catalysts for chemical reactions. These products are used as obstacles to the passage of the approved ions of heavy metals and are also used in electrochemistry as polymer composites [73] and [74].

A study has shown that the use of carbon nanotubes increases the rate of corrosion because these products are conductors of heat and electricity [75–77].

A research team has shown that carbon nanotubes can be used as corrosion inhibitors of ferrous and non-ferrous alloys [78].

A research team has shown that polyaminobenzene sulfonic acid and carboxylic acid can be used as corrosion inhibitors of carbon steel immersed in a 3.5% solution of NaCl. The team also showed that these organic products can also be used as corrosion inhibitors for stainless steel immersed in a 3% solution of NaCl chloride [79].

Polyaniline can be used as a mild steel corrosion inhibitor with a 3.5% solution of sodium chloride [80].

Carbonaceous polyaniline nanotubes immersed in amine-treated epoxy-modified acrylic resin give good corrosion inhibition in neutral, acidic or alkaline solutions [81].

The molecules of the thin film of filamentous nanocarbon graphite inhibit the corrosion of copper in seawater, the same was observed when using polyurethane used as a corrosion inhibitor of stainless steel in 3% of NaCl. The use of hydroxyapatite composite inhibits corrosion of stainless steel in simulated liquid and nanocomposites of silver-substituted hydroxyapatite also inhibit corrosion of stainless steel in SBF. Poly(o-phenylenediamine) produces 3.5% NaCl corrosion in the steel. Semiconductor quantum dots can be used as zinc plate corrosion inhibitors in 3.5% NaCl solution, or 1 M HCl solution [82–92].

Graphene and GO as Corrosion Inhibitors

Graphene and graphene oxide disperse into the polymer and thus improve its corrosion resistance. Their use as inhibitors has some defects:

- The use of graphene with the polymer limits its dispersibility and decreases its anti-corrosiveness.
- The agglomeration of graphene in the polymer reduces its dispersibility and decreases its anti-corrosiveness.

The use of small quantities of graphene produces a large diffusion of corrosive products. Graphene increases corrosion inhibition by increasing:

- The hydrophobia of inhibitors,
- Adhesion of inhibitors,
- The effectiveness of inhibitors [93].

Chemically Modified Graphene Oxide as Aqueous Phase Corrosion Inhibitors

Diaminopyridine and Diazopyridine inhibit corrosion of mild steel in 1M HCl solution by slowing the rate of anodic corrosion [94].

Biocorrosion Inhibition by Nanostructured Allotropic Carbon

The US Environmental Protection Agency (EPA) defines a material as antimicrobial if it kills 99.9% of bacteria after 2 hours [95].

Several articles have studied the bacterial action of graphene oxide [96–98].

One-ply graphene leaves have one atom thick, while graphene oxide has surfaces in the order of one square centimetre. Thus, the bacterial actions are different in both cases [99,100].

One study has shown that graphene blocks bacterial activity and biocorrosion on the surface of copper, which acts as a protective membrane preventing bacteria from coming into contact with metal [101].

One study investigated the interaction between bacteria and graphene-coated Cu atoms. they found that graphenes inhibit copper corrosion [102].

Researchers used graphene oxide to reduce sulphate-reducing bacteria. During this inhibition, an electrophoretic deposition of graphene oxide is formed on the surface of the metals, which gives it the role of biocorrosion inhibitor [103].

Conclusion

It has been shown that carbon allotropes, such as graphene, graphene oxide, carbon nanotubes and their derivatives, have great corrosion inhibiting power because they have large surfaces, they also have a great filling power and a great hydrophobicity. While these materials have defects such as:

- Their random distribution in the polymer.
- These materials are linked by intermolecular interactions of the Van der Waals type.
- These products can readily agglomerate in the polymer, resulting in non-uniform distribution;
- These molecules can also stack with the polymer.
- These products have a small solubility in polar solutions.

The distribution of these molecules in the polymer can be increased by agitation or ultrasonic or ball grinding.

The disadvantage of carbon allotropes can be overcome by melding them with organic compounds.

These materials offer great protection because of their large size. Several articles have shown the inhibitory effect of graphene, graphene oxide, carbon nanotubes and their derivatives on the corrosion of different metals and alloys in different solutions.

The large filling capacity and the high hydrophobia around these less dispersible materials in polymers because of the great intermolecular interaction forces and the great agglomeration capacity. The dispersibility of these materials increases with the addition of organic compounds. Their dispersion can be increased physically by using magnetic stirring or mixing methods by ultrasonic or ball grinding. There is little literature on carbon allotropes as aqueous phase corrosion inhibitors. Then you have to chemically modify them with organic compounds. Modified carbon allotropes have high solubility in polar electrolytes and can therefore be used as aqueous phase corrosion inhibitors.

Future research should be directed towards the synthesis of solutions-soluble carbon allotropes derivatives. This inhibition can be improved by using a synergistic effect.

REFERENCES

1. Georgakilas, V., Perman, J. A., Tucek, J., Zboril, R. Broad Family of Carbon Nanoallotropes: Classification, Chemistry, and Applications of Fullerenes, Carbon Dots, Nanotubes, Graphene, Nanodiamonds, and Combined Superstructures. *Chem. Rev.*, 115, 4744–4822 (2015)
2. Yang, C.H., Wei, L., Guan, L., Guo, Y. Polymer Nanocomposites for Energy Storage, Energy Saving and. *Anticorrosion Mater. Chem. A*, 3 (29) 14929–14941 (2015).
3. Marie, F. *Libert et Olivier Bildstein chemical news - october-november - 2015 - n° 400-401.*
4. Luo, Z., Song, K., Li, G., Yang, L. Hydrogen Evolution Ability of Selected Pure Metals and Galvanic Corrosion Behavior between the Metals and Magnesium. *Journal of Electrochemical Science and Technology*, 11 (4) 323–329 (2020).
5. Decker, J. B., Rollins, K. M., Ellsworth, J. C. Corrosion Rate Evaluation and Prediction for Piles Based on Long-Term Field Performance. *Journal of Geotechnical and Geoenvironmental Engineering*, 143 (3) (2008).
6. Liu, Y., Li, X. Effect of Oxygen Differential Aeration on Iron Corrosion Mechanism. *International Journal of Electrochemical Science*, 16 (2021).
7. Mishra, P., Yavas, D., Alshehri, A., Shrotriya, P., Bastawros, A., Hebert, K. R. Vacancy diffusion assisted intergranular corrosion model in low alloy steel. *Acta Materialia*, 220 (November 2021).
8. Baroux, B., CORROSION OF METALS, Passivity and localized corrosion, Dunod (2014).
9. Hu, J., Zhang, L., Ma, A., Mao, P., Zheng, Y. Effect of Cavitation Intensity on the Cavitation Erosion Behavior of 316L Stainless Steel in 3.5 wt.% NaCl Solution, *Metals Journal* (2022).
10. Brandl, E., Malke, R., Beck, T., Wanner, A., Pirater, T., Stress corrosion cracking and selective corrosion of copper-zinc alloys for drinking water installations. *Materials and Corrosion*, 73 (8) 1171–1336 (2009).
11. Cui, C., Ma, R., Martínez-Pañeda, E. Phase field formulation for solution stress corrosion cracking. *Journal of Solid Mechanics and Physics*, 147 (2021).
12. Laleh, M., Hughes, A. E., Hughes, A. E., Gibson, I., Tan, M. Y. A critical review of the corrosion characteristics of additively manufactured stainless steels. *International Materials Reviews*, 66 (2022).
13. Dwivedi, S. K., Vishwakarma, M., Hydrogen fracture in different materials: a review. *International Hydrogen Energy Review*, 43 (46) 21603–21616 (2018).
14. Hemmerling, K. J., Weitzler, L., Bauer, T. W., Padgett, D. E., Wright, T. M. Shrinkage and corrosion of metal liners from dual mobility modular constructions. *The Bone & Joint Journal*, 103-B (7) 1238–1246 (2021).
15. Mehta, H., Kaur, G., Chaudhary, G. R., Batra, U. Evaluation of biocorrosion and antifouling properties of gold metallosurfactant monolayer on galvanised steel in simulated marine environments inoculated with halophiles. *Corrosion Science Journal*, 179 (2021).
16. Noyel Victoria, S., Sharma, A., Manivanna, R. Corrosion of metals induced by microbial activity - Mechanism and control options. *Journal of the Indian Chemical Society*, 98 (6) (2021).
17. Wang, J., Guo, X., Xue, J. Biofilm-grown microplastics as vectors of pollutants in aquatic environments. *American Chemical Society*, 55 (19) 12780–12790 (2021).
18. Feliu, S. Electrochemical Impedance Spectroscopy for the Measurement of the Corrosion Rate of Magnesium Alloys: Brief Review and Challenges. *Metals Journal*, 10, 775 (2020).
19. Xu, D., Jia, R., Li, Y., Gu, T. Progress in the treatment of problematic industrial biofilms. *World Journal of Microbiology and Biotechnology*, 33, Article number: 97, (2017).
20. Al-Otaibi, M. S., Al-Mayouf, A. M., Khan, M., Mousa, A. A., Al-Mazroa, S. A., Alkhathlan, H. Z. Corrosion inhibitory action of some plant extracts on the corrosion of mild steel in acidic media. *Arabian Journal of Chemistry*, 7, 340–346 (2014).
21. Fiala, A., Chibani, A., Darchen, A., Boulkamh, A., Djebbar, K. Investigations of the inhibition of copper corrosion in nitric acid solutions by ketene dithioacetal derivatives. *Applied Surface Science*, 253, 9347–9356 (2007).

22. Peter, A., Obot, I. B., Sharma, S. K. Use of natural gums as green corrosion inhibitors: an overview. *Int. J. Ind. Chem.*, 6, 153–164 (2015).
23. Peter, A., Sharma, S. K. *Corrosion Protection Using Graphene and Biopolymers* (2021).
24. Zeng, Q., Pan, C., Li, C., Fei, T., Ding, X., Du, X., Wang, Q. On-line monitoring of corrosion behaviour in molten metal using laser-induced degradation spectroscopy. *Spectrochimica Acta Part B: Atomic Spectroscopy*, 142, 68–73 (2018).
25. Shi, X., Nguyen, T., Suo, Z., Liu, Y. Effect of Nanoparticles on the Anticorrosion and Mechanical Properties of epoxy Coating. *Surf. Coat. Technol.*, 204 (3) 237–245 (2009).
26. Sørensen, P.A., Kiil, S., Johanasan, K.D., Weinell, C. Anticorrosive Coatings: A Review. *J. Coat. Technol. Res.* 6 (2) 135–176 (2009).
27. Pourhashem, S., Ghasemy, E., Rashidi, A., Vaezi, M. R. A review on application of carbon nanostructures as nanofiller in corrosion-resistant organic coatings. *J. Coat. Technol. Res.* (2019).
28. Hunter, M. E. Rust Inhibitors. in *Encyclopedia of Lubricants and Lubrication* (ed. Mang, T.) 1681–1702 (Springer Berlin Heidelberg, 2014). doi:10.1007/978-3-642-22647-2_184.
29. Xiao, H. P., Liu, S. H., Xu, Q., Zhang, H. Carbon quantum dots: An innovative additive for water lubrication. *Sci. China Technol. Sci.* 62, 587–596 (2019).
30. Wolk, A. et al. A Novel Lubricant Based on Covalent Functionalized Graphene Oxide Quantum Dots. *Sci. Rep.*, 8, 1–9 (2018).
31. Hu, Y.S., Lu, Y. One-pot pyrolysis preparation of carbon dots as eco-friendly nanoadditives of water based lubricants. *Carbon N. Y.*, 152, 511–520 (2019).
32. Cui, M., Ren, S., Zhao, H., Wang, L., Xue, Q. Novel nitrogen doped carbon dots for corrosion inhibition of carbon steel in 1 M HCl solution. *Appl. Surf. Sci.*, 443, 145–156 (2018).
33. Cui, M., Ren, S., Xue, Q., Zhao, H., Wang, L. Carbon dots as new eco-friendly and effective corrosion inhibitor. *J. Alloys Compd.*, 726, 680–692 (2017).
34. Al Jahdaly, B. A., Elsadek, M. F., Ahmed, B. M., Farahat, M. F., Taher, M. M., Khalil, A. M., Outstanding Graphene Quantum Dots from Carbon Source for Biomedical and Corrosion Inhibition Applications. *Sustainability Journal*, 13, 2127 (2021).
35. Bunch, J.S., Verbridge, S., Alden, J.S. Impermeable atomic membranes from graphene sheets. *Nano Letters*, 8, 2458–2462 (2008).
36. Sun, P.Z., Yang, Q. Limits on gas impermeability of graphene. *Nature*, 579, 229–232 (2020).
37. Berman, D., Erdemir, A., Sumant, A. V. Graphene: a new emerging lubricant. *Mater. Today*, 17, 31–42 (2014).
38. Oudbashi, O., Wanhill, R. Long-Term Embrittlement of Ancient Copper and Silver Alloys. *Heritage*, 4, 2287–2319 (2021).
39. Liu, L., Wei, Q. Recent advances in friction and lubrication of graphene and other 2D materials: Mechanisms and applications. *Friction*, 7, 199–216 (2019).
40. Tang, W., Huang, Z., Wang, B. Synthesis of ionic liquid functionalized graphene oxides and their tribological property under water lubrication. *Fullerenes Nanotub. Carbon Nanostructures*, 26, 175–183 (2018).
41. Min, C. Fluorinated graphene oxide nanosheet: A highly efficient water-based lubricated additive. *Tribol. Int.*, 140, 105867 (2019).
42. Upadhyay, R. K., Kumar, A. Boundary lubrication properties and contact mechanism of carbon/MoS 2 based nanolubricants under steel/steel contact. *Colloids Interface Sci. Commun.*, 31, 100186 (2019).
43. Wu, L., Gu, L., Xie, Z., Zhang, C., Song, B. Improved tribological properties of Si3N4/GCr15 sliding pairs with few layer graphene as oil additives. *Ceram. Int.*, 43, 14218–14224 (2017).
44. Paul, G., Hirani, H., Kuila, T., Murmu, N. C. Nanolubricants dispersed with graphene and its derivatives: An assessment and review of the tribological performance. *Nanoscale*, 11, 3458–3483 (2019).
45. Zhao, X., Kwon, A.T., Arenillas, D. Enhanced Mechanical Properties of Graphene Based Poly(vinyl alcohol) composite. *Macromolecules*, 43 (5) 2357–2363 (2010).
46. Christopher, G., Anbu Kulandainathan, M., Harichandran, G. Comparative Study of Effect of Corrosion on Mild Steel with Waterborne Polyurethane Dispersion Containing Graphene Oxide Versus Carbon Black Nanocomposites. *Prog. Org. Coat.*, 89, 199–211 (2015).
47. Wang, X., Spandidos, A. Covalent Functionalization of Graphene with Organosilane and Its Use as a Reinforcement in Epoxy Composites. *Compos. Sci. Technol.*, 72 (6) 737–743 (2012).
48. Wan, Y.J., Tang, L.C., Gang, L.X. Grafting of Epoxy Chains onto Graphene Oxide for Epoxy Composites with improved Mechanical and Thermal Properties Carbon, 69, 467–480 (2014).

49. Wan, Y.J., Tang, L.C. Improved Dispersion and Interface in the Graphene/Epoxy Composites via a facile Surfactant Assisted Process. *Compos. Sci. Technol.*, 82, 60–68 (2013).
50. Lin, Y., Gao, F., Wen, H. Improvement of Mechanical Properties and Anticorrosion Performance of Epoxy Coatings by the Introduction of Polyaniline/Graphene Composite. *Surf. Coat. Technol.*, 374, 1128–1138 (2018).
51. Ramezanzadeh, B., Haeri, Z. Enhancement of Barrier and Corrosion Protection Performance of an Epoxy Coating Through Wet Transfer of Amino Functionalized Graphene Oxide. *Corros. Sci.*, 103, 283–304 (2016).
52. Marchebois, H., Joiret, S. Zinc-Rich Powder Coatings Corrosion in Sea Water: Influence of Conductive Pigments. *Prog. Org. Coat.*, 45 (4) 415–421 (2002).
53. Marchebois, H., Saval, C. Electrochemical Behavior of ZincRich Powder Coatings in Artificial Sea Water. *Electrochim. Acta*, 49 (17) 2945–2954 (2004).
54. Ding, R., Li, W., Gui, T. Study of Water Permeation Dynamics and Anti-Corrosion Mechanism of Graphene.
55. Karmakar, A., Zhang, Q., Zhang, Y. Neurotoxicity of nanoscale materials *J Food Drug Anal*, 22, 147 (2014).
56. Verma, C., Quraish, M.A., Ebenso, E. E., Hussain, C. M., Nano y Wiley Select journal, 1–20 (2021).
57. Rizwan, M., Singh, M., Mitra, C. K., Morve, R. K. Ecofriendly Application of Nanomaterials: Nanobioremediation *J Nanopart*, 2014, 1 (2014).
58. Hussain, C. M. *Handbook of Nanomaterials for Industrial Applications*, Elsevier (2018).
59. Saji, V. S., Thomas, J., *Curr. Sci*, 51 (2007).
60. Saji, V. S., Cook, R. *Corrosion Protection and Control Using Nanomaterials*, Elsevier (2012).
61. Bruder, F. K., Hagen, R., Roelle, T., Weiser, M. S., Faecke, T. Angew Chem, *Int Ed*, 50, 4552 (2011).
62. Zhu, Q.-L., Xu, Q., *Chem*, 1, 220 (2016).
63. Speed, D., Westerhoff, P., Sierra-Alvarez, R., Draper, R., Pantano, P., Aravamudhan, S., Chen, K. L., Hristovski, K., Herckes, P., Bi, X. *Environ. Sci.: Nano*, 2, 227 (2015).
64. Zhang, L., Webster, T. J. *Nano Today*, 4, 66 (2009).
65. Pourhashem, S., Vaezi, M. R., Rashidi, A., Bagherzadeh, M. R. *Prog. Org. Coat.*, 111, 47 (2017).
66. Jyotheender, K. S., Srivastava, C. *Composites, Part B*, 175, 107145 (2019).
67. Verma, C., Olasunkanmi, L. O., Akpan, E. D., Quraishi, M., Dagdag, O., El Gouri, M., Sherif, E.-S. M., Ebenso, E. E. *React. Funct. Polym*, 104741 (2020).
68. Dagdag, O., Safi, Z., Hsissou, R., Erramli, H., El Bouchti, M., Wazzan, N., Guo, L., Verma, C., Ebenso, E., El Harfi, A. *Sci. Rep.*, 9, 1 (2019).
69. Verma, C., Ebenso, E., Bahadur, I., Obot, I., Quraishi, M. *J. Mol. Liq.*, 212, 209 (2015).
70. Shihab, M. S., Al-Doori, H. H. *J. Mol. Struct*, 1076, 658 (2014).
71. Berisha, A., Combellas, C., Kanoufi, F., Pinson, J., Podvorica, F. I. *Electrochim. Acta*, 56, 10762 (2011).
72. Peng, Y. G., Ji, J. L., Zhang, Y. L., Wan, H. X., Chen, D. J. *Environ. Prog. Sustainable Energy*, 33, 123 (2014).
73. Zare, E. N., Lakouraj, M. M., Ramezani, A. *New J. Chem.*, 40, 2521 (2016).
74. Baghayeri, M., Zare, E. N., Lakouraj, M. M. *Microchim. Acta*, 182, 771 (2015).
75. Dresselhaus, G., Dresselhaus, M. S., Saito, R. *Physical Properties of Carbon Nanotubes*, World scientific (1998), ISBN: 9781783262410, 1783262419.
76. Salvetat, J.-P., Bonard, J.-M., Thomson, N., Kulik, A., Forro, L., Benoit, W., Zuppiroli, L. *Appl. Phys. A*, 69, 255 (1999).
77. Lu, J. P. *Phys. Rev. Lett.*, 79, 1297 (1997).
78. Ioniţă, M., Prună, A. *Prog. Org. Coat.*, 72, 647 (2011).
79. Ganash, A. *J. Compos. Mater.*, 48, 2215 (2014).
80. Farag, A. A., Kabel, K. I., Elnaggar, E. M., Al-Gamal, A. G. *Corros. Rev.*, 35, 85 (2017).
81. Rajyalakshmi, T., Pasha, A., Khasim, S., Lakshmi, M., Murugendrappa, M., Badi, N. *J. Electron. Mater.*, 49, 341 (2020).
82. Jeong, N., Jwa, E., Kim, C., Choi, J. Y., Nam, J.-Y., Hwang, K. S., Han, J.- H., Kim, H.-K., Park, S.-C., Seo, Y. S. *Chem. Eng. J.*, 314, 69 (2017).
83. Wei, H., Ding, D., Wei, S., Guo, Z. *J. Mater. Chem. A*, 1, 10805 (2013).
84. Sivaraj, D., Vijayalakshmi, K. *J. Alloys Compd.*, 777, 1340 (2019).
85. Sivaraj, D., Vijayalakshmi, K. *Ultrason. Sonochem.*, 59, 104730 (2019).

86. Zare, E. N., Lakouraj, M. M., Ghasemi, S., Moosavi, E. *RSC Adv.*, 5, 68788 (2015).
87. Prasannakumar, R., Chukwuike, V., Bhakyaraj, K., Mohan, S., Barik, R. *Appl. Surf. Sci.*, 507, 145073 (2020).
88. Cubides, Y., Castaneda, H. *Corros. Sci.*, 109, 145 (2016).
89. Kandasamy, K., Surendhiran, S., Khadar, Y. S., Rajasingh, P. *Mater. Today: Proc.*, 26, 3503 (2020).
90. Zhang, Y., Zhu, H., Zhuang, C., Chen, S., Wang, L., Dong, L., Yin, Y. *Mater. Chem. Phys.*, 179, 80 (2016).
91. Palaniappan, N., Cole, I., Kuznetsov, A., Thomas, K. J., *RSC Adv.*, 9, 38677 (2019).
92. Zare, E. N., Lakouraj, M. M., Moosavi, E., *Compos. Interfaces*, 23, 571 (2016).
93. Compton, O. C., Kim, S., Pierre, C., Torkelson, J. M., Nguyen, S. T. *Adv. Mater.*, 22, 4759 (2010).
94. Ansari, K., Chauhan, D. S., Quraishi, M., Adesina, A., Saleh, T. A. *RSC Adv.*, 10, 17739 (2020).
95. Zhou, F., Li, Z., Shenoy, G. J., Liand, L., Liu, H. Enhanced Room-Temperature Corrosion of Copper in the Presence of Graphene. *ACS Nano*, 7, 6939–6947 (2013).
96. Hans, M., Erbe, A., Mathews, S., Chen, Y., Solioz, M., Mü cklich, F. Role of Copper Oxides in Contact Killing of Bacteria. *Langmuir*, 29, 16160–16166 (2013)
97. Yu, L., Zhang, Y., Zhang, B., Liu, J., Zhang, H., Song, C. Preparation and Characterization of HPEI-GO/PES Ultrafiltration Membrane with Antifouling and Antibacterial Properties. *J. Membr. Sci.*, 447, 452–462 (2013)
98. Yu, L., Zhang, Y., Zhang, B., Liu, J. Enhanced Antibacterial Activity of Silver Nanoparticles/Halloysite Nanotubes/Graphene Nanocomposites with Sandwich-Like Structure. *Sci. Rep.*, 4, 4551–4555 (2014).
99. Zhang, W., Lee, S., McNear, K. L., Chung, T. F., Lee, S., Lee, K., Crist, S. A., Ratliff, T. L., Zhong, Z., Chen, Y. P., Yang, C. Use of Graphene as Protection Film in Biological Environments. *Sci. Rep.*, 4, 4097–4105 (2014)
100. Grass, G., Rensing, C., Solioz, M. Metallic Copper as an Antimicrobial Surface. *Appl. Environ. Microbiol.*, 77, 1541–1547 (2011).
101. Li, J., Wang, G., Zhu, H., Zhang, M., Zheng, X., Di, Z., Liu, X., Wang, X. Antibacterial Activity of Large-Area Monolayer Graphene Film Manipulated by Charge Transfer. *Sci. Rep.*, 4, 4359–4366 (2014).
102. Parra, C., Montero-Silva, F., Henríquez, R., Flores, M., Garín, C., Ramírez, C., Moreno, M., Correa, J., Seeger, M., Haberlë P. Suppressing Bacterial Interaction with Copper Surfaces through Graphene and Hexagonal-Boron Nitride Coatings. *Am. Chem. Soc.* (2015).
103. Song, T. S., Tan, W. M., Xie, J. Bioreduction of graphene oxide using sulphate-reducing bacteria and its implication on anti-biocorrosion. *Nanosci. Nanotechnol. J.* (2018).

3 Carbon Allotropes in Corrosion Prevention

Shveta Sharma and Richika Ganjoo
Department of Chemistry, Faculty of Technology and Science, Lovely Professional University, Phagwara, Punjab, India

Ashish Kumar
Department of Chemistry, Department of Science and Technology, NCE Bihar Engineering University, Government of Bihar, India

CONTENTS

3.1 INTRODUCTION

Corrosion is a damaging phenomenon in which metals deteriorate as a result of chemical reactions with their environment. Corrosion is projected to rise as a result of increased industrialisation and the usage of metallic materials in a variety of industries. Corrosion is a major cause of economic losses in many industrial processes like acid cleaning, acid descaling, acid pickling and oil-well acid that use extremely acidic fluids [1–3]. There are substantial expenses associated with testing for corrosion and repairing or restoring damaged structures, as well as an increased danger to public safety, which calls for developing novel materials that act as corrosion inhibitors [4–6]. Literature suggested the use of organic compounds as corrosion-mitigating substances due to the presence of various functional groups in the compounds, which helped the organic compounds to adsorb either physically, chemically, or both types of adsorptions together on the metallic surface [7–10]. However, lots of shortcomings are also present while using these compounds. The major problem is the toxic nature of organic compounds. Solubility problems and dissociation of compounds at elevated temperatures also interfere with inhibition efficiency. Therefore, this problem of inhibition efficiency of organic compounds can be solved if we replace the organic compounds with a polymeric material because polymers have a large surface area, which will ultimately cover the metallic surface more as compared to organic compounds and result in enhanced inhibition efficacy. Further, the polymeric material is associated with solubility problems in polar solvents; therefore, current research is oriented towards the usage of material in nano dimensions [11–13]. Anticorrosive nano compounds are generally organic in nature, and long chains of carbons are present in their structure.

In this chapter, we will discuss the various carbon allotropes employed as corrosion inhibitors when coated superficially on metal. These anticorrosive coatings make extensive use of nano-sized carbon allotropes, including graphene (G), graphene oxide (GO), single- and multi-walled carbon

DOI: 10.1201/9781003323976-4

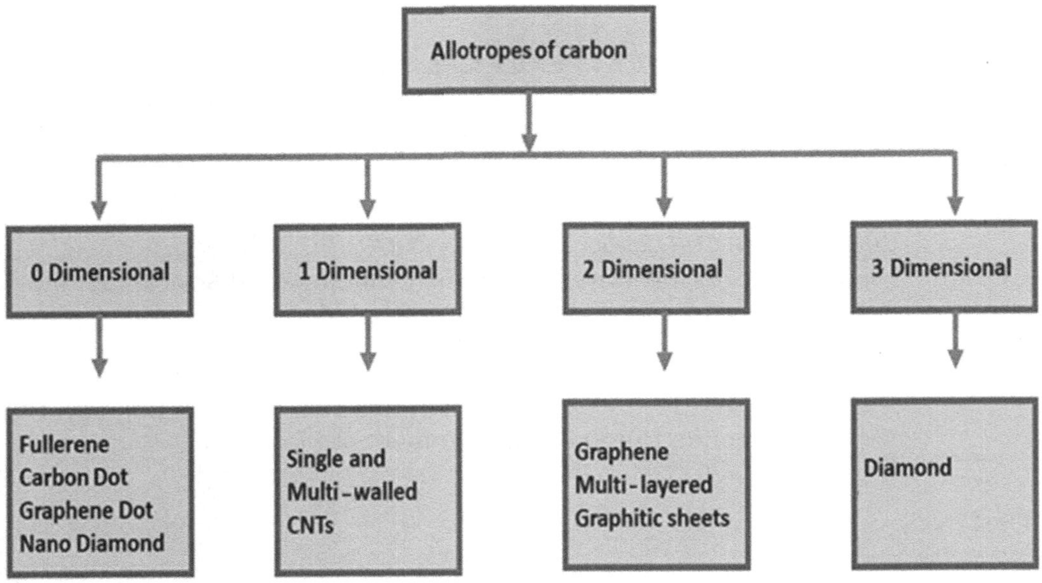

FIGURE 3.1 Carbon allotropes.

nanotubes, and their chemically modified derivatives. Coating is widely accepted as the best technique to protect a material that has been impacted by the external environment. Figure 3.1 displays the main types of carbon allotropes. Anticorrosive coatings are categorised based on how well they protect against corrosion. Cathodic protection, anodic passivation, and electrolytic inhibition are also there, along with active corrosion inhibitors, for the protection of metal [14]. Nanomaterials, whenever used as coatings, are allowed to interact with various polymeric materials to form composites and are used as filler in the polymeric material to increase their corrosion mitigation capacity.

3.2 COATINGS OF VARIOUS CARBON ALLOTROPES

The use of carbon allotropes as corrosion inhibitors, specifically carbon nanotubes (CNTs) and GO and their derivatives, is receiving a lot of interest in the modern world. These carbon nanoparticles and their derivatives have remarkable features, including outstanding mechanical resilience, high thermal and chemical resistance, high surface-to-volume ratio, and high dispersion, including excellent capacity to interact with metallic surfaces. They also can easily fill the coating's small pores and flaws to lengthen the electrolyte channel. Therefore, they are preferred by researchers for studying corrosion mitigation of metals.

3.2.1 Role of Carbon Nanotubes in Corrosion Inhibition

In 1952, Radushkevich and Lukyanovich identified CNTs; in 1972, Oberin discovered single-walled CNTs, and Iijima is widely regarded as the one responsible for the discovery of CNTs in recent times. For identifying multi-walled carbon nanotubes (MWCNTs), there are many methods in the literature for the preparation of CNTs like the Arc discharge method, Laser ablation method, chemical vapour deposition, flame synthesis method, etc. [15]. Polymer composites made from CNTs in particular are being studied as potential anticorrosive materials for both ferrous and non-ferrous metals because of their unique and intriguing features.

Zachariah and Liu [16] synthesised, characterised, and then investigated the composite made up of polybenzoxazine and MWCNTs; for the composite formation, radical transfer, and addition, reactions were used. Further, the characterisation was done with the help of FTIR, XPS, thermogravimetric analysis and Raman spectroscopies. Anticorrosive efficiency of the coating was analysed with an electrochemical technique by employing a CH instrument. Cyclic voltammetry was performed with a scanning rate of $10mVs^{-1}$ in the potential range −1.0 to 1.0V. Electrochemical analysis was performed in the frequency 0.01 Hz and 100 kHz. Results revealed that inhibition efficacy of investigated composite coatings came out to be 99.91%, with a significant reduction in the corrosion rate.

Kumar and Gasem [17] studied polyaniline (PANI) and functionalised CNT to explain the effectiveness of the composite in protecting mild steel surfaces from corrosion. Their structure and morphology were analysed using several methods, including Raman spectroscopy, ATR-IR, and FESEM analysis. Results revealed that mechanical strength was enhanced, along with the increased hydrophobic character of the composite formed (elucidated with contact angle study). The behaviour of coating towards the corrosion was studied with electrochemical measurements. This composite coating reinforced with f-CNTs created a more effective protective barrier layer to reduce the corrosion of metal substrates.

Oliveira et al. [18] worked on the composite made up of Ni–P and MCNTs with the elecrtroless deposition method and studied the property changes in the coating when MCNTs were added. The main techniques employed for this purpose were X-ray diffractometry to detect the crystallinity, surface study with scanning electron microscopy (SEM)-EDX, scratch test for adhesive power, and corrosion study with electrochemical methods. X-ray diffraction study was conducted for Ni-P without MWCNT and with MWCNT. Crystallites of Ni_3P, along with crystalline planes (100 and 200) of Ni, were present in the Ni-P coating, while on adding the MWCNTs, there were visible changes in the crystallinity of coating and plane Ni(200) reduced and plane Ni(100) became more oriented. There was a significant decrease in the crystalline behaviour on the addition of MWCNTs. SEM was employed to elaborate on the surface changes that occurred on adding the MWCNTs to Ni-P. Without the addition of CNTs, the structure of Ni-P was more nodular, but after the addition of CNTs, the surface became smoother with small nodules because of the nucleation, resulting in a lesser corrosion rate. A scratch test was performed for coatings, revealing that a very small and superficial scratch was there when the coating had CNTs, which confirmed the firmness of the coating. An electrochemical impedance study was performed, and the Nyquist graph plotted (given in Figure 3.2) confirmed the small impedance of the uncoated sample as compared to that of the coated sample, confirming the protection of the test sample due to the coating. Similarly, a potentiodynamic polarisation study was conducted, and a Tafel plot was drawn (given in Figure 3.3). Decreased corrosion current values were visible in the plot. At last, the SECM investigation revealed high current values when no coating was there and low current values when CNTs were being coated as composite. The results showed that the electrochemical activity of the coated samples was decreased, whereas the strength of the composite was boosted thanks to the use of MWCNTs.

Coatings are very effective in saving the test metal from corrosion, and that was proved by applying various techniques. For example, electrochemical analysis proved that after coating the material with carbon allotropes, the corrosion current was significantly reduced. Here, the effects of coatings are discussed by considering coatings of two elements, Ni and P, and then another coating of Ni-P but with varying concentrations of MWCNTs. Figure 3.4a demonstrates the corrosion mechanism of the traditional coating consisting of Ni and P, which happens when the anodic dissolution of the steel substrate occurs through intrinsic coating flaws such as micro cracks, releasing Fe^{2+} ions into the electrolyte. But the whole case changes when MWCNTs were added to the conventional coatings (demonstrated in Figure 3.4b). The corrosion current showed a dip, and the corrosion rate reduced, which slowed down the rate at which the steel substrate was being dissolved by the anodic reaction. The blocking effect of the MWCNT particles resulted in a

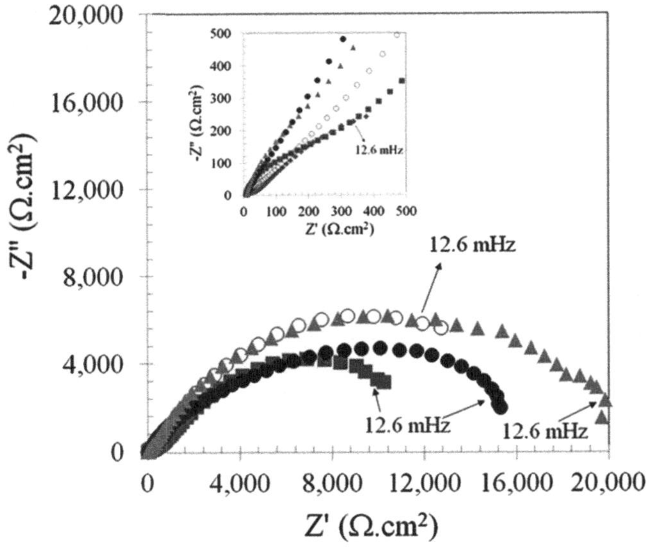

FIGURE 3.2 Nyquist plot for composite without and with MWCNTs [18].

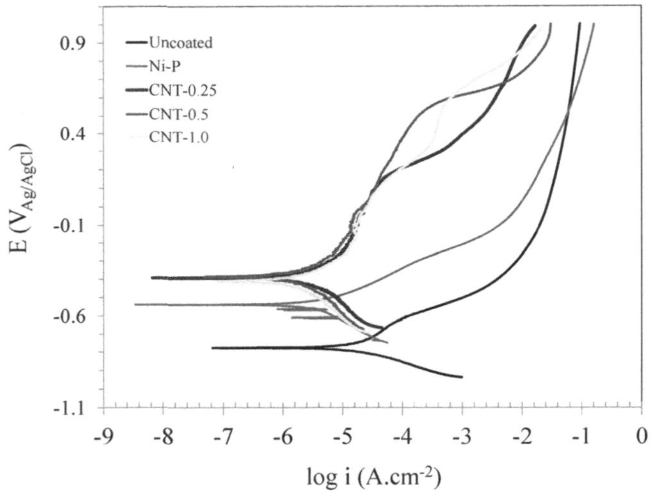

FIGURE 3.3 Tafel exploration without and with MWCNTs in composite film [18].

gradual reduction in the amount of Fe^{2+} ions that were released into the environment. Because of the increased amount of MWCNT in the composite film, the quantity of Fe^{2+} ions that were discharged into the solution was further reduced (Figures 3.4c and 3.4d).

Composites made up of polyurethane and multi-walled carbon nanotube were studied by Wei et al. [19] for polyurethane and MWCNT composite for their anticorrosive properties in NaCl solution for stee samples. Composite was fabricated by the surface-initiated polymerisation, and characterisation was done with SEM and FTIR methods. Further, the thermogravimetric method was used to check the thermal stability of the composite, and electrochemical changes were evaluated by electrochemical analysis. Chemical structure investigation revealed that the proper formation of polymer around the CNTs was there, resulting in exceptional bonding between polyurethane and CNTs. The thermal stability test revealed that after adding the MWCNTs to the

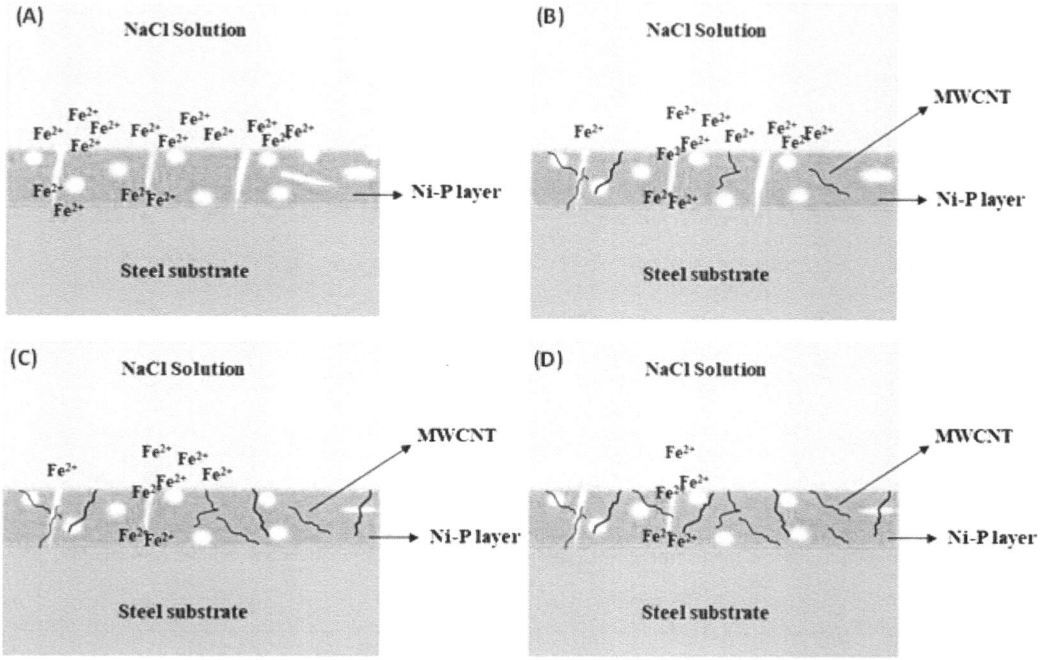

FIGURE 3.4 Pictorial presentation of mechanism [18].

polymeric matrix, the stability of the composite was enhanced with increased decompositional temperature. Tafel plots illustrated by PDP gave a positive corrosion potential and predicted a decreased corrosion current. According to the findings, a formidable physical barrier was formed by the composite of polyurethane and MWCNT against the attack of a corrosive medium.

Calabrese et al. [20] investigated coating made up of N-propyl-trimethoxy-silane and MWCNTs for the protection of Al metal in sodium chloride solution. Surface morphology was studied by SEM instrument, where three types of samples by varying the amount of CNTs were studied (given in Figure 3.5). Figure 3.5a shows that the first sample, AS3-CNT2, which was taken into consideration, does not have a regular structure, and CNT structures on the surface were clearly visible there. In some places, few depressions or pits were also present. Therefore, it was clear from the first picture that the surface was not homogeneous and smooth. A uniform morphology with the CNT nanofiller in the form of well-distributed aggregates was obtained by increasing the CNT content (AS3-CNT4 in Figure 3.5b). As a result of the constituents' solidifications during the curing process, the top of the coating was comprised of several pits and cracks of varying sizes but

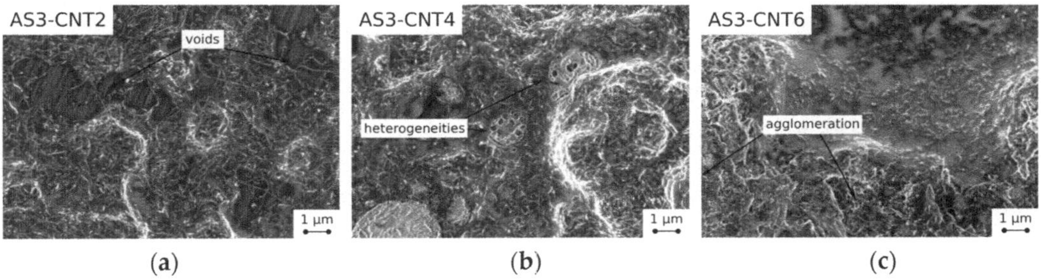

FIGURE 3.5 (a–c) SEM images of samples with varying concentrations of CNTs [20].

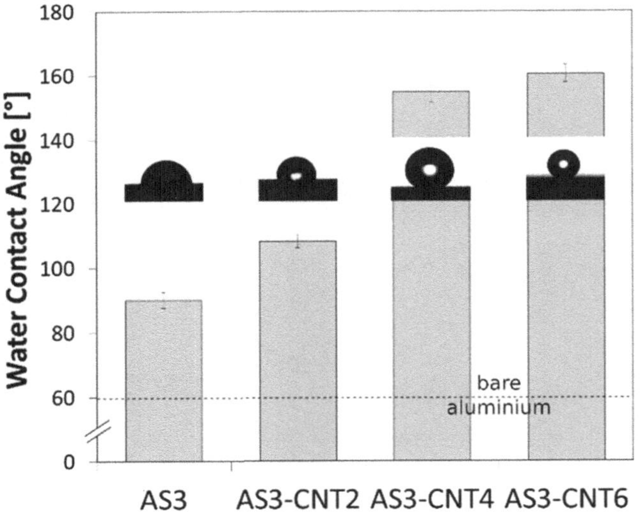

FIGURE 3.6 Contact angle measurements of composite with varying concentrations of MWCNTs [20].

TABLE 3.1
Investigations on Carbon Nanotube Composites as a Corrosion-Resistant Coating

S.No.	Name	Metal	Electrolyte
1	Polybenzoxazine/ MWCNT	Cold roll steel	3.5% NaCl
2	Polyaniline/ f-CNTs	MS	3.5% NaCl
3	Ni-P/ MCNTs	API 5L X80 Pipeline steel	3.5% NaCl
4	Polyurethane/ MCNTs	SS	3% NaCl
5	Silane/ CNTs	Al	3.5% NaCl

comparatively was smoother. In the end, a structure that had the greatest quantity of CNT, designated AS3-CNT6 in Figure 3.5c, revealed a structure that was interlinked and comprised of CNT wires that were randomly intertwined with the silane matrix. Additionally seen are several enormous aggregation zones, which suggest the presence of significant connections, and this absence of large cracks or other flaws in the coating is indicative of good adherence.

A further water contact angle study was also incorporated, mentioned in Figure 3.6, to check the hydrophobic nature of the composite coating. For this study, pure metallic samples without any coating, and with a coating (by increasing the amount of CNTs), were considered. Eventually, the water wettability went from 60° to 90°, proving that the presence of the silane layer improved the hydrophobic behaviour of the surface. With the increased percentage of CNTs, the contact angle also increased, and the maximum was 160.3°. Various investigations for CNTs with other materials are summarised in Table 3.1 [16–20].

3.2.2 Role of Graphene in Corrosion Inhibition

Anticorrosive coatings for a variety of metals and alloys, graphene, and their composites are seeing widespread usage because these nanoparticles have exceptional mechanical, thermal, electrical, and corrosion-reducing capabilities. The main benefit to adding graphene to the polymeric matrix is that graphene sheets create a high-energy barrier that prevents atomic oxygen from migrating from the top of the graphene sheets to the reactive polymer surface underneath. This protects the polymer from atomic oxygen (AO) erosion, which can cause the polymer to erode over time, and also saves the metallic surface under study from further oxidation [21–23]. Epoxy resin coating that is based on graphene nanosheets (GNS) is believed to be an excellent type of anticorrosive material, and lots of research is orienting toward the usage of epoxy and graphene. Graphene fillers with very low loading have the potential to match or exceed the performance of large quantities of traditional composite fillers, and the majority of research to date has focused primarily on improving the mechanical, electrical, or flame retardancy properties of epoxy resin by adding graphene fillers new [24–31].

Zhang et al. [21] synthesised, characterised, and then investigated the epoxy-graphene composite for corrosion inhibition. The scheme for the preparation of the composite is illustrated in Figure 3.7. First, graphene stabilised with polyvinylpyrrolidone was synthesised from graphite. The second step was to incorporate PVP-rGO aqueous solution weight-wise into the epoxy resin in the water-borne epoxy latex by using a magnetic stirrer. Then, that mixture was casted into the Zn plate and cured. The coating was characterised with Raman spectra and SEM techniques, and corrosion-reducing tendency for Zn metal was investigated by employing the Tafel exploration method.

Prasai et al. [22] further studied graphene, in the form of ultrathin films, to prevent rust and corrosion in metals. Here, the graphene on copper and nickel was allowed to grow, or multi-layer graphene was transferred onto them physically, and then electrochemical techniques were used to examine the corrosion inhibition of these metals. The first thing was to investigate how the application of graphene as a protective layer affects the chemical processes that take place throughout the process of corrosion. This was investigated through cyclic voltammetry. Measurements using cyclic voltammetry have shown that the graphene coating is very good at preventing the oxidation of metals and the reduction of oxygen. Further, quantitatively the corrosion rate was calculated by Tafel analysis, and the voltage used was −150 to + 150 mV with a scan rate of 0.005 mV/s. Interestingly, graphene can provide the same level of corrosion reduction as typical organic coatings that are much thicker than graphene. Results also claimed that although the graphene itself is not affected, the metal that lies underneath it is corroding at fissures in the graphene coating. According to these findings, the pace at which copper films coated with graphene created using chemical vapour deposition erode in an aerated solution of Na_2SO_4 is seven times slower than the rate at which naked copper corrodes in the same solution. In addition, nickel that has had a graphene film formed on it corrodes at a rate that is 20 times lower than bare nickel, whereas nickel surfaces that have been covered with four layers of graphene that have been physically transferred corrode at a rate that is only four times lower. According to these studies, graphene has the distinction of being the most ultrathin corrosion-resistant layer.

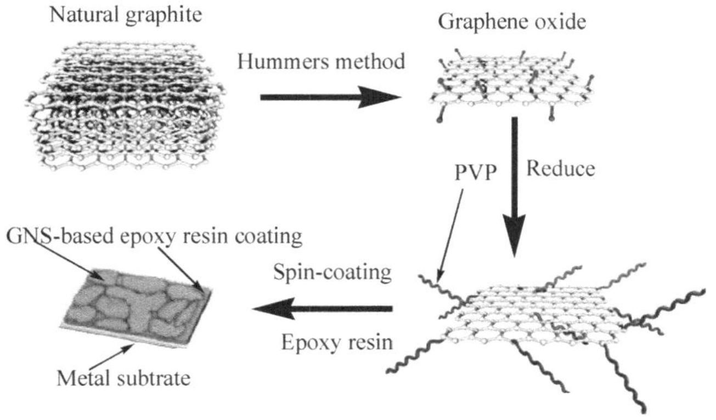

FIGURE 3.7 Synthesis of graphene-epoxy coating [21].

Nayak et al. [23] investigated the protection of the Ni surface from graphene deposited on the Ni surface by using the chemical vapour deposition method. Graphene inhibits the formation of oxide on the surface of the material and protects the metal from the effects of a reactive environment. Coated samples were characterised by various methods like X-ray diffraction, optical micrograph, Raman spectroscopy, X-ray photoelectron spectroscopy, and even after heating the specimen to a maximum temperature of 500° Celsius in the air for three hours; the results revealed that the surface was effectively protected against oxidation. Additionally, it has been found that graphene offers a good level of resistance against H_2O_2. Results depicted the superior performance of graphene as a layer of protection. Because it is chemically inert to oxidising gas and liquid solutions, it can be used in a diverse range of environments.

Yu et al. [32] demonstrated the formation of polymer-graphene composite to coat the metal to check the effectiveness of the coating in saving the metal from corrosion. In this study, two layers of graphene were kept between the three layers of polymer, and then the modified coating was compared with the pure polymeric coating. Further, the long-term effect of the graphene-inserted polymeric coating was investigated with electrochemical tests. Electrochemical analysis was performed over a frequency range of 100,000–0.01 Hz, and PDP was performed with a scan rate of 1 mV/s over a potential variation of −500 to + 500 mV vs open circuit potential. Electrochemical impedance spectroscopy (also known as EIS) is a strong instrument that may be used to analyse the corrosion behaviour of coatings. It reveals information about the reactions of coated metals as well as the barrier qualities of anticorrosive coatings. There is a correlation between greater anticorrosion performance and a higher magnitude of impedance at low frequencies. EIS study also revealed that the impedance value was 500 times higher than the impedance calculated for an uncoated metallic sample and was almost 20 times higher than the metallic sample coated with pure polymer without the addition of graphene. Even when the analysis was performed for two layers of graphene, the impedance value was 5000 times higher than the plane metallic sample. PDP measurements performed were well in agreement with the EIS analysis. The graphene-induced coating was showing low corrosion current values, even when readings were taken after 120 days of immersion, which confirmed that the coating was not damaged over a longer period. SEM analysis was performed to understand the surface morphology of the coated sample. Surface changes were studied over 30 and 120 days of exposure of the sample to the test solution (3.5% NaCl). The uncoated sample was fully corroded even after one day of exposure; even the surface looked damaged for the polymeric-coated sample. When the sample was coated with graphene-incorporated polymeric composite, the surface appeared to be smooth with no damage.

The CVD graphene exhibits natural cracks and tears that allow molecules and other chemical species to easily move through them. When it was applied to the sample, corrosion would occur, and the metal would get damaged. Even when the polymeric matrix was coated, it failed to protect the sample for a longer period. But when graphene was added to the polymer (a single sheet of graphene), the composite formed protected the metallic sample, but the effect was not long-lasting. However, as time passes, the corrosive species that were absorbed by the topmost polymer film made their way via graphene flaws and tears, diffused through the bottom polymer layer, and eventually reached the metal surface, which began the process of its deterioration. When two sheets of graphene are added to a polymer coating, the rate at which corrosive species can diffuse through the coating can be significantly reduced and slowed down. The resulting coating does an excellent job of protecting aircraft aluminium for a period of up to four months when it is submerged in water simulating the environment of the ocean.

Tiwari and Raman [33] studied the multi-layer graphene coated over the Cu metal and checked the changes in the capability of graphene to protect the metal from corrosion when coated in multilayers. Single-layer coating of graphene and multi-layer coating of graphene is given in Figure 3.8. Further, the coating formed was characterised by using Raman spectra, SEM, and optical microscopy. Raman spectroscopy confirmed the presence of graphene on the top of the metallic sample. With the help of Raman spectra, graphene was presented in multi-layers over the sample

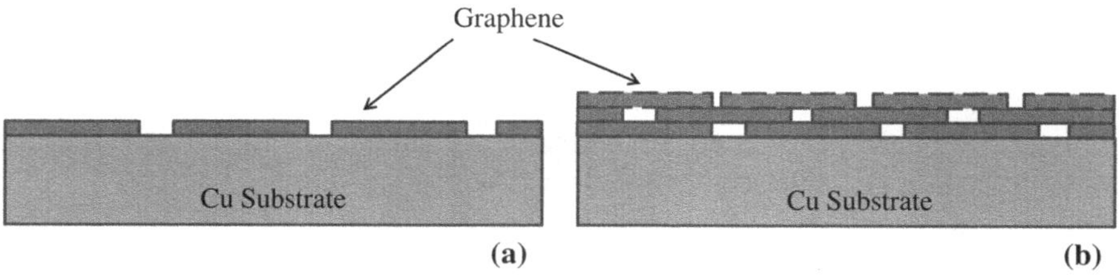

FIGURE 3.8 Single-layer and multi-layer coating of graphene on the Cu surface [33].

TABLE 3.2
Investigations on Graphene-Based Composites as a Corrosion-Resistant Coating

S.No.	Name of Carbon Allotrope	Metal	Electrolyte
1	Epoxy/ Graphene sheets	Zn	3.5% NaCl
2	Graphene	Cu/ Ni	Na_2SO_4
3	Graphene	Ni	H_2O_2
4	Polyvinyl butyral/ Graphene	Al	3.5%NaCl
5	Multi-layered Graphene	Cu	0.1 M NaCl

as the ratio of G and 2d was 1.42 in the present study, and the D peak also signifies that the voids were there on the graphene layers. In the current research, graphene-covered copper was analysed using SEM, which revealed dark spots, and these dark areas have been interpreted as either graphene granules or ripples. Corrosion inhibition property was interpreted with the help of EIS, and the value of corrosion resistance came out to be higher in the case of coated copper metal than the corrosion resistance value of the uncoated copper metal. The multi-layer graphene coating that was generated in this work gave significant corrosion resistance in a chloride solution. Furthermore, the corrosion resistance was maintained for long durations (around 400 hours), which can be attributed to the multi-layer graphene coating. Also, it is common knowledge that polymeric coatings are utilised in situations that involve the coupling of mechanical, thermal, and corrosive forces. Therefore, the ability of polymeric coatings to withstand mechanical and thermal stress is also an important consideration for their use in industry. When the coating barrier is destroyed either mechanically or thermally and corrosive species reach the metal surface, the corrosion process cannot be avoided. This can happen for several reasons. Therefore, the research into the mechanical and thermal properties of the polymer coating is particularly crucial for producing anticorrosive features. On the other hand, very few mechanical and thermal properties of polymer coatings have been evaluated and reported [34,35]. All the past investigations discussed in this chapter are summarised in Table 3.2 [21–23,32,33].

3.2.3 Role of Graphene Oxide in Corrosion Inhibition

In the field of energy, as well as in the development of hydrophobic coatings, GO and functionalised GO have been investigated. GO has a wide range of potential applications and key functions in the field of coating creation. GOs are distinguished by their unusual surface area and characteristics [36,37]. The effectiveness of composite coatings is highly dependent on the GO's dispersal and interaction. GOs are incompatible with coating matrices due to their high van der Waals force and surface hydrophilic oxygen-containing groups. Furthermore, these characteristics cause clustering, which in turn increases the number of small pores and fissures. The presence of a significant number of oxygen-containing groups, such as hydroxyl, carboxyl, and epoxide groups, on the surfaces of GO may make them useful candidates for use as reaction sites in the process of modification. As a result, surface modification of GO is an intriguing strategy for enhancing compatibility [38]. Palaniappan et al. [39] synthesised octylamine-functionalised GO and studied it for corrosion mitigation in a chloride environment. The coating was characterised by SEM, TEM, FTIR, Raman spectroscopy etc., and corrosion property was studied with the help of electrochemical investigation. The corrosion study showed that the physical adsorption of epoxy was not good, which resulted in increased corrosion. For the octylamine functionalised GO-coated magnesium, the surface of Mg was more protected, and a clear dip in the corrosion current value was visible as the coating with functionalised GO restricted the ions from penetrating into the surface. AFM study also confirmed the better inhibition of corrosion when functionalised GO was used. It was clear from AFM that because of an increase in the diffusion of corrosive ions into the epoxy coating, pitting corrosion was promoted on the epoxy-coated magnesium alloy. Due to the very hydrophobic surface, it is hypothesised that the pitting progression was regulated by the octylamine-functionalised GO sheet. Different locations for pitting progression were tested; however, no pitting initiation was discovered due to the hydrophobic surface.

Palaniappan et al. [40] reported the coating modified GO over the Ti6Al4V alloy and the capability of coating to inhibit corrosion by characterising it with various methods like XRD, TEM, FESEM, FTIR, UV, and Raman spectroscopy. The coating was also theoretically analysed with Gaussian software. Electronic transitions of GO and GO + Nd were studied with the help of Uv-IR. In GO + Nd, there were n-π transitions shown. With investigation using IR, the conventional stretching of GO for carboxyl group was altered, along with changes in other band values, which confirmed reduced GO in the decorated one. Structure elucidation was done with Raman spectra, which gives D, G and 2G bands in GO, and in the case of decorated GO, the same bands with a high G band and extra peak for Nd-hydroxide were there, which confirmed the reduction in GO amount. The X-ray diffraction study revealed various peaks increase of modified GO, which indicated Nd-hydroxide in the coating. Further, the surface of the metal was studied for GO and GO-Nd, which revealed the Nd was present in the form of powder on the GO surface, indicating the interaction of GO and Nd. An electrochemical study confirmed the shift of OCP values towards the lesser negative, which validated the protection of the alloy by the applied coating as it did not allow the electrolyte to get into the coating. EIS showed the increased double-layer capacitance and bigger semi-circular loops, confirming the physical adsorption of the coating on the alloy surface. DFT investigations revealed neodymium-modified GO is on the alloy surface, and chemical adsorption was held responsible for adsorption.

Zhang et al. [41] synthesised (2-(3,4-epoxycyclohexyl)ethyl triethoxysilane) to modify the GO to enhance the properties of epoxy resin coating to functionalised GO. The coatings were further investigated by FTIR, X-ray diffraction, X-ray photoelectron spectroscopy, SEM and TEM, electrochemical impedance spectroscopy, and a salt spray test. A spray test was performed for different coatings (illustrated in Figure 3.9) over a period of 400 hours. The sample deteriorated in the case of epoxy coating. The surface was damaged with blisters on it, and the number of micropores and microcracks in the coating grew as a result of the GO agglomeration process. The test solution penetrated the coating and damaged the sample. Coatings that were enhanced by ETEO-GO resulted in the creation of corrosion products with reduced levels. These results demonstrated that the barrier and protective capabilities of the coatings were significantly enhanced as a result of the surface modification of GO with ETEO. The amount of ETEO-GO in the solution reduced the amount of corrosion. The surface of the coatings made with EGO3 and EGO4 exhibited only a minor amount of corrosion products and no cracks, indicating that the coatings provided superior protection against corrosion. The result of all the findings confirmed that using GO that has been functionalised makes the epoxy composite coating more effective at preventing corrosion.

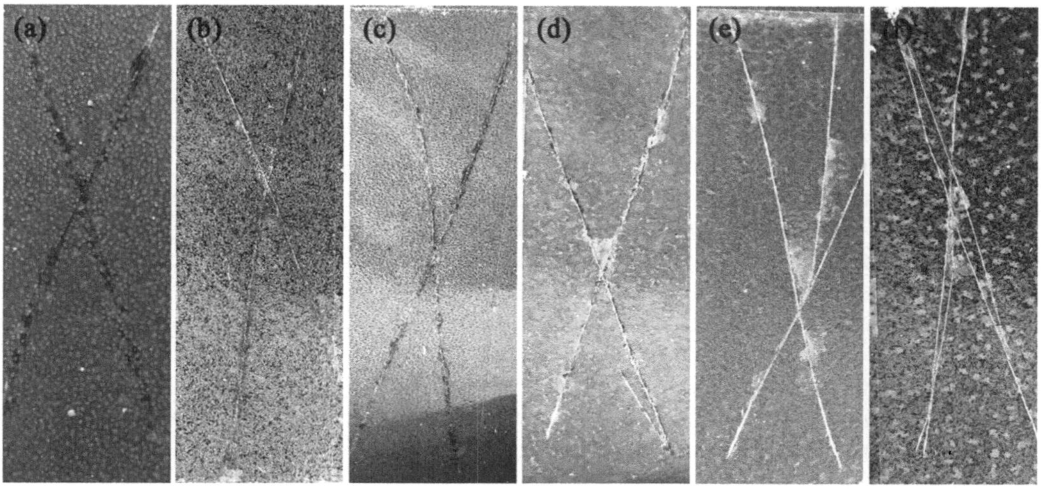

FIGURE 3.9 Spray test performed over the period of 400 hours [41].

Yu et al. [42]. studied vinyl polymer/graphene-based nanocomposites for their usage in corrosion protection. Various characterisation techniques like FTIR, UV–vis, and X-ray photoelectron spectroscopy were employed, along with several electrochemical studies. The FT-IR was performed for graphite, GO, and modified GO (p-GO and pv-GO), and variations in bands indicated the formation of functionalised GO. Again, the UV-vis spectra of GO, p-GO, and pv-GO were studied, as well as the variation in the spectral values, along with the difference in colours as indicated in the UV–vis study. The results of the UV–vis spectra demonstrated that the reactions occurred in a stepwise fashion during the modification of GO. Results revealed increased corrosion mitigation efficacy from 37.90% to 99.53% after the addition of 2 wt% modified GO in the PS polymeric framework. The corrosion potential and corrosion resistance values increased, and the corrosion current and corrosion rate decreased, confirming the anticorrosion protection.

Lu et al. [43] combined 3-aminobenzenesulfonic acid (m-ABSA) with oligoanilines to prepare GO-based sulphonated oligoaniline coatings. Aniline trimer (AT) and sulphonated aniline trimer were both subjected to FT-IR spectroscopy to demonstrate that distinct bonds and functional groups are present in each of these compounds. In addition, a scanning electron microscope (SEM) was used to confirm that GO and SAT had been combined, and a transmission electron microscope (TEM) was utilised to observe the surface morphology of the GO-SAT composite coating that was formed. The electrochemical investigation also proved the excellent protection of samples with modified GO. The main studies related to GO are summarised in Table 3.3 [39–43].

TABLE 3.3
Investigations on Graphene Oxide as a Corrosion-Resistant Coating

S.No.	Name of Carbon Allotrope	Metal	Electrolyte
1	Epoxy, Octyl amine / Graphene oxide	Mg	3.5% NaCl
2	Neodymium / Graphene oxide	Ti_6Al_4V	Na_2SO_4
3	2-(3,4-epoxycyclohexyl)ethyl triethoxysilane / Graphene oxide	Steel	NaCl
4	Polystyrene (PS) Functionalized- Graphene oxide	Fe sample	3.5% NaCl
5	Sulphonated oligoanilines/ Graphene oxide	Steel	3.5% NaCl

3.3 CONCLUSION

When applied as coatings, different allotropes of carbon were able to significantly decrease the amount of corrosion that various metals experienced. Carbon allotropes have the properties of larger surface area and hydrophobicity, which allow them to show anticorrosive properties and to serve as a suitable replacement for conventional corrosion-inhibiting molecules. Moreover, conventional anticorrosion coatings nearly always consist of heavy metals, which are both hazardous and damaging to the environment. Organic coatings are comparatively useful, but microscopic channels can form in coatings due to the presence of free volumes and flaws, which can be created during the coating. Therefore, to provide long-term protection, nanoparticles can be incorporated, and for this purpose, nanoparticles of carbon are one of the best substitutes. In the case of carbon allotropes, a more thorough investigation is required; for example, to get CNTs for the necessary applications, synthesis processes will need to be refined further, as the catalyst used in the synthesis process has an impact on CNT size. Cost is another factor that is involved in preparing CNTs; researchers should work on this.

REFERENCES

1. Haruna, K., Obot, I. B., Ankah, N. K., Sorour, A. A., & Saleh, T. A. 2018. Gelatin: A green corrosion inhibitor for carbon steel in oil well acidizing environment. *Journal of Molecular Liquids*, *264*, 515–525.
2. Singh, A., Ansari, K. R., Chauhan, D. S., Quraishi, M. A., Lgaz, H., & Chung, I. M. 2020. Comprehensive investigation of steel corrosion inhibition at macro/micro level by ecofriendly green corrosion inhibitor in 15% HCl medium. *Journal of Colloid and Interface Science*, *560*, 225–236.
3. Ituen, E., Mkpenie, V., & Dan, E. 2019. Surface protection of steel in oil well acidizing fluids using L-theanine-based corrosion inhibitor formulations: Experimental and theoretical evaluation. *Surfaces and Interfaces*, *16*, 29–42.
4. Trabanelli, G. 1991. Whitney award lecture: Inhibitors—An old remedy for a new challenge. *Corrosion*, *47*(6), 410–419.
5. Singh, A. K. 2012. Inhibition of mild steel corrosion in hydrochloric acid solution by 3-(4-((Z)-indolin-3-ylideneamino) phenylimino) indolin-2-one. *Industrial & Engineering Chemistry Research*, *51*(8), 3215–3223.
6. Sharma, S., Ganjoo, R., Saha, S. K., Kang, N., Thakur, A., Assad, H., & Kumar, A. 2022. Investigation of inhibitive performance of betahistine dihydrochloride on mild steel in 1 M HCl solution. *Journal of Molecular Liquids*, *347*, 118383.
7. Verma, C., Olasunkanmi, L. O., Ebenso, E. E., Quraishi, M. A., & Obot, I. B. 2016. Adsorption behavior of glucosamine-based, pyrimidine-fused heterocycles as green corrosion inhibitors for mild steel: Experimental and theoretical studies. *The Journal of Physical Chemistry C*, *120*(21), 11598–11611.
8. Verma, C., Olasunkanmi, L. O., Ebenso, E. E., Quraishi, M. A., & Obot, I. B. 2016. Adsorption behavior of glucosamine-based, pyrimidine-fused heterocycles as green corrosion inhibitors for mild steel: Experimental and theoretical studies. *The Journal of Physical Chemistry C*, *120*(21), 11598–11611.
9. Pancharatna, P. D., Lata, S., & Singh, G. 2019. Imidazolium based ionic liquid as an efficient and green corrosion constraint for mild steel at acidic pH levels. *Journal of Molecular Liquids*, *278*, 467–476.
10. Dahiya, S., Pahuja, P., Lgaz, H., Chung, I. M., & Lata, S. 2019. Advanced quantum chemical and electrochemical analysis of ravage drugs for corrosion inhibition of mild steel. *Journal of Adhesion Science and Technology*, *33*(10), 1066–1089.
11. Umoren, S. A., & Eduok, U. M. 2016. Application of carbohydrate polymers as corrosion inhibitors for metal substrates in different media: A review. *Carbohydrate Polymers*, *140*, 314–341.
12. Xiao, H., & Liu, S. 2017. 2D nanomaterials as lubricant additive: A review. *Materials & Design*, *135*, 319–332.
13. Baer, D. R., Amonette, J. E., Engelhard, M. H., Gaspar, D. J., Karakoti, A. S., Kuchibhatla, S., … & Wang, C. M. 2008. Characterization challenges for nanomaterials. *Surface and Interface Analysis: An*

International Journal Devoted to the Development and Application of Techniques for the Analysis of Surfaces, Interfaces and Thin Films, *40*(3–4), 529–537.
14. Othman, N. H., Ismail, M. C., Mustapha, M., Sallih, N., Kee, K. E., & Jaal, R. A. 2019. Graphene-based polymer nanocomposites as barrier coatings for corrosion protection. *Progress in Organic Coatings*, *135*, 82–99.
15. Rahman, G., Najaf, Z., Mehmood, A., Bilal, S., Shah, A. U. H. A., Mian, S. A., & Ali, G. 2019. An overview of the recent progress in the synthesis and applications of carbon nanotubes. *C*, *5*(1), 3.
16. Zachariah, S., & Liu, Y. L. 2020. Nanocomposites of polybenzoxazine-functionalized multiwalled carbon nanotubes and polybenzoxazine for anticorrosion application. *Composites Science and Technology*, *194*, 108169.
17. Kumar, A. M., & Gasem, Z. M. 2015. In situ electrochemical synthesis of polyaniline/f-MWCNT nanocomposite coatings on mild steel for corrosion protection in 3.5% NaCl solution. *Progress in Organic Coatings*, *78*, 387–394.
18. de Oliveira, M. C. L., Correa, O. V., da Silva, R. M. P., de Lima, N. B., de Oliveira, J. T. D., de Oliveira, L. A., & Antunes, R. A. 2021. Structural characterization, global and local electrochemical activity of electroless Ni–P-multiwalled carbon nanotube composite coatings on pipeline steel. *Metals*, *11*(6), 982.
19. Wei, H., Ding, D., Wei, S., & Guo, Z. 2013. Anticorrosive conductive polyurethane multiwalled carbon nanotube nanocomposites. *Journal of Materials Chemistry A*, *1*(36), 10805–10813.
20. Calabrese, L., Khaskoussi, A., & Proverbio, E. 2020. Wettability and anti-corrosion performances of carbon nanotube-silane composite coatings. *Fibers*, *8*(9), 57.
21. Zhang, Z., Zhang, W., Li, D., Sun, Y., Wang, Z., Hou, C., ... & Liu, Y. 2015. Mechanical and anticorrosive properties of graphene/epoxy resin composites coating prepared by in-situ method. *International Journal of Molecular Sciences*, *16*(1), 2239–2251.
22. Prasai, D., Tuberquia, J. C., Harl, R. R., Jennings, G. K., & Bolotin, K. I. 2012. Graphene: Corrosion-inhibiting coating. *ACS Nano*, *6*(2), 1102–1108.
23. Nayak, P. K., Hsu, C. J., Wang, S. C., Sung, J. C., & Huang, J. L. 2013. Graphene coated Ni films: A protective coating. *Thin Solid Films*, *529*, 312–316.
24. Liu, K., Chen, S., Luo, Y., Jia, D., Gao, H., Hu, G., & Liu, L. 2013. Edge-functionalized graphene as reinforcement of epoxy-based conductive composite for electrical interconnects. *Composites Science and Technology*, *88*, 84–91.
25. Qian, X., Song, L., Yu, B., Wang, B., Yuan, B., Shi, Y., ... & Yuen, R. K. 2013. Novel organic–inorganic flame retardants containing exfoliated graphene: Preparation and their performance on the flame retardancy of epoxy resins. *Journal of Materials Chemistry A*, *1*(23), 6822–6830.
26. Chatterjee, S., Wang, J. W., Kuo, W. S., Tai, N. H., Salzmann, C., Li, W. L., ... & Chu, B. T. T. 2012. Mechanical reinforcement and thermal conductivity in expanded graphene nanoplatelets reinforced epoxy composites. *Chemical Physics Letters*, *531*, 6–10.
27. Wang, X., Xing, W., Zhang, P., Song, L., Yang, H., & Hu, Y. 2012. Covalent functionalization of graphene with organosilane and its use as a reinforcement in epoxy composites. *Composites Science and Technology*, *72*(6), 737–743.
28. Prolongo, S. G., Jimenez-Suarez, A., Moriche, R., & Ureña, A. 2013. In situ processing of epoxy composites reinforced with graphene nanoplatelets. *Composites Science and Technology*, *86*, 185–191.
29. Shen, X. J., Liu, Y., Xiao, H. M., Feng, Q. P., Yu, Z. Z., & Fu, S. Y. 2012. The reinforcing effect of graphene nanosheets on the cryogenic mechanical properties of epoxy resins. *Composites Science and Technology*, *72*(13), 1581–1587.
30. Zandiatashbar, A., Picu, R. C., & Koratkar, N. 2012. Mechanical behavior of epoxy-graphene platelets nanocomposites. *ASME Journal of* Engineering Materials *and* Technology, *134*(3), 031011.
31. Wajid, A. S., Ahmed, H. T., Das, S., Irin, F., Jankowski, A. F., & Green, M. J. 2013. High-performance pristine graphene/epoxy composites with enhanced mechanical and electrical properties. *Macromolecular Materials and Engineering*, *298*(3), 339–347.
32. Yu, F., Camilli, L., Wang, T., Mackenzie, D. M., Curioni, M., Akid, R., & Bøggild, P. 2018. Complete long-term corrosion protection with chemical vapor deposited graphene. *Carbon*, *132*, 78–84.
33. Tiwari, A., & Singh Raman, R. K. 2017. Durable corrosion resistance of copper due to multi-layer graphene. *Materials*, *10*(10), 1112.
34. Mirabedini, S. M., & Kiamanesh, A. 2013. The effect of micro and nano-sized particles on mechanical and adhesion properties of a clear polyester powder coating. *Progress in Organic Coatings*, *76*(11), 1625–1632.

35. Verker, R., Grossman, E., & Eliaz, N. 2009. Erosion of POSS-polyimide films under hypervelocity impact and atomic oxygen: The role of mechanical properties at elevated temperatures. *Acta Materialia*, *57*(4), 1112–1119.
36. Stankovich, S., Dikin, D. A., Dommett, G. H., Kohlhaas, K. M., Zimney, E. J., Stach, E. A., ... & Ruoff, R. S. 2006. Graphene-based composite materials. *Nature*, *442*(7100), 282–286.
37. Liang, J., Huang, Y., Zhang, L., Wang, Y., Ma, Y., Guo, T., & Chen, Y. 2009. Molecular-level dispersion of graphene into poly (vinyl alcohol) and effective reinforcement of their nanocomposites. *Advanced Functional Materials*, *19*(14), 2297–2302.
38. Dreyer, D. R., Park, S., & Bielawski, C. W. 2010. RS Ruoff The chemistry of graphene oxide. *Chemical Society Reviews*, *39*, 228–240.
39. Palaniappan, N., Cole, I. S., & Kuznetsov, A. E. 2020. Experimental and computational studies of graphene oxide covalently functionalized by octylamine: Electrochemical stability, hydrogen evolution, and corrosion inhibition of the AZ13 Mg alloy in 3.5% NaCl. *RSC Advances*, *10*(19), 11426–11434.
40. Palaniappan, N., Cole, I. S., Caballero-Briones, F., Manickam, S., Lal, C., & Sathiskumar, J. 2019. Neodymium-decorated graphene oxide as a corrosion barrier layer on Ti6Al4V alloy in acidic medium. *RSC Advances*, *9*(15), 8537–8545.
41. Zhang, C., Dai, X., Wang, Y., Sun, G., Li, P., Qu, L., ... & Dou, Y. 2019. Preparation and corrosion resistance of ETEO modified graphene oxide/epoxy resin coating. *Coatings*, *9*(1), 46.
42. Yu, Y. H., Lin, Y. Y., Lin, C. H., Chan, C. C., & Huang, Y. C. 2014. High-performance polystyrene/graphene-based nanocomposites with excellent anti-corrosion properties. *Polymer Chemistry*, *5*(2), 535–550.
43. Lu, H., Zhang, S., Li, W., Cui, Y., & Yang, T. 2017. Synthesis of graphene oxide-based sulfonated oligoanilines coatings for synergistically enhanced corrosion protection in 3.5% NaCl solution. *ACS Applied Materials & Interfaces*, *9*(4), 4034–4043.

4 Carbon Allotropes

Fundamental, Synthesis, Characterization, and Properties Functionalization

Swati Sahu and Sanjay Tiwari
Institute of Renewable Energy Technology & Management, School of Studies in Electronics & Photonics, Pt. Ravishankar Shukla University, Raipur, India

CONTENTS

4.1 INTRODUCTION

Carbon atoms' remarkable ability to create strong covalent bonds with other carbon atoms in various hybridization states (sp, sp^2, sp^3) or nonmetallic elements allows them to build a wide range of structures, from small molecules to long chains. This trait explains why organic chemistry and biochemistry are so important in life. Over two centuries ago, carbon was first discovered in organic molecules and biomolecules, as well as natural carbon materials such as various varieties of amorphous carbon, diamond, and graphite. Even though diamond and graphite are both made entirely of carbon atoms, their characteristics are vastly different. Diamond is the hardest known substance and a

DOI: 10.1201/9781003323976-5

transparent electrical insulator. Graphite, on the other hand, is a black opaque soft material with exceptional electrical conductivity. The way the carbon atoms are bonded in each case causes these discrepancies. Diamond is made up of tetrahedral sp^3 carbon atoms that come together to create massive crystals. On the other hand, graphite is composed of layered graphene monolayers held together by van der waals forces. These graphene monolayers are made up of closely packed sp^2 carbon atoms in a two-dimensional hexagonal lattice.

The development of a variety of novel materials with surprising properties and broad potential applications in technology has partially filled the formerly empty gap between organic molecules and natural carbon materials in recent years. The C_{60} molecule, also referred to as fullerene, was the first of these carbon nanostructures to be discovered, and it was first described in 1985 [1,2]. Many other fullerenes, including C_{20}, C_{70}, and even larger species, were later discovered, but C_{60} is by far the most studied to date. Each C_{60} molecule is made up of 60 sp^2 carbon atoms organized in a spherical (truncated icosahedral) shape made up of hexagons and pentagons. Fullerenes are the tiniest efficient carbon nanostructures reported, and they exist at the intersection of molecules and nanomaterials. Given its solubility in organic solvents, C_{60}, for example, may be considered a big spherical organic molecule (toluene in particular). Iijima's discovery of carbon nanotubes (CNTs) six years later was a huge step forward in the development of carbon nanomaterials [3–5].

Following the discovery of these two fundamental materials, fullerenes and CNTs, new carbon nanostructures with distinct shapes, like single-walled carbon nanohorns (SWNHs), onion carbon spheres, and bamboo-like nanotubes, were developed. Although discovering these materials was less groundbreaking than that of CNTs, their existence is significant since it further highlights carbon's potential to construct novel nanostructures that were unimaginable only a few years ago. Graphene, the building block of graphite, is the most newly isolated carbon nanostructure. Although its presence had been anticipated decades before [6] and experimentally found in 1962 by Boehm et al. [7], Andre Geim and Konstantin Novoselov [8] were the first to isolate and analyze it in 2004. Graphene is a family of nanostructures that are made up of a single graphene monolayer or a few graphene monolayers. Kang et al. studied the various graphyne structures that have been postulated over the last few decades, including α-graphyne, β-graphyne, and γ-graphyne, to name a few [9]. These structures feature various atom layouts, symmetries, and ratios of sp- and sp^2-hybridized atoms. Later, Haley showed how the carbon allotrope graphyne, which is made up equally of sp- and sp^2-hybridized carbon atoms, was prepared using the alkyne metathesis. The carbon sheets can be exfoliated like graphene and are stacked in an ABC-staggered interlayer configuration [10].

Because they are predominantly made up of sp^2 carbon atoms organized in a hexagonal network, all these carbon allotropes belong to the same family. They all have certain common features as a result of their shared structure, but they also have substantial distinctions because of their diverse sizes and forms. Electrical conductivity, mechanical strength, chemical reactivity, and optical characteristics are all similar. The most significant distinction is their dispersibility in organic solvents: C_{60} is the sole readily soluble nanostructure, although graphene can be dispersed in a limited number of organic solvents.

In terms of their principles, synthesis, behavior, functionalization, and potential applications, carbon allotropes—particularly 0-D, 1-D, and 2-D carbon nanostructures—have been thoroughly investigated. However, several recent studies have described the integration of various carbon nanostructures to produce carbon superstructures or all-carbon composite structures.

4.2 FUNDAMENTAL AND CLASSIFICATION OF CARBON ALLOTROPES

Using their morphological characteristics, carbon allotropes can also be categorized, and different classifications of carbon allotropes are depicted in Figure 4.1. In such a classification system, the first group would include nanostructures like fullerene, CNTs, and nanohorns that have empty inner regions. The inside spaces of the hollow nanostructures can host visitors' molecules, metals, atoms, and other

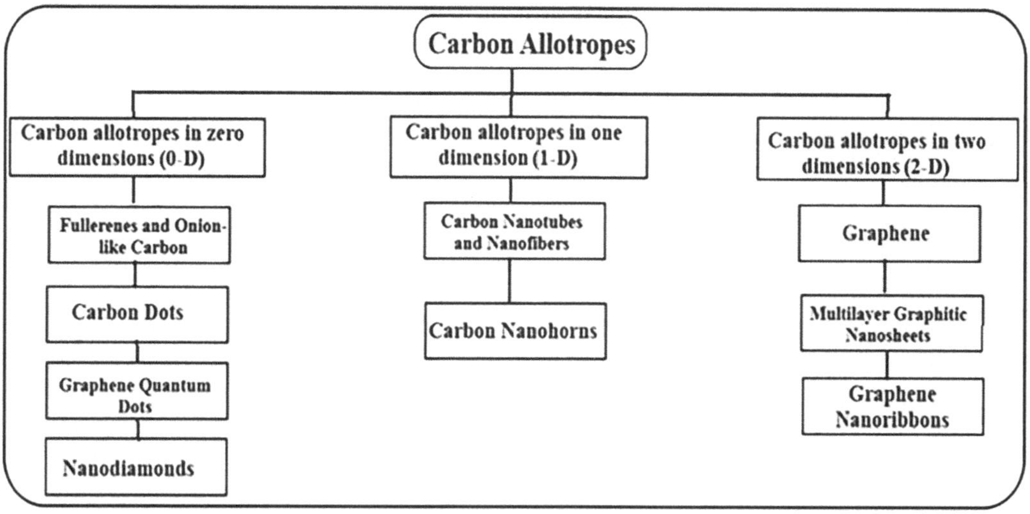

FIGURE 4.1 Classification of carbon allotropes.

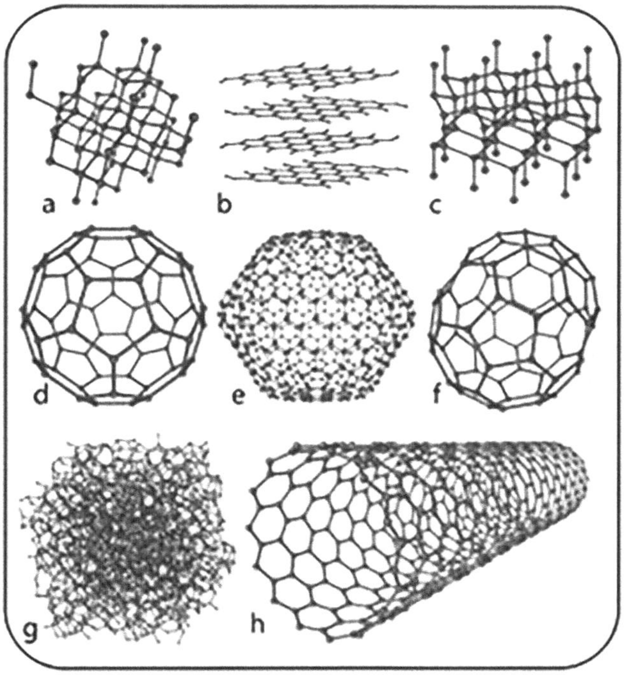

FIGURE 4.2 Eight carbon allotropes are represented structurally as follows: (a) diamond, (b) graphite, (c) lonsdaleite, (d) C_{60} buckminsterfullerene, (e) C_{540} fullerite, (f) C_{70} fullerene, (g) amorphous carbon, and (h) zig-zag single-walled carbon nanotube [11].

nanostructures. They can even produce microhabitats that have microhabitatsular reactions in rare circumstances. This idea assigns a new second group to strong nanostructures without internal spaces, such as diamonds, C-dots, and onion carbon spheres. Graphene might be a part of this category because it does not have any internal vacancies. The structure of eight carbon allotropes is shown in Figure 4.2.

4.2.1 Carbon Allotropes in Zero Dimensions (0-D)

4.2.1.1 Fullerenes and Onion Carbon

After fullerenes were discovered, scientists began studying carbon nanostructures. Fullerenes are closed, hollow structures formed of sp^2-hybridized carbon atoms grouped in 12 pentagons and, depending on the total number of carbon atoms, a certain number of hexagons. N hexagons will be present in a fullerene with 20+2n carbon atoms. The closed forms of fullerenes define the number of pentagons, which is always 12 in those with complete structures. Basic spherical C_{60} has an exterior diameter of 0.71 nm and numerous chemical properties in common with organic molecules. The tiniest carbon nanostructure, however, can be thought of as a standard 0-D carbon allotrope. The C_{60} molecule's ability to inhibit the development of double bonds in the pentagonal rings has led to widespread usage of the term "non-super aromatic." Although numerous similar configurations with more or fewer C atoms (e.g., C_{70}, C_{76}, C_{82}, and C_{84}) can be found in experimentally made fullerene soot, C_{60} is the most prevalent, has drawn the most attention from researchers, and has been the most frequently employed. Different types of fullerenes are shown in Figure 4.3. Its shape is an icosahedron with 20 hexagons and 12 pentagons that have been twisted in half. Its carbon atoms all went through sp^2 hybridization. However, because the arrangement of carbon atoms is pyramidalized rather than planar, there is a "pseudo"-sp^3-bonding. Sp^2 carbons must contain the essential component. Consequently, C_{60} and other bigger fullerenes can be thought of as carbon. A nanoallotrope is made up of the sp^2 and sp^3 molecules together. Multi-layer fullerene, commonly known as carbon onions, is one of the newest fullerene family members among carbon nanomaterials. Carbon onions are multilateral graphic layers that are almost spherical. These layers can develop crystalline quasispherical or amorphous layers due to a range of holes and flaws produced by various pentagonal and hexagonal carbon rings. Researchers are focusing on developing sensors for the detection of compounds that are significant to both biology and nonbiology due to their outstanding thermal stability and electrical conductivity. Carbon onions are being used as the main building block in these sensors.

4.2.1.2 Carbon Dots

C-dots are carbon nanoparticles with a diameter of 2–10 nm, a high oxygen content, and different volumetric ratios of graphitic and turbostratic carbon. C-dots are typically amorphous in form and comprise primarily sp^3-hybridized carbon. C-dots' most distinctive and important attribute is their moderately high photoluminescence (PL), which is influenced by their size, excitation wavelength, and surface functionalization. Due to their excellent and controllable PL, high quantum yield (QY), low toxicity, small size, notable biocompatibility, and plentiful low-cost sources, C-dots, a new rising star in the carbon family, have drawn significant attention.

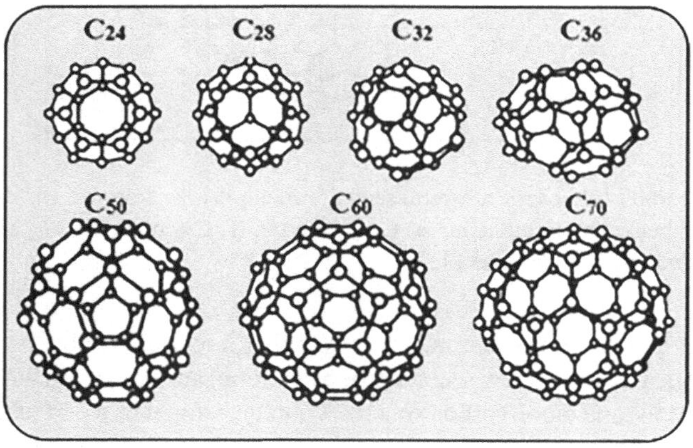

FIGURE 4.3 Fullerene subtypes.

These applications include biomedicine, photocatalyst, optical devices, and anti-counterfeiting [12–20].

4.2.1.3 Graphene Quantum Dots (GQDs)

GQDs are the outcome of severing a graphene monolayer into little fragments (discs) with thicknesses of a few nanometers (2–20 nm). GQDs are crystalline and made primarily of sp^2-hybridized carbon [21]. However, they often consist of a few more stacked graphene monolayers based on the topologies of their graphene nanosheet progenitors. They arose from the fact that quantum confinement-related edge effects in graphene nanosheets with dimensions less than 100 nm become increasingly evident as the sheets get smaller, especially when the sizes drop beyond 10 nm [21]. GQDs have nonzero band gaps owing to quantum confinement as well as edge effects, whereas huge graphene nanosheets have a band gap of zero width, limiting their applicability in electrical and optoelectronic applications. As a result, they have particularly fascinating optical and electrical properties [22–24].

4.2.1.4 Diamonds

Diamonds [25] are sp^3 carbon nanoparticles with dimensions more than 0–2 nm but less than 20 nm, which are made up of crystal domains with a diamondoid-like structure. They cannot be dissolved and are often made using top-down techniques like jet milling or abrasion of micro diamonds [26,27]. Such nanostructures act like bulk diamonds when their diameters are greater than 20 nm. On the other hand, diamondoids, which are sp^3 carbon nanostructures with a diameter of less than 1 nm, are a naturally occurring phenomenon in petroleum deposits [28]. These tiny diamondoids typically have hydrogen or other non-carbon atoms connected to their surface-bound, sp^3-hybridized carbon atoms.

4.2.2 Carbon Allotropes in One Dimension (1-D)

4.2.2.1 Carbon Nanotubes and Nanofibers

Many years before graphene was isolated, CNTs were identified and studied [3–5]. A broad category of tubular nanostructures with comparable structures and forms are referred to as "carbon nanotubes," a general phrase. They must ideally be constructed using a hexagonal lattice of sp^2 carbon atoms, like graphene. Furthermore, the graphene sheet's edges are bonded in nanotubes to create a cylinder with a large aspect ratio. The most basic CNT is sealed at both ends and has a single graphenic wall. Single-walled carbon nanotubes (SWNTs) are long and have an empty interior region with a dimension of around 0.4–2 nm. Depending on the number of graphene layers in the walls of the cylindrical structure, CNTs can either be multi-walled (MWNTs) or double-walled (DWNTs) [29,30]. Chirality, or the angle between the hexagons and the nanotube axis, is another crucial characteristic of CNTs in addition to diameter and length. Carbon atoms can be organized in a variety of configurations around the nanotube's circumference depending on their chirality; the most frequent examples are armchairs, zig-zag, and chiral patterns, as shown in Figure 4.4.

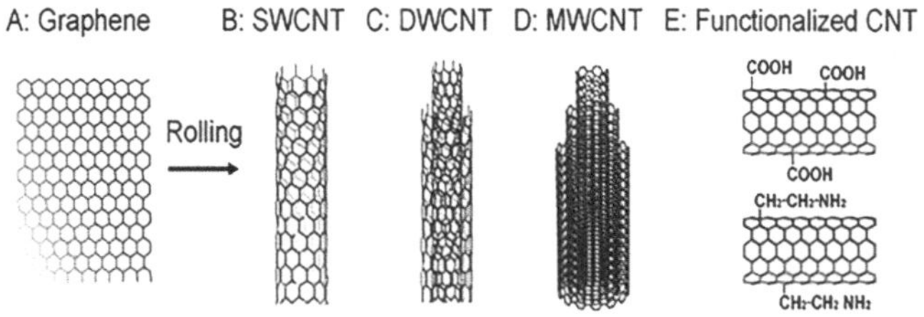

FIGURE 4.4 The distinction between graphene, single-walled carbon nanotubes (SWCNTs), double-walled carbon nanotubes (DWCNTs), and multi-walled carbon nanotubes (MWCNTs) [31].

4.2.2.2 Carbon Nanohorns

While investigating the creation of CNTs, Iijima discovered SWNHs in 1999 [32]. Nanohorns are structures consisting of a single graphene layer that resembles tubules or cones. They typically form enormous, spherical aggregates with a diameter of 80–100 nm that resemble dahlia flowers. The dimension of a single nanohorn is 1–2 nm at the tips and 4–5 nm at the base of the cone [33]. Nearly 0.4 nm separates adjacent SWNHs wall from the wall. Additional SWNH aggregation types with characteristics of buds and seeds have been seen [34].

4.2.3 Carbon Allotropes in Two Dimensions (2-D)

4.2.3.1 Graphene

Since it makes up the majority of natural graphite, graphene is an exceedingly abundant substance. Covalent bonds connect each carbon atom in a graphene sheet to three adjacent carbon atoms, forming a sturdy honeycomb lattice (hcb). The strongest substance currently understood is graphene. The half-full band that gives graphene its aromatic properties is formed by interactions between the unhybridized p-orbitals of the carbon atoms, which are aligned perpendicular to the planar structure of the graphene sheet.

4.2.3.2 Multi-Layer Graphitic Nanosheets

Two to ten graphene monolayers make up a multi-layer graphitic nanosheet (MGN) [35]. They have characteristics in common with graphene. In organic solvents, MGNs can be distributed to create stable, transparent suspensions.

4.2.3.3 Graphene Nanoribbons

An alternate kind of graphene that is currently garnering major research attention is thin ribbons of graphene monolayers. The thin, elongated graphene monolayer strips that can be made by "unzipping" CNTs have received the majority of attention in investigations on these materials [36,37]. A common description of graphene nanoribbons is that they are one-dimensional, sp^2-hybridized carbon strips with defined edges and carbon atoms that are not 3-coordinated in nature. Based on the edge termination, three different forms of graphene nanoribbons are usually known: (a) armchair, (b) zig-zag, (c), and chiral nanoribbons. Specifically, the number of dimer C-C lines (N_a) across the armchair graphene nanoribbon is used to indicate its width, whereas the count of zig-zag chains (N_z) across the nanoribbon is used to express the width of the graphene nanoribbon having zig-zag edges [38–41] (Figure 4.5).

Edge reconstruction could happen because the carbon atoms at the edges are not bonded saturated. The armchair graphene nanoribbon's edge pattern is stable because of the strong dangling bonds, but edge reconstructions are anticipated for zig-zag graphene nanoribbons at high temperatures. It is typical to use hydrogen saturation to stabilize the edge structure. Other edge profiles incorporating pentagonal as well as heptagonal carbon rings have been seen, although those edge reconstructions are extremely uncommon [38]. In some cases, graphene nanoribbons may have a variety of edges. Hybrid graphene nanoribbons are the name given to formations like these, which most typically have heterojunctions of an armchair and zig-zag patterns. Graphene nanoribbons can exhibit bilayered or few-layered patterns, similar to graphene; if several layers of finite graphene strips are placed together, the design is known as a graphitic nanoribbon [38].

4.3 SYNTHESIS AND CHARACTERIZATION OF CARBON ALLOTROPES

There is no disputing that carbon nanomaterials offer intriguing properties that have insured their employment in numerous applications. However, structural variability and imprecise manufacturing brought on by the synthesis process have made it highly challenging to utilize these

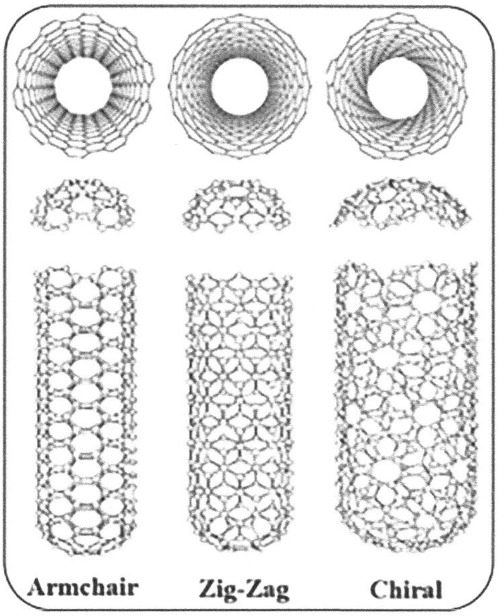

FIGURE 4.5 Chirality variations in carbon nanotubes.

materials in a variety of applications. But since new synthesis methods have been developed and current ones are continually improved, large quantities of high-quality carbon-based nanomaterials have been produced while using less time and energy. Depending on how the carbon atoms are positioned adjacent to one another, they may produce compounds with a variety of properties due to their allotropic features. The classification of carbon allotropes can be done using their morphological characteristics. These carbon allotropes' atomic arrangements and interactions with other substances have a big impact on how they behave. In terms of their optical, electrical, mechanical, and surface area-to-volume ratio properties, carbon-based nanomaterials outperform conventional materials. Carbon allotropes have been characterized using several spectroscopic and spectrometric techniques that are often used in organic chemistry, including UV-vis, NMR, FTIR, Raman, and mass spectrometry. For instance, among other techniques, fullerenes have been produced by arc discharge, chemical vapor deposition (CVD), and laser ablation [42]. CNTs have been produced via arc discharge [43], laser ablation [44], CVD [45], solid-state pyrolysis [46], plasma torch [47], and liquid electrolysis [48]. A diagram of the CVD synthesis process is presented in Figure 4.6. Graphene has been produced by exfoliation [48], CVD [49], arc discharge [50], laser ablation (pulsed laser deposition) [51], unzipping of CNTs [52], Hummer's method [53], and chemical-based techniques [54]. In this part, the three most frequently used synthesis techniques – arc discharge, laser ablation, and CVD – are covered in detail.

Arc Discharge: The arc discharge method has been used extensively to create carbon nanomaterials ever since it was first developed [2]. The approach has undergone important changes targeted at accomplishing certain goals. The alterations include using DC power instead of AC power, using relatively low AC instead of high AC, using electrodes made of highly refined demineralized coal, and using tapered equipment for collecting carbon soot [55]. The method uses high-purity graphite electrodes and is easy to assemble, as shown in Figure 4.6. It also has the potential to provide reasonably high yields (anode and cathode). In essence, the graphite anode is coated with a metal catalyst depending on the type of nanomaterials that are wanted. By adjusting the anode position, one can control or regulate the gap that exists between the anode and cathode. High temperatures (over 17,000°C), low pressures (between 30 and 130 Torr), and an inert gas environment are all maintained during synthesis, with reaction times ranging from 50–60 seconds to 2–10 minutes [56]. Reactor temperature, catalyst makeup, gas pressure (hydrogen), and carbon

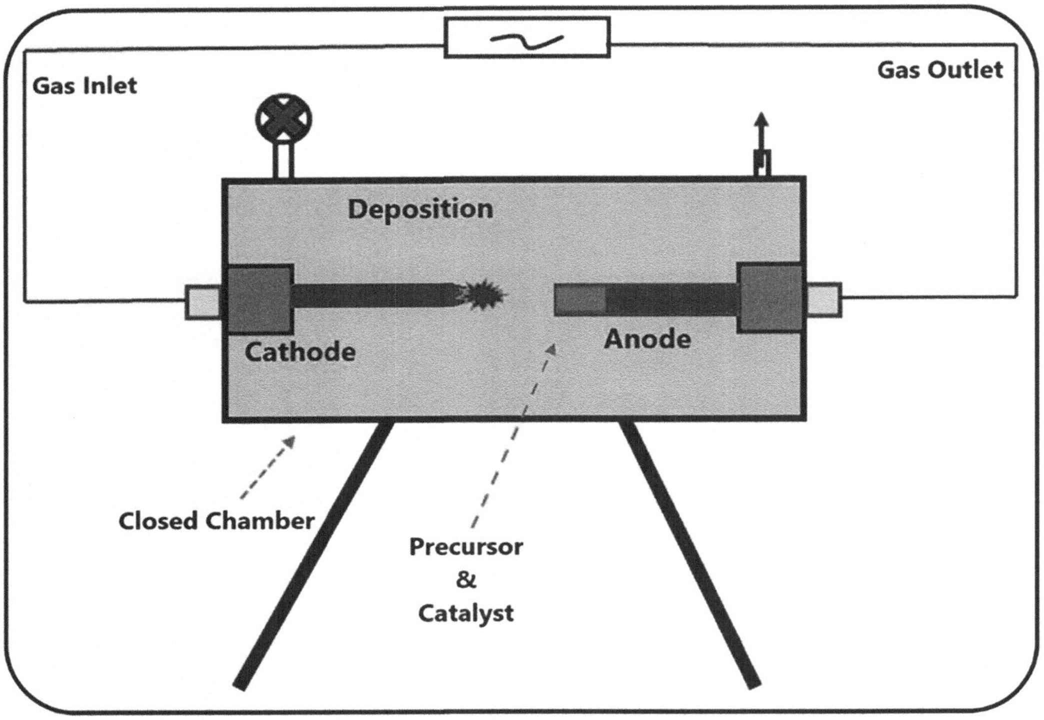

FIGURE 4.6 Method of arc discharge synthesis.

vapor concentration are a few examples of physical and chemical parameters that have an impact on the arc discharge process.

Laser Ablation: The laser ablation method, depicted in Figure 4.7, was created and improved by Guo et al. [57] before being used by other researchers [58,59]. As opposed to arc discharge, which employs liquid carbon sources, laser ablation uses solid carbon sources and operates at significantly lower temperatures. The set-up is inspired by the Geo et al. investigation's [57] usage

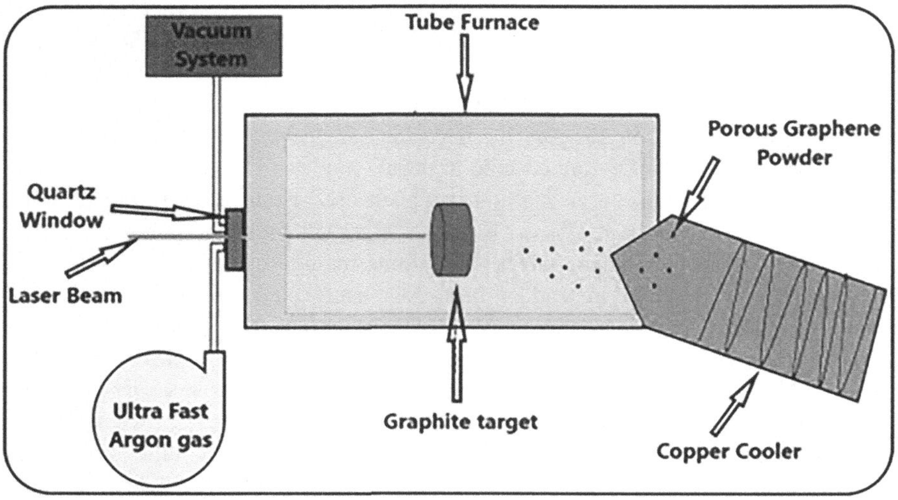

FIGURE 4.7 Schematic depiction of laser ablation.

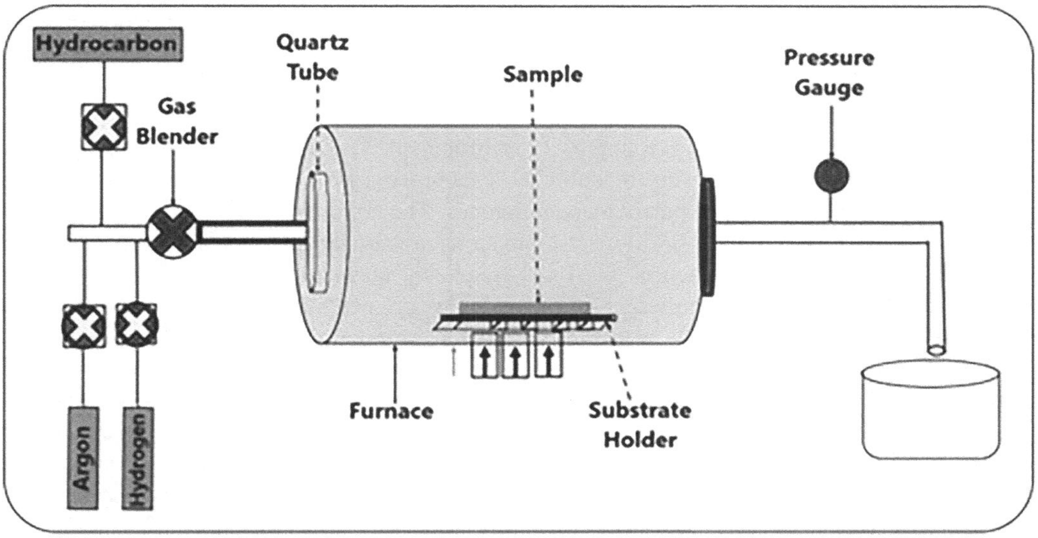

FIGURE 4.8 A diagram illustrates the CVD synthesis method.

of the metal-graphite inside of a furnace that was heated to 1.2×10^3°C and assaulted with a 532 nm laser beam. Under computer control, smooth laser beam scanning is performed to produce uniform vaporization. Soot is created by laser vaporization and then moved from the high temperature of the furnace to a lower cooling collector constructed of copper using argon gas. With this technique, carbon nanomaterials including carbon quantum dots, graphene, diamond, and carbon nanoparticles may all be made in a straightforward one-step procedure.

Chemical Vapor Deposition: A feedstock containing hydrocarbons or carbon monoxide must be broken down by the employment of a metal catalyst in CVD. It should be mentioned that polycyclic aromatic hydrocarbons, ethylene, benzene, and methane are the main carbon sources used in CVD. This approach is applied in two different ways: the horizontal furnace configuration and the vertical furnace configuration. The horizontal arrangement (Figure 4.8) is a common comparison to other allotropies of carbon nanomaterials and is used in the manufacturing of CNTs. To begin the synthesis process, the catalyst is first placed on ceramic or quartz pots or crucibles before being inserted into the quartz tube. The catalyst bed is subjected to a reaction mixture of inert gas and hydrocarbon that occurs between 5×10^2°C and 1.10×10^4°C. Once the system has reached room temperature, it is cooled [56]. Significantly, a wide range of CVD methods has been created, namely oxygen-assisted CVD, radio-frequency water-assisted CVD, CVD (RF-CVD), hot filament (HFCVD), microwave plasma (MPECVD), and thermal or plasma (PE) oxygen-assisted-catalytic chemical deposition (CCVD) [60].

4.3.1 Carbon Allotrope Synthesis in Zero Dimensions (0-D)

4.3.1.1 Synthesis of Fullerenes and Onion Carbon

By vaporizing graphite using arc and plasma discharges or laser irradiation, C_{60}, and other fullerenes can be created. Naphthalene pyrolysis and hydrocarbon combustion are two alternatives. The latter method was developed by Howard et al. and is helpful for the mass synthesis of fullerenes for commercial purposes [61]. All these techniques produce crude products with a minor fraction of fullerenes, with C_{60} being the most abundant. Simple extraction with benzene or toluene is used to separate the fullerene mixture from the initially generated soot, and then column or liquid chromatography is used to separate the required components. Despite using cheap and plentiful raw materials like graphite, the main drawbacks of most fullerene preparation methods include their low yields and the difficulty in isolating and purifying the desired products, which significantly raises the cost and environmental impact of fullerene production [42,62].

The carbon source vaporization approach is inefficient because the yield is so low (less than 1%). A carbon source, including graphite, is laser-irradiated. Temperatures of 1300°C and pressures up to 1 kbar are required for this reaction [2]. This process yields C_{60}, C_{70}, and trace amounts of C_{84} fullerenes. It is difficult to purify these fullerenes. The commercial purpose of a single type of fullerene is challenging in the hostile environment necessary for fullerene synthesis. There are several additional techniques, such as the radio-frequency plasma technique or the arc plasma approach, that can be used to synthesize fullerenes on a commercial scale. The disadvantages of using laser irradiation are present in even commercial production. Pallavolu et al. developed porous nitrogen-doped carbon nano-onions (N-CNO) using a different ingenious, straightforward, inexpensive, one-step pyrolysis method. Numerous physical and electroanalytical investigations were used to describe the produced material. The N-CNO material's graphitic character was confirmed by the D, G, and 2-D bands from Raman and the pyridinic-, pyrrolic-, graphitic-, and pyridinic-N-O from X-ray photoelectron spectroscopy. When N-CNO-containing materials were tested as grown at a current density of 1A/g, the findings showed a specific capacitance of 234 Fg^{-1}. Additionally, 98% of the N-CNO was retained after 5000 extended charge/discharge cycles. The operating voltages of the created asymmetric devices were increased to 1.6 V in 2 M KOH solution, resulting in significant specific capacitances of 79 F g1 and greater energy densities of 27.2 Wh kg^{-1} at 1 Ag^{-1}, as shown in Figure 4.9 [63].

First, onion carbon structures were created by bombarding CNTs with strong electron beams. Carbon soot is heated under vacuum at 2100–2250 °C to produce macroscopic quantities of onion carbon structures. The resulting nanostructures contain several non-spherical forms and 28 graphenic shells with sizes of 310 nm. By using an arc discharge between graphite electrodes submerged in water, Sano et al. [64,65] and Alexandrou et al. [66] generated high-quality spherical onion carbon structures with mean diameters of 25–30 nm on a milligram scale. A considerable number of spherical onion carbon structures with diameters between 10 and 20 nm have also been created by Cabioc'h et al. [67,68] by implanting carbon ions into metal matrixes like silver.

4.3.1.2 Synthesis of Carbon Dots

By accident, C-dots were found by Xu et al. [69] while purifying SWNTs made utilizing the arc discharge process using gel electrophoresis. In recent years, a wide range of procedures and methods for producing C-dots have been developed, reflecting the growing interest in carbon nanostructures

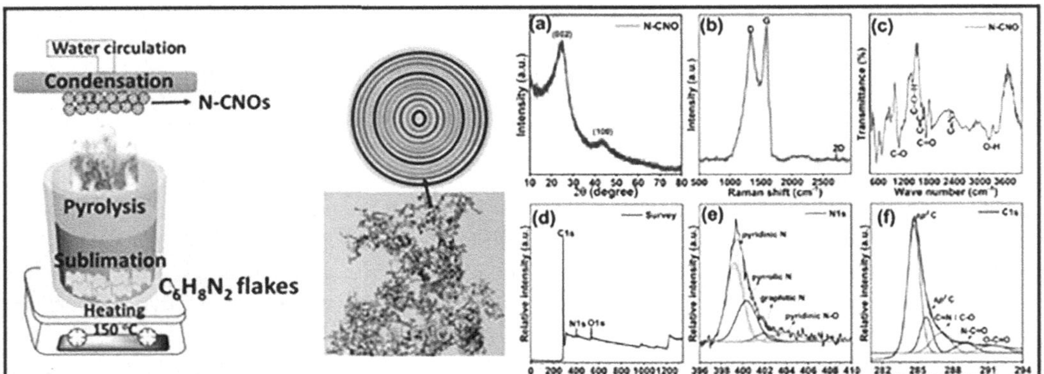

FIGURE 4.9 (a) Graphical presentation of synthesis of highly efficient nitrogen-doped carbon nano-onions, (b) Raman spectra and its band structure of nitrogen-doped carbon nano-onions [63].

and their interesting features. Li et al. [70] have provided a detailed description of the synthesis, characteristics, and uses of C-dots. They are normally created utilizing a top-down method that relies on laser ablation of properly prepared mixes of cement and graphite powder. The C-dots' fundamental structure is created by laser ablation, which is followed by an oxidative acid treatment that enriches the surfaces of the newly generated dots with reactive oxygen groups. Thereafter, organic molecules and oligomers are trapped on the surface of the dots via a surface-passivation phase. A typical diameter of 5 nm [71] or so characterizes the produced nanoparticles. A straightforward electrochemical device with two graphite electrodes in clean water is used in an intriguing alternative technique for C-dot synthesis [72]. The surface of the dots is then passively treated, trapping organic molecules and oligomers there. The resultant nanoparticles are roughly 5 nm in diameter on average. The use of a straightforward electrochemical set-up with two graphite electrodes in pure water is an intriguing alternative method for C-dot production. C-dots have been made using graphite electrodes or MWNTs as the carbon source using similar electrochemical techniques.

Other techniques for producing C-dots are based on the thermal breakdown or carbonization of organic molecules in extremely hot solvents, such as "water in oil" emulsions or powders, the pyrolysis of polymer–silica nanocomposites, the ultrasonic carbonization, the solvothermal, the hydrothermal, the microwave, the treatment of carbohydrates, polymers, organic compounds, organosilanes, or natural products. The two procedures that are most frequently employed to create C-dots are hydrothermal and microwave [73–77]. A very recent alternative technique for making C-dots relies on the microwave-assisted pyrolysis of various chemical molecules that serve as carbon precursors. Glucose and liquid ammonia combinations can be ultrasonically processed to create C-dots as well [78]. This technique yields highly scattered C-dots with lattice spacings of approximately 3.43 and sizes of about 10 nm, which are indicative of a graphitic structure. By chemically reducing carbon tetrachloride in reverse micelles of tetraoctylammonium bromide, Linehan et al. recently demonstrated a straightforward solution phase synthesis of highly luminous C-dots [79].

4.3.1.3 Synthesis of Graphene Quantum Dots

Two different methodologies have served as the foundation for the development of numerous GQD preparation techniques [80–83]. In the "top-down" method, graphitic structures such as carbon black, coal, graphene oxide, graphite fibers, MWNTs, SWNTs, graphite [84–86], or graphene are sliced using cutting techniques to create GQDs [21]. On the other hand, in the "bottom-up" method, GQDs are put together from big aromatic molecules, as presented in Figure 4.10 [80].

High-resolution electron-beam lithography was used to carve graphene as one of the earliest techniques for GQD manufacturing. This process calls for extremely specialized equipment and has limited yields [21]. GQDs can also be produced by hydrothermally treating micrometer-sized partly reduced graphene oxide nanosheets. This technique involves oxidizing reduced graphene oxide nanosheets using potent acids and heating them to 200 °C, which significantly shrinks their size. A dialysis membrane is used to separate the resulting ultra-small GQDs, producing flat nanoparticles with a diameter of approximately 10 nm and one-third graphenic layers. This method's process of cutting is thought to involve the creation of rows of epoxides and carbonyls on the graphene oxide surface during oxidation, followed by a stage in which the graphene sheets are separated into numerous smaller pieces [87].

Coal may be employed as a readily available and affordable starting material for the effective synthesis of GQDs, according to a recent study by Ye et al. [88]. Graphite has also been electrolyzed to create GQDs. For instance, the Zhang et al. method uses an anode made of a graphite rod submerged in an alkaline solution to electrolyze GQDs [84]. After reduction with hydrazine and electrolysis, GQDs are produced that are light brown and can be centrifuged to separate them from the remaining graphite. Using a dialysis membrane, the electrolyte and leftover hydrazine are exchanged for pure water. These techniques result in GQDs with 5–10 nm sizes.

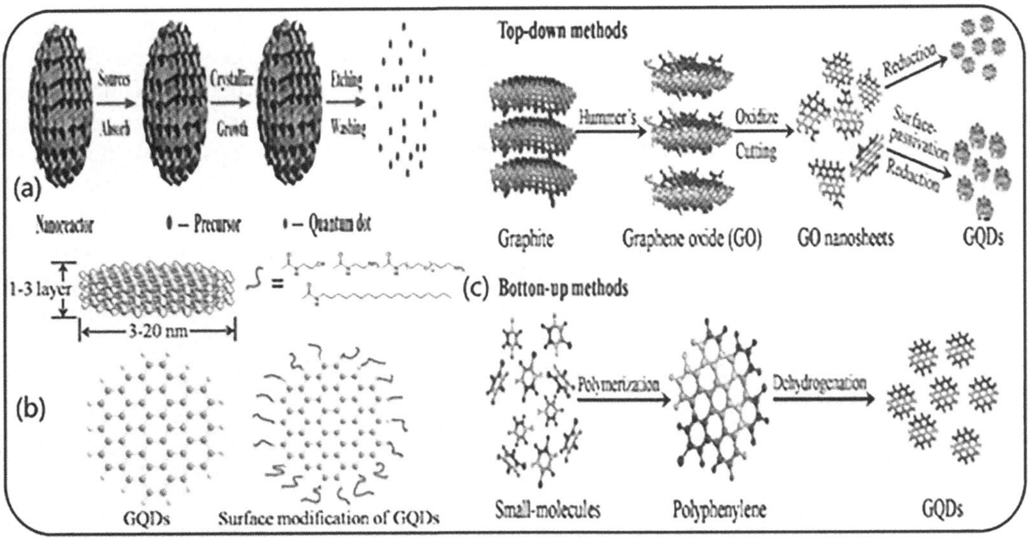

FIGURE 4.10 (a) Representation of confined reaction in nanoreactors for the synthesis of quantum dots. (b) GQDs and structural models during surface-passivation reagent functionalization. (c) Diagram showing the top-down and bottom-up approaches to GQD synthesis [80].

Lee et al. reported on a unique strategy for synthesizing hexagonal-shaped size-controlled single-crystalline GQDs as well as a breakthrough in the use of catalytic solution chemistry to create 2-D and 3-D graphene single crystal structures from d-glucose. Through the use of size-controlled single-crystalline GQDs, we were able to determine the critical part edge states play in luminescence by examining the relationship between their crystalline size and exciton lifespan. Additionally, light-emitting diodes that use blue-emissive single-crystalline GQDs as an emitter provide stable deep-blue emission regardless of voltage and doping level (see Figure 4.11) [89].

4.3.1.4 Synthesis of Diamonds

Diamonds are produced using several techniques, and these processes affect their size, shape, and quality. Detonating an explosive mixture of carbon-containing substances like trinitrotoluene and

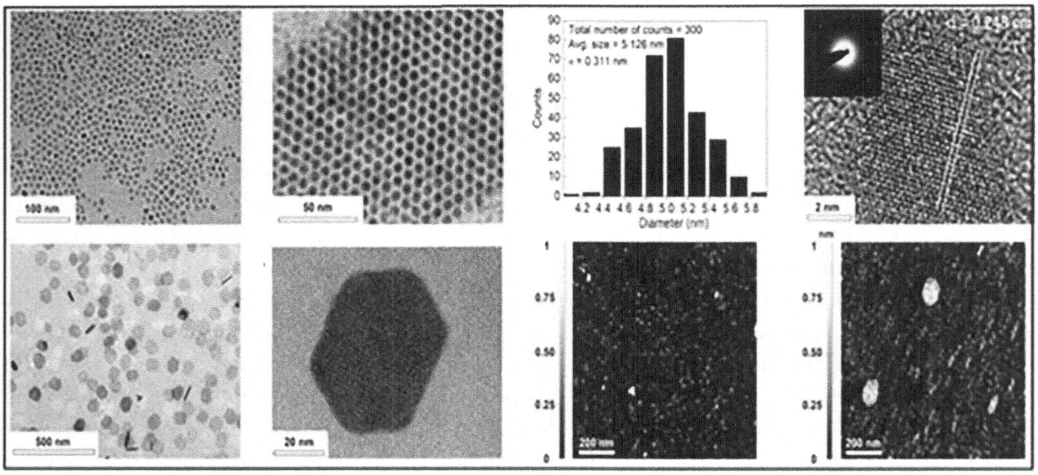

FIGURE 4.11 TEM images, AFM images, and topographical images of 5 and 70 nm size GQD [89].

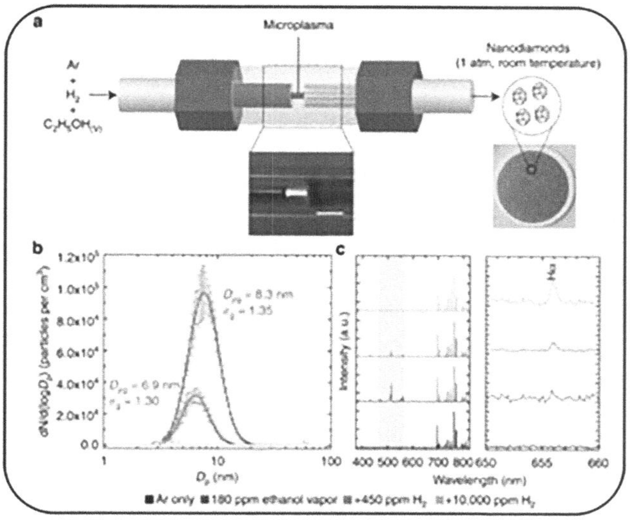

FIGURE 4.12 A mixture of Ar, H_2 gas and ethanol vapor is constantly dissociated in an atmospheric pressure microplasma to synthesize carbon nanoparticles in the gas phase: (a) A commercial scanning mobility particle sizing (SMPS) device was used to classify aerosols in situ to monitor particle production. (b) OES tracks the dissociation of ethanol vapors. (c) The C_2 band has been associated with the creation of nanodiamonds [97].

hexogen (1,3,5-trinitro-1,3,5-triazinane) is the most well-known technique for mass-producing diamonds [90–92]. Up to 75% of the resulting soot is typically composed of diamonds with diameters of 4–5 nm and a somewhat narrow size distribution. The diamonds have a tendency to group into larger agglomerates, which makes them difficult to handle or filter because they are not dispersible in water or organic solvents. The dispersion of the diamonds is further complicated by the fact that these agglomerates frequently have a layer of graphitic material on them [93].

Using ceramic microbead milling or ultrasonication, these agglomerates can be mechanically eliminated. Diamonds treated in this way become extremely hydrophilic, making it simple to disperse them in polar solvents like water or DMSO [94,95]. From diamond microcrystals, it has also been possible to create nanodiamonds with a decent yield using a high-energy ball milling process [27,96]. Based on the dissociation of ethanol utilizing a microplasma process in an inert atmosphere, Kumar et al. established a straightforward and affordable alternative approach for generating high-purity nanodiamonds, as shown in Figure 4.12 [97].

4.3.2 Carbon Allotrope Synthesis in One Dimension (1-D)

4.3.2.1 Synthesis of Carbon Nanotubes and Nanofibers

Although they have the same basic structure as graphene, the synthesis of CNTs is very different from that of graphene. SWNTs are similar to cylindrical graphene sheets. Several tried-and-true methods can be used to create CNTs [98]. The easiest approach uses an arc discharge among two carbon electrodes inside of a container filled with inert gas [99]. The amount, purity, and quality of the nanotubes created by this procedure depend on the kind and pressure of the inert gas. Other volatile organic molecules can be added to the chamber in conjunction with gases like He, CH_4, and H_2 to change the properties of the resulting nanotubes. MWNTs are produced by arc discharges without the use of a catalyst [100–102]. SWNTs can, however, be created if the electrodes are made of graphite that has catalytic nanoparticles of metals including Ni, Fe, Co, Pt, and Rh or their alloys. The first two methods were developed by Iijima, Ichihashi, and Bethune et al., who each used Co nanoparticles as catalysts [3–5].

Other methods of producing DWNTs include pulsed arc discharges with a Y/Ni alloy as the catalyst and the arc discharge method in the presence of H_2 or Ar utilizing a catalyst mixture of Ni, Co, Fe, and S. By abusing a graphite target with a pulsed laser in a chamber with a regulated environment [103], CNTs have been created. Nd: YAG or CO_2 lasers are used for ablation, and the graphite target often contains catalytic nanoparticles of Ni or Co or a combination of the two [104].

The CVD process over a metal catalyst is the most effective way to make CNTs currently available. The metal catalyst is often composed of nanoparticles of Fe, Co, or Ni, or an alloy of these metals, and the carbon vapor is typically one of the following: methane, ethane, acetylene, ethylene, an H_2/CH_4 combination, or ethanol. Following catalytic nucleation, the carbon vapors flow through the catalytic nanoparticles and are thermally or plasma irradiated decomposed, supplying carbon atoms for the creation of CNTs [105–107]. The composition and morphology of the catalytic nanoparticles, the carbon source, the kind of substrate, and the decomposition temperature are the most crucial factors affecting the quality, yield, purity, and characteristics of the final CVD product [106]. Dinh et al. investigated the utilization of in situ miniemulsion polymerizations as well as various physical mixing techniques for the creation of nanocomposite films, including MWNTs and statistical copolymers made of styrene and n-butyl acrylate. Simple drop-casting of the produced nanocomposite latex allowed for the fabrication of nanocomposite films at room temperature. The latex stability and physical characteristics of the resultant nanocomposite films are shown to be significantly influenced by the production process as well as compositional variables like the kind of surfactant. Miniemulsion polymerized films showed excellent stretchability, as shown in Figure 4.13 [108].

4.3.2.2 Synthesis of Carbon Nanohorns

By using an arc discharge technique or a CO_2 laser to vaporize a graphite target at room temperature, SWNHs are formed [109–112]. As opposed to SWNH aggregates created by CO_2 laser ablation, it has been noted that when using the arc discharge technique, SWNH aggregate size is frequently lower. Unlike CNTs, SWNHs can be made in the absence of a metal catalyst with excellent yields and purity. This indicates that SWNHs are typically totally free of metal contaminants and relatively easy to produce. The kind and pressure of the ambient gas that is employed in the laser ablation have been discovered to have an impact on the structure of individual SWNHs and the size of aggregates [33].

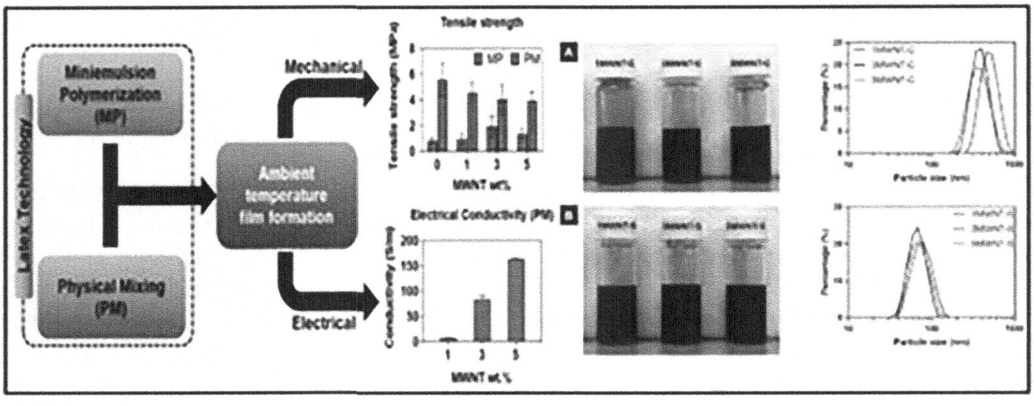

FIGURE 4.13 Synthesized from surfactant-free emulsion polymerization (SF) and miniemulsion polymerization (MP) [108].

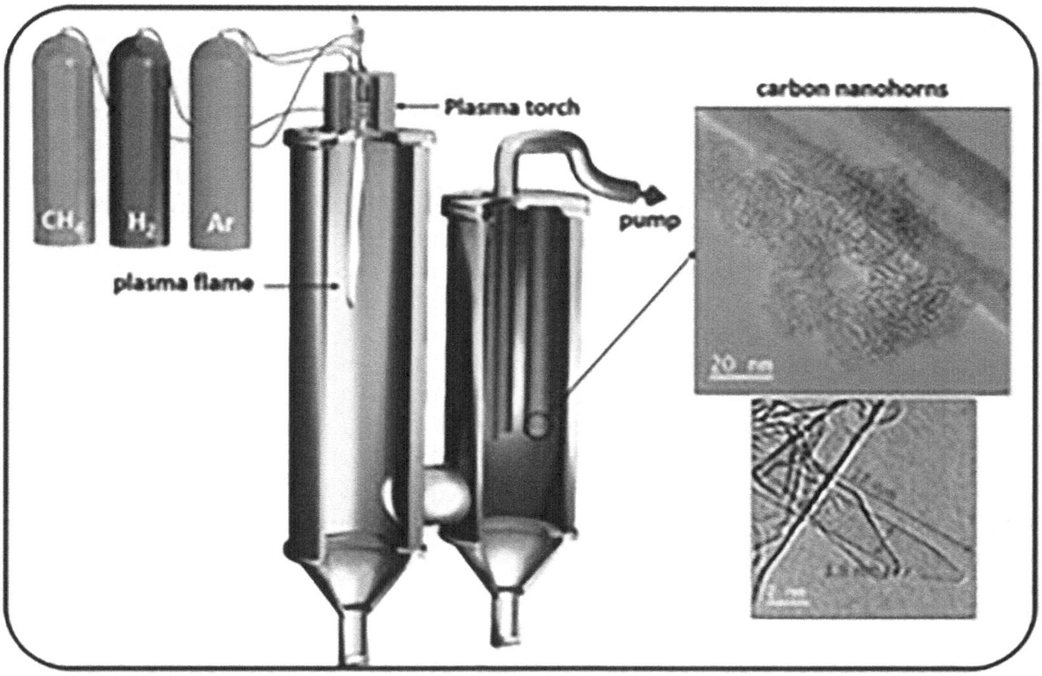

FIGURE 4.14 Carbon nanohorns are formed using inductively coupled plasma [113].

Casteignau et al. employed inductively coupled plasma (ICP) and gaseous precursors to show a new method for producing carbon nanohorns (CNHs) in a reactor. According to this research, pressure has the most influence on the selectivity of carbon nanostructures: pressures below 53.3 kPa favor the growth of carbon nanoflakes, whereas pressures of 66.7 kPa and above encourage the production of CNHs. The reaction also depends on the ratio of methane to hydrogen as well as the overall concentration of $CH_4 + H_2$ inside the plasma flame. By injecting H_2 and CH_4 in a 1:2 ratio at an output pressure of 82.7 kPa at a rate of 20 g/h, CNHs are preferentially produced [113] (Figure 4.14).

4.3.3 Carbon Allotrope Synthesis in Two Dimensions (2-D)

4.3.3.1 Synthesis of Graphene

By liquid exfoliating graphite with ultrasonication in particular organic solvents with surface tensions of under 40 mJ/m^2, graphene monolayers can be isolated [114]. Many other solvents, including N-methyl-2-pyrrolidone, perfluorinated aromatic solvents, pyridine, dimethylformamide, and o-dichlorobenzene, have been used to prepare fine dispersions of graphene. Surfactant molecules are sometimes utilized to keep graphene in liquid solutions stable [115–118]. Cai et al. have evaluated the various processes for exfoliating graphite [119]. This method often produces single graphene monolayers as well as several multi-layered carbon nanosheets with ten or fewer monolayers in the scattered carbon material. Compared to what is possible with Novoselov's micromechanical approach, these technologies allow for the creation of remarkably large amounts of graphene. The reduction of graphene oxide can also be used to create graphene nanoplatelets. According to studies by Brodie [120], Staudenmaier [121], and Hummers and Offeman [122,123], among others, graphene oxide is created by oxidizing graphite with potent oxidizing agents. Epoxy, carboxy, and hydroxy groups are all added to the graphenic surface via oxidation. Graphene nanoplatelets can be created by partially reducing graphene oxide nanoplatelets, which are easily exfoliated in water. For this aim, a wide range of reducing substances have been employed, including hydrazines, hydrides, hydroquinone, and titanium oxide under ultraviolet radiation [124]. Heating in an inert atmosphere is another way to reduce graphene oxide.

Other methods of producing graphene include those based on CVD, the intercalative expansion of graphite, thermal annealing of SiC, and others. Graphenic nanosheet production using an interesting bottom-up technique was recently published by Turchanin et al. [125]. In this procedure, biphenyl molecules are self-assembled on a substrate, cross-linked by electron irradiation to form atomically thin carbon nanosheets, and then annealed to form graphite. Following annealing at 900 °C, graphitization dramatically lowers the sheets' resistivity and increases their mechanical strength. By adjusting the size of the exposed surface and using nanolithographic processes, this technology has the main advantage of enabling the creation of nanosheets with predetermined shapes and sizes, ranging from the macroscopic to the nanoscopic. Choi et al. [126] discuss three cutting-edge mass manufacturing techniques, as well as their shortcomings and prospective upgrades. According to a study by Abdelkhalek et al., functionalized silica nanoparticles on graphene oxide nanosheets (FGS) were created using a quick, inexpensive, and efficient synthesis method for the adsorption of trypan blue (TB) and bisphenol A. (BPA). The effective chemical activation of rice husk ash (RHA) with KOH at a high autoclaving temperature to produce FGS has been demonstrated, as shown in Figure 4.15 [127].

4.3.3.2 Synthesis of Multi-Layer Graphitic Nanosheets

When graphite is ultrasonically exfoliated in liquid form in a variety of solvents and when synthesizing graphene via the CVD method, MGNs and graphene monolayers are both formed [128–130]. MGNs are sometimes dismissed as simple leftovers of graphene manufacturing due to the widespread interest in graphene monolayers. They might, nevertheless, be helpful in a variety of situations. Low yields are the primary drawback of ultrasonic techniques for the manufacturing of MGN. The air pyrolysis of organic compounds like betaine is a different method of making

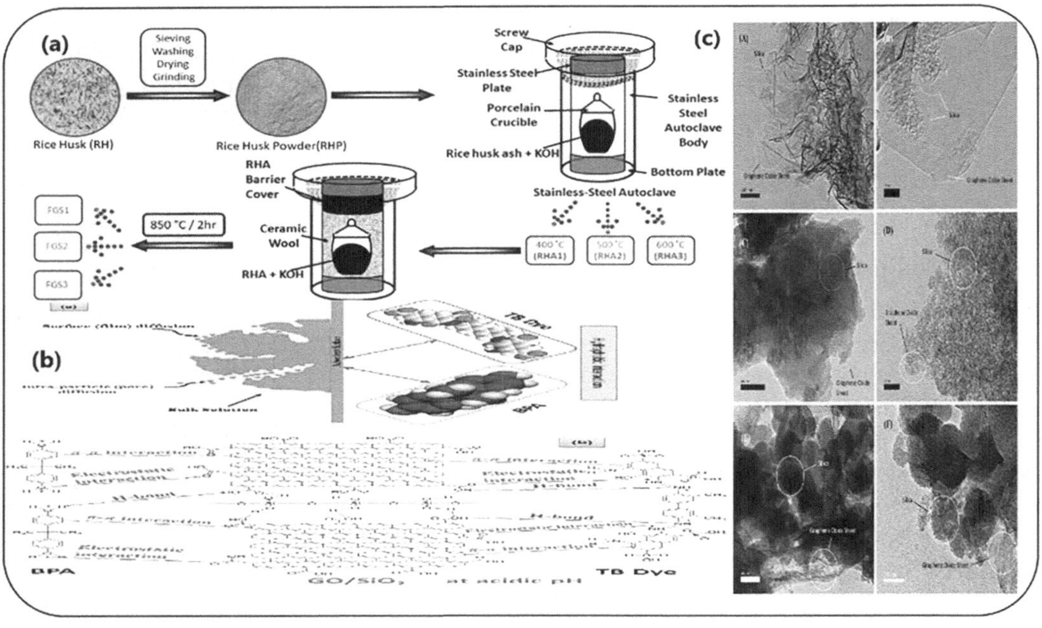

FIGURE 4.15 (a) System set-up diagram for the fabrication of GO/SiO_2 nanocomposites, (b) Pore filling and hydrophobic interaction, π-π interaction and H-bonding mechanism, (c) TEM images of GO/SiO_2 nanocomposites [127].

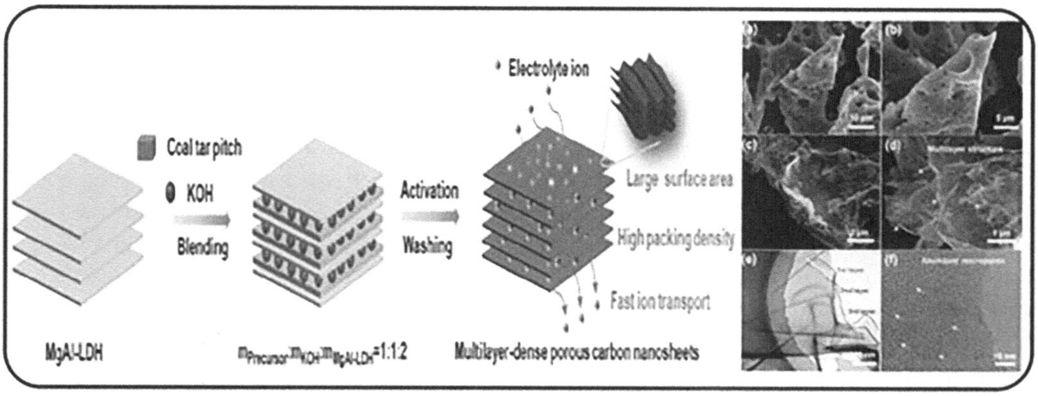

FIGURE 4.16 (a) 2-D space-confined method for making multi-layer-dense porous carbon nanosheets (MPCN), (b) SEM, TEM, and HR-TEM images of the MgAl-LDH [132].

MGNs. This straightforward, cost-effective, and easily scalable process yields ultrathin carbon nanosheets with thicknesses of much less than 10 nm. Glucose can be hydrothermally carbonized at the air-liquid interface using sodium dodecyl sulfate as a template, or carbon tetrachloride can be treated solvothermal to create carbon nanosheets [131]. Che et al. presented a simple 2-D space-confined approach for fabricating multi-layer-dense porous carbon nanosheets (MPCN) by exploiting the stacking effect of rapid-drying MgAl-layered double hydroxides (MgAl-LDH). In comparison to the standard activation technique, the inclusion of MgAl-LDH improves sample yield and packing density of porous carbon materials (see Figure 4.16) [132].

Using a hybrid covering of amorphous alumina and carbon, which was simultaneously embedded in a porous framework of in situ exfoliated graphene/graphite nanosheets (GGNs), Wang et al. demonstrated how ball-milled silicon particles of submicron size were strategically changed. The composite displays improved electrochemical performance, including higher rate capability and high cycle stability (see Figure 4.17) [133].

4.3.3.3 Synthesis of Graphene Nanoribbons

As was already indicated, MWNTs are frequently chemically unzipped during treatment with sulfuric acid and potassium permanganate, leading to the formation of graphene nanoribbons (GNRs). The intensely oxidizing conditions lead the graphitic layers of the tube to separate one at a time, releasing them as GNRs that are enriched with oxygen at the edges and the bulk surface in

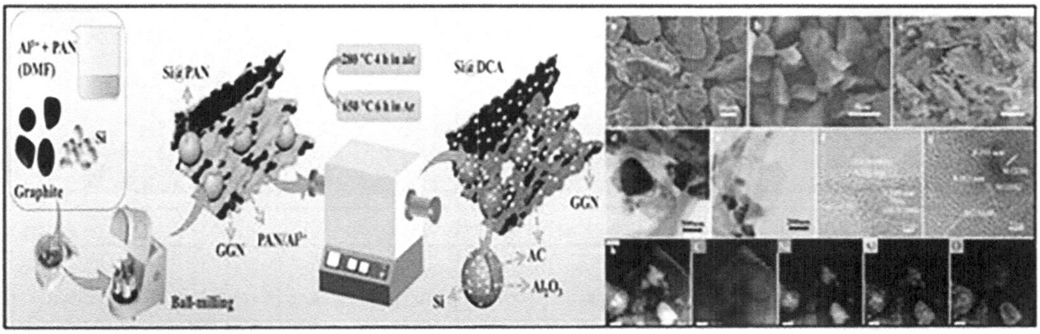

FIGURE 4.17 Schematic representation of the preparation of graphene/graphite nanosheets using ball-milled silicon with amorphous alumina and carbon hybrid coating embedded [133].

the form of carbonyl, carboxyl, and hydroxy groups. These GNRs have low electrical conductivity as a result of the aromatic system's disruption at the oxidized sites. Using hydrazine for reduction or annealing under H_2 [37], however, can significantly enhance it. Additionally, Jiao et al. have demonstrated that MWNTs may also be calcined in air and then sonicated to produce graphene nanoribbons [134,135]. The electrical unwrapping of MWNTs or high-pressure treatment of MWNTs, lithographic methods, methods based on organic synthesis, and the CVD technique are only a few other ways of creating carbon nanoribbons that have been documented. In the review study by Terrones et al., methods for making graphene nanoribbons by unrolling or unzipping CNTs were described in great detail [38]. It turns out that the degree of nanotube crystallinity and uniformity in the length, diameter, and several layers of CNTs are what control the appropriate edge structure and crystallinity of graphene nanoribbons made from those tubes. Scaling-up of catalytic cutting offers the possibility of producing graphene nanoribbons with high yields.

With the use of exfoliation, short sonication, centrifugation to remove big particles, and smooth-edged GNR preparation, Li et al. created a technique for graphite. Using electrochemical exfoliation with ionic-liquid assistance [136], Lu et al. produced GNRs from graphite electrodes as well as graphene nanoparticles and graphene [137]. Finally, using a bottom-up strategy, polycyclic aromatic hydrocarbons may be condensed to create graphene nanoribbons. Dobner et al. proposed a unique diffusion-controlled on-surface methodology for graphene nanoribbons (GNR) based on two types of precursor molecules that utilize different disparities in the top mobilities of the precursors. This method is a step forward in the controlled manufacturing of complicated GNR heterostructures, and it should be relevant to a synthesis of a wide range of GNR heterojunctions (see Figure 4.18) [138]. Way et al. demonstrated that when combined with growth processes that selectively extend seeds in a single direction, molecular-scale carbon seeds may be used to commence the CVD synthesis of graphene to produce one-dimensional graphene nanoribbons that are smaller than 5 nm (see Figure 4.19) [139].

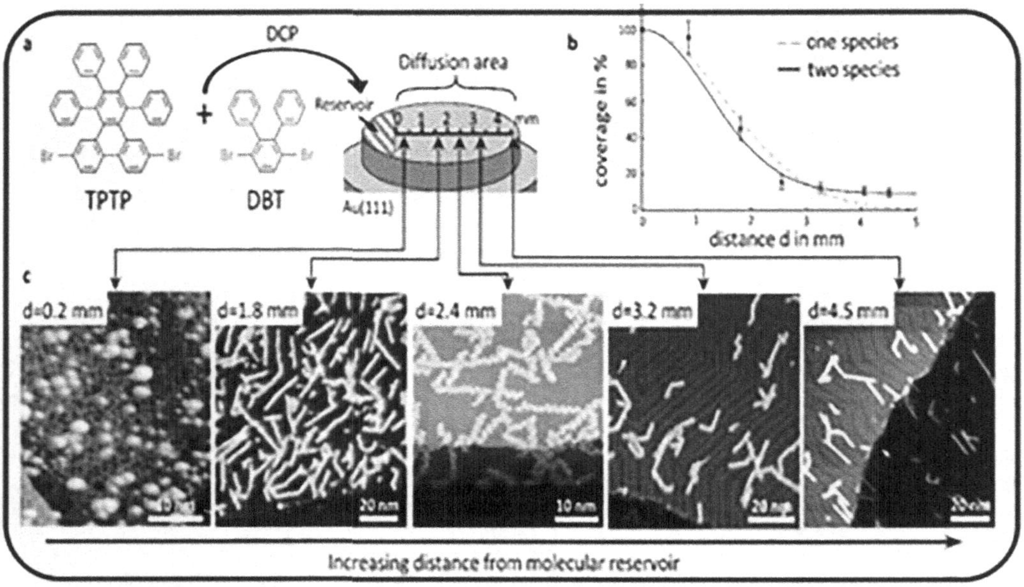

FIGURE 4.18 The technique of deposition and experimental findings of density gradient-driven GNR self-assembly (a) Precursor monomer structure and deposition principle graphic, (b) GNR surface coverage, (c) STM images [138].

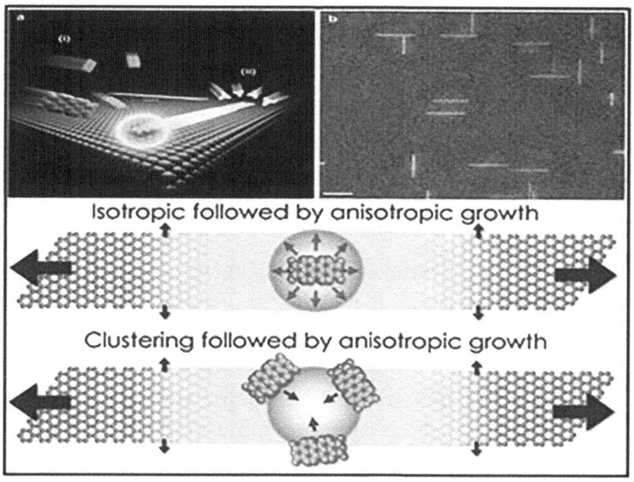

FIGURE 4.19 (a) Nanoribbon fabrication from seeds generated from molecules, (b) Two potential nanoribbon starting paths from PAH-derived seeds [139].

4.4 PROPERTIES FUNCTIONALIZATION OF CARBON ALLOTROPES

Graphene and CNTs are the most popular graphitic carbon nanoallotropes due to their exceptional mechanical and electrical capabilities, and especially because they can transmit these beneficial qualities to composite materials on a macroscopic scale, creating sophisticated hybrid systems. When electrons occupy accessible bands up to the Fermi level (i.e., when they are delocalized electrons), they behave as though they are imprisoned inside a potential well in a confined dimension. In contrast, in a nonconfined dimension, electrons are represented as a Fermi gas with a band structure. Only one dimension is restricted in 2-D nanomaterials like graphene, whereas the other two are unconfined. Two dimensions are constrained while the third is unconstrained in 1-D nanomaterials like nanotubes. All three dimensions are constrained in 0-D nanomaterials, such as dots and nanodiamonds. Constrained and delocalized effects coexist in 1-D and 2-D nanomaterials, resulting in heterogeneous behaviors with continuous and discontinuous densities of states. Consequently, intriguing electronic features are created by the growth of size-dependent band gaps, peaks, and/or singularities with odd electronic states. The density of states in 0-D nanomaterials is discontinuous, with large gaps that are notably visible in optical characteristics (e.g., PL). In addition, peaks in the density of states that are typically spin polarized and appear as localized magnetic moments may be caused by surface flaws, unsaturated bonds, or edge profiles.

The 2s, $2p_x$, $2p_y$, and $2p_z$ orbitals are occupied by the valence electrons in carbon and are essential for the formation of covalent bonds in carbon nanomaterials. As the binding energy of the chemical bonds is minimal compared to the energy difference between the higher energy 2p levels and the lower energy 2s levels in carbon, the electronic wave functions for these four valence electrons can easily mix. As a consequence, sp, sp^2, and sp^3 bonding states are produced depending on the degree of hybridization by altering the occupancy of two 2s and three 2p atomic orbitals. Different structural arrangements are caused by different bonding states: sp bonding favors chain structures, sp^2 bonding results in planar structures, and sp^3 bonding encourages tetragonal forms.

4.4.1 Fundamental Properties of C_{60}

Due to its molecular structure and small size, pure C_{60} lacks several of the fundamental properties of other carbon nanostructures, such as conductivity and mechanical strength. Although C_{60} lacks these characteristics, its structural shape nonetheless makes it an extremely potent radical scavenger. Due to its fast reactions with all forms of free radicals, its spherical shape, and its electron-deficient nature with 30 C=C bonds, C_{60} has been dubbed a "radical sponge." Another crucial quality of C_{60} is its ability to serve as an electron acceptor in the donor-acceptor components of energy conversion systems. This capability is correlated with its high electron affinity and low reorganization energy [140–146].

4.4.2 Photoluminescence Properties of Carbon Dots and Graphene Quantum Dots

The most fascinating property of C-dots is PL. This feature, along with their size, water dispersibility, and extremely low toxicity, makes them versatile in their potential applications, especially in biology and nanomedicine. Both the C-dots generated by the acid treatment method and the laser ablation method do not exhibit PL. PL is only visible when the dots have achieved surface passivation in both cases. This is done by adding oxygen-containing functional groups during the acidic treatment stage, which covalently bond various organic molecules to the surfaces of the dots [147].

4.4.3 Properties of Diamonds

A type of sp^3 carbon with a diamond crystal structure and particles that are generally 5 nm in size is referred to as a diamond. The spherical shape of their particles is consistent with the idea that they were created from liquid droplets [148]. Diamonds have been said to be stable if their surfaces are stabilized by terminating with functional groups (different oxygen-containing groups in most cases) or made to adopt sp^2 hybridization [149]. The core-shell structure of diamonds is consequently frequently described as consisting of a diamond core (sp^3 carbon atoms), a middle layer (sp^2+x carbon atoms), and a graphitic outer shell (sp^2 carbon atoms) with dangling bonds terminated by functional groups. The majority of impurities in diamonds take the form of substituted non-carbon atoms [148].

4.4.4 Mechanical Properties of Graphene, CNTs, and Carbon Nanofibers

Graphene nanoplatelets and CNTs are the strongest and most flexible materials now in use because of the powerful covalent C-C bonds that exist between neighboring sp^2 carbon atoms. The astonishing superplasticity of SWNTs – which caused them to grow approximately 280% longer and 15 times thinner before breaking – is accounted for by the formation and mobility of kinks in the structure [150,151]. It was discovered that graphene has Young's modulus of 1.0 TPa [152]. CNTs and graphene offer a variety of nonlinear mechanical qualities, including adaptability, buckling instability, and fracture strength. They have been discovered to be strong, enduring tremendous stresses without displaying signs of brittleness, flexibility, or atomic rearrangements.

4.4.5 Electronic and Other Physical Properties of Carbon Nanohorns, CNTs, Nanoribbons, and Graphene

The density of states in graphene is zero at the Fermi energy level, which qualifies it as a zero band gap semiconductor or semimetal. Additionally, the ambipolar property of graphene allows its charge carriers to change from electrons to holes when an applied electric field is present [153,154]. In addition, because of its strong C–C covalent bonds and phonon scattering, graphene has been shown to display exceptional thermal conductivity. It has been discovered that the number of layers and the relative locations of the atoms in the neighboring layers have a significant impact on the electronic characteristics of graphene. Conduction and valence bands noticeably overlap when there are more graphene sheets added, which leads to the appearance of more charge carriers.

Due to the graphene nanoribbons' finite size (i.e., width), unique quantum phenomena are anticipated to develop as a result of the confinement effect. According to theory, the width and edge topology of graphene nanoribbons have a major impact on their electrical characteristics. According to the band gap width, armchair graphene nanoribbons may be divided into three groups; as the width rises, the band gap shrinks until it reaches a zero value, which is a hallmark of 2-D graphene. Nanoribbons made of armchair graphene are only semiconducting. If armchair graphene nanoribbons have a multi-layer configuration, the band gaps narrow as both metallic and semiconductor activity emerges [155]. The band structure produced by zig-zag graphene nanoribbons with edges is considerably changed by the edge states, which introduce a flat band around the Fermi level that extends along the nanoribbon edge [156,157]. The extreme degeneracy of the edge states causes them to be spin-polarized. The band gap for zig-zag nanoribbons narrows with width like that of armchair nanoribbons. In contrast to their armchair counterparts, single-layer zig-zag graphene nanoribbons can turn metallic when their width reaches a certain level [38]. Additionally, the edge states are linked to the creation of magnetic moments, leaving zig-zag graphene nanoribbons with exceptional magnetic characteristics. CNTs, or SWNTs or MWNTs, which may be thought of as cylindrically fixed graphene monolayers or multi-layers, share many of the same features as graphene. CNTs may be either metals or semiconductors because of how the graphene monolayer is rolled up to form the cylinder, which determines the conductivity of the material. In other words, the diameter and chirality of CNTs substantially influence their electrical properties.

4.4.6 Magnetic Properties of Carbon Allotropes

It is well known that the valence state of carbon may be represented as a linear combination (mixing) of the 2s and $2p^3$ orbitals, producing an sp^3- or sp^2-hybridized orbital depending on the number of bonds formed among the carbon atoms (four single bonds in the case of sp^3 hybridization, and three bonds (one double bond and two single bonds) for sp^2 hybridization). Another sign of sp hybridization in carbon is the presence of one single bond and one triple bond. When no electrons are left unpaired, carbon acts in a diamagnetic way as a result of bonding. Diamond's magnetic susceptibility is governed by three terms: two diamagnetic contributions from the core and valence electrons, as well as a paramagnetic factor that results from magnetic dipole transitions between the valence and conduction bands and is symmetrically influenced by the chemical bond. Due to the delocalization of electrons travelling through graphitic rings in the plane, the diamagnetic term from valence electrons in graphite is greatly increased. Graphite exhibits the biggest diamagnetic susceptibility of all the documented materials, excluding superconductors, due to its enormous and anisotropic magnetic susceptibility [158].

Fullerenes and CNTs, two more carbon nanoallotropes, are similarly magnetically unattractive. A fullerene structure, which includes both pentagons and hexagons, often controls its short-range magnetic characteristics [158]. CNTs exhibit diamagnetic behavior regardless of their structure (single- or multi-walled); for MWCNTs, the diamagnetic susceptibility is extremely similar to that of incorrectly oriented graphite [159]. According to reports, the magnetic behavior of nanohorns, which are CNTs that are closed at one end, is comparable to that of pure fullerenes and is defined by a magnetic susceptibility that includes both paramagnetic and diamagnetic components [160].

4.4.7 Chemical Reactivity Properties

Carbon allotropes are often regarded as stable materials due to their low reactivity toward chemicals during chemical processing, such as gases or acids. With an electron-poor polyene molecule, C_{60} functions as an unbounded conjugated system. This shows that, unlike other polyenes, its carbon atoms are only bonded to other carbon atoms and do not create covalent links with hydrogen atoms. Due to their broad systems, fullerenes can engage in addition and redox activities.

Despite having a rather stable structure, the spherical form of C_{60} causes its C=C bonds to deviate from planarity, leading to some angle strain in the molecule. This strain is responsible for a considerable percentage of its chemical reactivity. Therefore, C_{60} can be used in the majority of sp^2 carbon-based chemical processes. Because the structural elements of the different carbon nanoallotropes, including CNTs, SWNHs, graphenes, GQDs, and onion carbon structures, are identical, so too are their chemical reactivities.

4.5 SUMMARY AND FUTURE PERSPECTIVES

As the first comprehensive attempt to date, this chapter discussed the systematization of the multiple predicted carbon nanostructures, the existing carbon nanoallotropes, and the structures produced by combining the nanoallotropes. For this reason, we divided these structures into groups primarily based on the dimensionality of the structures (0-D, 1-D, and 2-D) and the hybridization state of the carbon atoms that make it up (sp, sp^2, and sp^3). Additionally, we discussed their traits, the functionalization with their most significant attributes, and existing synthesis techniques.

The technology could undergo a revolution thanks to nanoscale carbon, which could also improve durability and precision while allowing humans to reach previously inconceivable depths. Carbon allotropes are the technology of the future. Despite the success of the synthesis methods, the results are not satisfactory when it comes to the efficient application of carbon allotropes because existing technologies are a little behind what is needed for optimal application.

Carbon is without a doubt the element with the broadest and most diverse range of nanoallotropes. It is feasible to fabricate a broad variety of distinctive configurations utilizing different combinations of sp^2- and sp^3-hybridized carbon atoms, as well as the morphological modifications that sp^2 graphene sheets undergo and the results of their size limitations. Additionally, it is feasible to create allotropic hybrids, mixed nanoarchitectures, and new superstructures with a range of dimensionalities, traits, and potential applications by combining these nanostructures in different ways. Many additional carbon nanoallotropes and composite structures have been projected to exist (especially at high pressures), in addition to the large variety of carbon nanoallotropes with composite structures that have been characterized experimentally. These materials are anticipated to have exceptional characteristics and exciting potential uses.The authors declare that we have no conflict of interest. No funding was received to assist with the preparation of this manuscript.

REFERENCES

1. Kroto, Harold W., James R. Heath, Sean C. O'Brien, Robert F. Curl, and Richard E. Smalley. "C60: Buckminsterfullerene." *Nature* 318, no. 6042 (1985): 162–163.
2. Krätschmer, Wolfgang, Lowell D. Lamb, K.H.D.R. Fostiropoulos, and Donald R. Huffman. "Solid C60: A new form of carbon." *Nature* 347, no. 6291 (1990): 354–358.
3. Iijima, Sumio. "Helical microtubules of graphitic carbon." *Nature* 354, no. 6348 (1991): 56–58.
4. Iijima, Sumio, and Toshinari Ichihashi. "Single-shell carbon nanotubes of 1-nm diameter." *Nature* 363, no. 6430 (1993): 603–605.
5. Bethune, D.S., Ch H. Kiang, M.S. De Vries, G. Gorman, R._ Savoy, J. Vazquez, and R. Beyers. "Cobalt-catalysed growth of carbon nanotubes with single-atomic-layer walls." *Nature* 363, no. 6430 (1993): 605–607.
6. Wallace, Philip Richard. "The band theory of graphite." *Physical Review* 71, no. 9 (1947): 622.
7. Boehm, Hans P. "Chemical identification of surface groups." In *Advances in catalysis*, vol. 16, pp. 179–274. Academic Press, 1966.
8. Novoselov, Kostya S., Andre K. Geim, Sergei V. Morozov, De-eng Jiang, Yanshui Zhang, Sergey V. Dubonos, Irina V. Grigorieva, and Alexandr A. Firsov. "Electric field effect in atomically thin carbon films." *Science* 306, no. 5696 (2004): 666–669.
9. Kang, Jun, Zhongming Wei, and Jingbo Li. "Graphyne and its family: Recent theoretical advances." *ACS Applied Materials & Interfaces* 11, no. 3 (2018): 2692–2706.

10. Miller, David C., Soumitra V. Athavale, and Frances H. Arnold. "Combining chemistry and protein engineering for new-to-nature biocatalysis." *Nature Synthesis* 1, no. 1 (2022): 18–23.
11. Wikipedia. Allotropes of carbon [Internet]. Available from: http://en.wikipedia.org/wiki/Allotropes_of_carbon#mediaviewer/File:Eight_Allotropes_of_Carbon.png
12. Yan, Yibo, Jun Gong, Jie Chen, Zhiping Zeng, Wei Huang, Kanyi Pu, Jiyang Liu, and Peng Chen. "Recent advances on graphene quantum dots: from chemistry and physics to applications." *Advanced Materials* 31, no. 21 (2019): 1808283.
13. Pan, Dengyu, Jingchun Zhang, Zhen Li, Chao Wu, Xiumei Yan, and Minghong Wu. "Observation of pH-, solvent-, spin-, and excitation-dependent blue photoluminescence from carbon nanoparticles." *Chemical Communications* 46, no. 21 (2010): 3681–3683.
14. Jiang, Kai, Yuhui Wang, Xiaolu Gao, Congzhong Cai, and Hengwei Lin. "Facile, quick, and gram-scale synthesis of ultralong-lifetime room-temperature-phosphorescent carbon dots by microwave irradiation." *Angewandte Chemie International Edition* 57, no. 21 (2018): 6216–6220.
15. Hu, Chao, Mingyu Li, Jieshan Qiu, and Ya-Ping Sun. "Design and fabrication of carbon dots for energy conversion and storage." *Chemical Society Reviews* 48, no. 8 (2019): 2315–2337.
16. Liu, Meng Li, Bin Bin Chen, Chun Mei Li, and Cheng Zhi Huang. "Carbon dots: synthesis, formation mechanism, fluorescence origin and sensing applications." *Green Chemistry* 21, no. 3 (2019): 449–471.
17. Zhu, S., Y. Song, X. Zhao, J. Shao, J. Zhang, and B. Yang. "The photoluminescence mechanism in carbon dots (graphene quantum dots, carbon nanodots, and polymer dots): Current state and future perspective." *Nano Res.*, 8 (2015) (2): 355–381.
18. Dave, Khyati, and Vincent G. Gomes. "Carbon quantum dot-based composites for energy storage and electrocatalysis: Mechanism, applications and future prospects." *Nano Energy* 66 (2019): 104093.
19. Zhang, Zhengting, Guiyun Yi, Peng Li, Xiuxiu Zhang, Haiyang Fan, Yulong Zhang, Xiaodong Wang, and Chuanxiang Zhang. "A minireview on doped carbon dots for photocatalytic and electrocatalytic applications." *Nanoscale* 12, no. 26 (2020): 13899–13906.
20. Yuan, Fanglong, Ya-Kun Wang, Geetu Sharma, Yitong Dong, Xiaopeng Zheng, Peicheng Li, Andrew Johnston, et al. "Bright high-colour-purity deep-blue carbon dot light-emitting diodes via efficient edge amination." *Nature Photonics* 14, no. 3 (2020): 171–176.
21. Ponomarenko, Leonid A., Fredrik Schedin, Mikhail I. Katsnelson, Ruyan Yang, Ernie W. Hill, Kostya S. Novoselov, and Andre K. Geim. "Chaotic Dirac billiard in graphene quantum dots." *Science* 320, no. 5874 (2008): 356–358.
22. Yan, Xin, Xiao Cui, Binsong Li, and Liang-shi Li. "Large, solution-processable graphene quantum dots as light absorbers for photovoltaics." *Nano Letters* 10, no. 5 (2010): 1869–1873.
23. Neubeck, Soeren, Leonid A. Ponomarenko, Frank Freitag, A.J.M. Giesbers, Ulrich Zeitler, Sergey V. Morozov, Peter Blake, Andre K. Geim, and Kostya S. Novoselov. "From one electron to one hole: Quasiparticle counting in graphene quantum dots determined by electrochemical and plasma etching." *Small* 6, no. 14 (2010): 1469–1473.
24. Li, Liang-shi, and Xin Yan. "Colloidal graphene quantum dots." *The Journal of Physical Chemistry Letters* 1, no. 17 (2010): 2572–2576.
25. Williams, Oliver Aneurin. "Nanocrystalline diamond." *Diamond and Related Materials* 20, no. 5-6 (2011): 621–640.
26. Niwase, K., T. Tanaka, Y. Kakimoto, K.N. Ishihara, and P.H. Shingu. "Raman spectra of graphite and diamond mechanically milled with agate or stainless steel ball-mill." *Materials Transactions, JIM* 36, no. 2 (1995): 282–288.
27. Boudou, Jean-Paul, Patrick A. Curmi, Fedor Jelezko, Joerg Wrachtrup, Pascal Aubert, Mohamed Sennour, Gopalakrischnan Balasubramanian, Rolf Reuter, Alain Thorel, and Eric Gaffet. "High yield fabrication of fluorescent nanodiamonds." *Nanotechnology* 20, no. 23 (2009): 235602.
28. Dahl, J.E., S.G. Liu, and R.M.K. Carlson. "Isolation and structure of higher diamondoids, nanometer-sized diamond molecules." *Science* 299, no. 5603 (2003): 96–99.
29. Kim, Yoong Ahm, Hiroyuki Muramatsu, Takuya Hayashi, Morinobu Endo, Mauricio Terrones, and Mildred S. Dresselhaus. "Fabrication of high-purity, double-walled carbon nanotube buckypaper." *Chemical Vapor Deposition* 12, no. 6 (2006): 327–330.
30. Jorio, Ado, and Gene Dresselhaus. *Carbon nanotubes: Advanced topics in the synthesis, structure, properties and applications*. Edited by Mildred S. Dresselhaus. Vol. 111. Berlin: Springer, 2008.
31. Jackson, Petra, Nicklas Raun Jacobsen, Anders Baun, Renie Birkedal, Dana Kühnel, Keld Alstrup Jensen, Ulla Vogel, and Håkan Wallin. "Bioaccumulation and ecotoxicity of carbon nanotubes." *Chemistry Central Journal* 7, no. 1 (2013): 1–21.

32. Iijima, Sumio, M. Yudasaka, R. Yamada, S. Bandow, K. Suenaga, F. Kokai, and K. Takahashi. "Nano-aggregates of single-walled graphitic carbon nano-horns." *Chemical Physics Letters* 309, no. 3-4 (1999): 165–170.
33. Jorio, Ado, and Gene Dresselhaus. *Carbon nanotubes: Advanced topics in the synthesis, structure, properties and applications*. Edited by Mildred S. Dresselhaus. Vol. 111. Berlin: Springer, 2008.
34. Jorio, Ado, and Gene Dresselhaus. *Carbon nanotubes: Advanced topics in the synthesis, structure, properties and applications*. Edited by Mildred S. Dresselhaus. Vol. 111. Berlin: Springer, 2008.
35. Bourlinos, Athanasios B., Theodore A. Steriotis, Radek Zboril, Vasilios Georgakilas, and Athanasios Stubos. "Direct synthesis of carbon nanosheets by the solid-state pyrolysis of betaine." *Journal of Materials Science* 44, no. 5 (2009): 1407–1411.
36. Sinitskii, Alexander, Ayrat Dimiev, Dmitry V. Kosynkin, and James M. Tour. "Graphene nanoribbon devices produced by oxidative unzipping of carbon nanotubes." *ACS Nano* 4, no. 9 (2010): 5405–5413.
37. Kosynkin, Dmitry V., Amanda L. Higginbotham, Alexander Sinitskii, Jay R. Lomeda, Ayrat Dimiev, B. Katherine Price, and James M. Tour. "Longitudinal unzipping of carbon nanotubes to form graphene nanoribbons." *Nature* 458, no. 7240 (2009): 872–876.
38. Terrones, Mauricio, Andrés R. Botello-Méndez, Jessica Campos-Delgado, Florentino López-Urías, Yadira I. Vega-Çantú, Fernando J. Rodríguez-Macías, Ana Laura Elías, et al. "Graphene and graphite nanoribbons: Morphology, properties, synthesis, defects and applications." *Nano Today* 5, no. 4 (2010): 351–372.
39. Chen, Long, Yenny Hernandez, Xinliang Feng, and Klaus Müllen. "From nanographene and graphene nanoribbons to graphene sheets: Chemical synthesis." *Angewandte Chemie International Edition* 51, no. 31 (2012): 7640–7654.
40. James, Dustin K., and James M. Tour. "Graphene: Powder, flakes, ribbons, and sheets." *Accounts of Chemical Research* 46, no. 10 (2013): 2307–2318.
41. James, Dustin K., and James M. Tour. "The chemical synthesis of graphene nanoribbons—A tutorial review." *Macromolecular Chemistry and Physics* 213, no. 10–11 (2012): 1033–1050.
42. Shetti, Nagaraj P., Amit Mishra, Soumen Basu, and Tejraj M. Aminabhavi. "Versatile fullerenes as sensor materials." *Materials Today Chemistry* 20 (2021): 100454.
43. Guo, Junjie, Xiaomin Wang, Yanli Yao, Xiaowei Yang, Xuguang Liu, and Bingshe Xu. "Structure of nanocarbons prepared by arc discharge in water." *Materials Chemistry and Physics* 105, no. 2–3 (2007): 175–178.
44. Ikegami, Tomoaki, Futoshi Nakanishi, Makoto Uchiyama, and Kenji Ebihara. "Optical measurement in carbon nanotubes formation by pulsed laser ablation." *Thin Solid Films* 457, no. 1 (2004): 7–11.
45. Zhu, Yanjuan, Tianjin Lin, Qiuxiang Liu, Yulian Chen, Guofu Zhang, Huifang Xiong, and Haiyan Zhang. "The effect of nickel content of composite catalysts synthesized by hydrothermal method on the preparation of carbon nanotubes." *Materials Science and Engineering: B* 127, no. 2–3 (2006): 198–202.
46. Kucukayan, Gokce, Rasim Ovali, Serim Ilday, Beril Baykal, Hilmi Yurdakul, Servet Turan, Oguz Gulseren, and Erman Bengu. "An experimental and theoretical examination of the effect of sulfur on the pyrolytically grown carbon nanotubes from sucrose-based solid state precursors." *Carbon* 49, no. 2 (2011): 508–517.
47. Onyancha, Robert Birundu, Uyiosa Osagie Aigbe, Kingsley Eghonghon Ukhurebor, and Perpetua Wanjiru Muchiri. "Facile synthesis and applications of carbon nanotubes in heavy-metal remediation and biomedical fields: A comprehensive review." *Journal of Molecular Structure* 1238 (2021): 130462.
48. Niyogi, Sandip, Elena Bekyarova, Mikhail E. Itkis, Jared L. McWilliams, Mark A. Hamon, and Robert C. Haddon. "Solution properties of graphite and graphene." *Journal of the American Chemical Society* 128, no. 24 (2006): 7720–7721.
49. Kataria, Satender, Stefan Wagner, Jasper Ruhkopf, Aamit Gahoi, Himadri Pandey, Rainer Bornemann, Sam Vaziri, Anderson D. Smith, M. Ostling, and Max C. Lemme. "Chemical vapor deposited graphene: From synthesis to applications." *Physica status solidi (a)* 211, no. 11 (2014): 2439–2449.
50. Vittori Antisari, Marco, Daniele Mirabile Gattia, L. Brandão, Renzo Marazzi, and Amelia Montone. "Carbon nanostructures produced by an AC arc discharge." *Materials Science Forum* 638 (2010): 1766–1771. Trans Tech Publications.
51. Yang, Zhibin, and Jianhua Hao. "Progress in pulsed laser deposited two-dimensional layered materials for device applications." *Journal of Materials Chemistry C* 4, no. 38 (2016): 8859–8878.

52. Cano-Marquez, Abraham G., Fernando J. Rodriguez-Macias, Jessica Campos-Delgado, Claudia G. Espinosa-González, Ferdinando Tristán-López, Daniel Ramírez-González, David A. Cullen, David J. Smith, Mauricio Terrones, and Yadira I. Vega-Cantú. "Ex-MWNTs: Graphene sheets and ribbons produced by lithium intercalation and exfoliation of carbon nanotubes." *Nano Letters* 9, no. 4 (2009): 1527–1533.
53. Zaaba, N.I., K.L. Foo, U. Hashim, S.J. Tan, Wei-Wen Liu, and C.H. Voon. "Synthesis of graphene oxide using modified hummers method: Solvent influence." *Procedia Engineering* 184 (2017): 469–477.
54. Horiuchi, Shigeo, Takuya Gotou, Masahiro Fujiwara, Toru Asaka, Tadahiro Yokosawa, and Yoshio Matsui. "Single graphene sheet detected in a carbon nanofilm." *Applied Physics Letters* 84, no. 13 (2004): 2403–2405.
55. Kyesmen, Pannan Isa, Audu Onoja, and Alexander Nwabueze Amah. "Fullerenes synthesis by combined resistive heating and arc discharge techniques." *SpringerPlus* 5, no. 1 (2016): 1–7.
56. Szabó, Andrea, Caterina Perri, Anita Csató, Girolamo Giordano, Danilo Vuono, and János B. Nagy. "Synthesis methods of carbon nanotubes and related materials." *Materials* 3, no. 5 (2010): 3092–3140.
57. Guo, Ting, Pavel Nikolaev, Andrew G. Rinzler, David Tomanek, Daniel T. Colbert, and Richard E. Smalley. "Self-assembly of tubular fullerenes." *The Journal of Physical Chemistry* 99, no. 27 (1995): 10694–10697.
58. Thess, Andreas, Roland Lee, Pavel Nikolaev, Hongjie Dai, Pierre Petit, Jerome Robert, Chunhui Xu, et al. "Crystalline ropes of metallic carbon nanotubes." *Science* 273, no. 5274 (1996): 483–487.
59. Rao, Apparao M., E. Richter, Shunji Bandow, Bruce Chase, P.C. Eklund, K.A. Williams, S. Fang, et al. "Diameter-selective Raman scattering from vibrational modes in carbon nanotubes." *Science* 275, no. 5297 (1997): 187–191.
60. Eatemadi, Ali, Hadis Daraee, Hamzeh Karimkhanloo, Mohammad Kouhi, Nosratollah Zarghami, Abolfazl Akbarzadeh, Mozhgan Abasi, Younes Hanifehpour, and Sang Woo Joo. "Carbon nanotubes: Properties, synthesis, purification, and medical applications." *Nanoscale research letters* 9, no. 1 (2014): 1–13.
61. Howard, Jack B., J. Thomas McKinnon, Yakov Makarovsky, Arthur L. Lafleur, and M. Elaine Johnson. "Fullerenes C60 and C70 in flames." *Nature* 352, no. 6331 (1991): 139–141.
62. Anctil, Annick, Callie W. Babbitt, Ryne P. Raffaelle, and Brian J. Landi. "Material and energy intensity of fullerene production." *Environmental Science & Technology* 45, no. 6 (2011): 2353–2359.
63. Pallavolu, Mohan Reddy, Yedluri Anil Kumar, M. Dhananjaya, Bandar Ali Al-Asbahi, Adem Sreedhar, and Sang W. Joo. "Design and synthesis of highly efficient Nitrogen-doped carbon nano-onions for asymmetric supercapacitors." *Journal of Alloys and Compounds* (2022): 165609.
64. Sano, Noboru, Houqi Wang, Manish Chhowalla, Ioannis Alexandrou, and Gehan A.J. Amaratunga. "Synthesis of carbon'onions' in water." *Nature* 414, no. 6863 (2001): 506–507.
65. Sano, N., H. Wang, I. Alexandrou, M. Chhowalla, K.B.K. Teo, G.A.J. Amaratunga, and K. Iimura. "Properties of carbon onions produced by an arc discharge in water." *Journal of Applied Physics* 92, no. 5 (2002): 2783–2788.
66. Alexandrou, Ioannis, Hanguang Wang, Nobuyuki Sano, and Gehan A.J. Amaratunga. "Structure of carbon onions and nanotubes formed by arc in liquids." *The Journal of Chemical Physics* 120, no. 2 (2004): 1055–1058.
67. Cabioc'h, Thierry, Jean-Christophe Girard, Michel Jaouen, Marie Françoise Denanot, and Gilles Hug. "Carbon onions thin film formation and characterization." *EPL (Europhysics Letters)* 38, no. 6 (1997): 471.
68. Cabioc'h, Thierry, Michel Jaouen, Marie Françoise Denanot, and P. Bechet. "Influence of the implantation parameters on the microstructure of carbon onions produced by carbon ion implantation." *Applied Physics Letters* 73, no. 21 (1998): 3096–3098.
69. Xu, Xiaoyou, Robert Ray, Yunlong Gu, Harry J. Ploehn, Latha Gearheart, Kyle Raker, and Walter A. Scrivens. "Electrophoretic analysis and purification of fluorescent single-walled carbon nanotube fragments." *Journal of the American Chemical Society* 126, no. 40 (2004): 12736–12737.
70. Li, Haitao, Zhenhui Kang, Yang Liu, and Shuit-Tong Lee. "Carbon nanodots: Synthesis, properties and applications." *Journal of Materials Chemistry* 22, no. 46 (2012): 24230–24253.
71. Sun, Ya-Ping, Bing Zhou, Yi Lin, Wei Wang, K.A. Shiral Fernando, Pankaj Pathak, Mohammed Jaouad Meziani, et al. "Quantum-sized carbon dots for bright and colorful photoluminescence." *Journal of the American Chemical Society* 128, no. 24 (2006): 7756–7757.

72. Ming, Hai, Zheng Ma, Yang Liu, Keming Pan, Hang Yu, Fang Wang, and Zhenhui Kang. "Large scale electrochemical synthesis of high quality carbon nanodots and their photocatalytic property." *Dalton Transactions* 41, no. 31 (2012): 9526–9531.
73. Liu, Jing, Xinling Liu, Hongjie Luo, and Yanfeng Gao. "One-step preparation of nitrogen-doped and surface-passivated carbon quantum dots with high quantum yield and excellent optical properties." *RSC Advances* 4, no. 15 (2014): 7648–7654.
74. Chen, Po-Cheng, Ya-Na Chen, Pin-Che Hsu, Chung-Chien Shih, and Huan-Tsung Chang. "Photoluminescent organosilane-functionalized carbon dots as temperature probes." *Chemical Communications* 49, no. 16 (2013): 1639–1641.
75. Zhou, Li, Benzhao He, and Jiachang Huang. "Amphibious fluorescent carbon dots: One-step green synthesis and application for light-emitting polymer nanocomposites." *Chemical Communications* 49, no. 73 (2013): 8078–8080.
76. Zhou, Jin, Xiaoyue Shan, Juanjuan Ma, Yamin Gu, Zhaosheng Qian, Jianrong Chen, and Hui Feng. "Facile synthesis of P-doped carbon quantum dots with highly efficient photoluminescence." *RSC Advances* 4, no. 11 (2014): 5465–5468.
77. Liu, Yi, Ning Xiao, Ningqiang Gong, Hao Wang, Xin Shi, Wei Gu, and Ling Ye. "One-step microwave-assisted polyol synthesis of green luminescent carbon dots as optical nanoprobes." *Carbon* 68 (2014): 258–264.
78. Ma, Zheng, Hai Ming, Hui Huang, Yang Liu, and Zhenhui Kang. "One-step ultrasonic synthesis of fluorescent N-doped carbon dots from glucose and their visible-light sensitive photocatalytic ability." *New Journal of Chemistry* 36, no. 4 (2012): 861–864.
79. Linehan, K., and H. Doyle. "Efficient one-pot synthesis of highly monodisperse carbon quantum dots." *RSC Advances* 4 (2014): 18–21.
80. Shen, Jianhua, Yihua Zhu, Xiaoling Yang, and Chunzhong Li. "Graphene quantum dots: Emergent nanolights for bioimaging, sensors, catalysis and photovoltaic devices." *Chemical Communications* 48, no. 31 (2012): 3686–3699.
81. Bacon, Mitchell, Siobhan J. Bradley, and Thomas Nann. "Graphene quantum dots." *Particle & Particle Systems Characterization* 31, no. 4 (2014): 415–428.
82. Li, Lingling, Gehui Wu, Guohai Yang, Juan Peng, Jianwei Zhao, and Jun-Jie Zhu. "Focusing on luminescent graphene quantum dots: Current status and future perspectives." *Nanoscale* 5, no. 10 (2013): 4015–4039.
83. Liao, Hsueh-Chung, Chun-Chih Ho, Chun-Yu Chang, Meng-Huan Jao, Seth B. Darling, and Wei-Fang Su. "Additives for morphology control in high-efficiency organic solar cells." *Materials Today* 16, no. 9 (2013): 326–336.
84. Zhang, Mo, Linling Bai, Weihu Shang, Wenjing Xie, Hong Ma, Yingyi Fu, Decai Fang, et al. "Facile synthesis of water-soluble, highly fluorescent graphene quantum dots as a robust biological label for stem cells." *Journal of Materials Chemistry* 22, no. 15 (2012): 7461–7467.
85. Markovic, Zoran M., Biljana Z. Ristic, Katarina M. Arsikin, Djordje G. Klisic, Ljubica M. Harhaji-Trajkovic, Biljana M. Todorovic-Markovic, Dejan P. Kepic, et al. "Graphene quantum dots as autophagy-inducing photodynamic agents." *Biomaterials* 33, no. 29 (2012): 7084–7092.
86. Sun, Yiqing, Shiqi Wang, Chun Li, Peihui Luo, Lei Tao, Yen Wei, and Gaoquan Shi. "Large scale preparation of graphene quantum dots from graphite with tunable fluorescence properties." *Physical Chemistry Chemical Physics* 15, no. 24 (2013): 9907–9913.
87. Pan, Dengyu, Jingchun Zhang, Zhen Li, and Minghong Wu. "Hydrothermal route for cutting graphene sheets into blue-luminescent graphene quantum dots." *Advanced Materials* 22, no. 6 (2010): 734–738.
88. Ye, Ruquan, Changsheng Xiang, Jian Lin, Zhiwei Peng, Kewei Huang, Zheng Yan, Nathan P. Cook, et al. "Coal as an abundant source of graphene quantum dots." *Nature communications* 4, no. 1 (2013): 1–7.
89. Lee, Seok Hwan, Do Youb Kim, Jaemin Lee, Seul Bee Lee, Hyun Han, Young Yun Kim, Sung Cik Mun, Sang Hyuk Im, Tae-Ho Kim, and O. Ok Park. "Synthesis of single-crystalline hexagonal graphene quantum dots from solution chemistry." *Nano Letters* 19, no. 8 (2019): 5437–5442.
90. Shenderova, Olga A., and Dieter M. Gruen. *Ultrananocrystalline diamond: Synthesis, properties and applications*. William Andrew, 2012.
91. Dolmatov, Valerii Yu. "Detonation synthesis ultradispersed diamonds: Properties and applications." *Russian Chemical Reviews* 70, no. 7 (2001): 607–626.
92. Danilenko, Viacheslav Vasilyovich. "On the history of the discovery of nanodiamond synthesis." *Physics of the Solid State* 46, no. 4 (2004): 595–599.

93. Krueger, Anke. “Diamond nanoparticles: Jewels for chemistry and physics.” *Advanced Materials* 20, no. 12 (2008): 2445–2449.
94. Krüger, Anke, Fumiaki Kataoka, M.A.A. Ozawa, T. Fujino, Y. Suzuki, Aleksandr E. Aleksenskii, A. Ya Vul, and Eiji Ōsawa. “Unusually tight aggregation in detonation nanodiamond: Identification and disintegration.” *Carbon* 43, no. 8 (2005): 1722–1730.
95. Ozawa, Masaki, Masayasu Inaguma, Makoto Takahashi, Fumiaki Kataoka, Anke Krueger, and E. Ōsawa. “Preparation and behavior of brownish, clear nanodiamond colloids.” *Advanced Materials* 19, no. 9 (2007): 1201–1206.
96. Mahfouz, Remi, Daniel L. Floyd, Wei Peng, Jennifer T. Choy, Marko Loncar, and Osman M. Bakr. “Size-controlled fluorescent nanodiamonds: A facile method of fabrication and color-center counting.” *Nanoscale* 5, no. 23 (2013): 11776–11782.
97. Kumar, Ajay, Pin Ann Lin, Albert Xue, Boyi Hao, Yoke Khin Yap, and R. Mohan Sankaran. “Formation of nanodiamonds at near-ambient conditions via microplasma dissociation of ethanol vapour.” *Nature Communications* 4, no. 1 (2013): 1–9.
98. Prasek, Jan, Jana Drbohlavova, Jana Chomoucka, Jaromir Hubalek, Ondrej Jasek, Vojtech Adam, and Rene Kizek. “Methods for carbon nanotubes synthesis.” *Journal of Materials Chemistry* 21, no. 40 (2011): 15872–15884.
99. Journet, Catherine, W.K. Maser, Patrick Bernier, Annick Loiseau, M. Lamy de La Chapelle, dl S. Lefrant, Philippe Deniard, R. Lee, and J.E. Fischer. “Large-scale production of single-walled carbon nanotubes by the electric-arc technique.” *Nature* 388, no. 6644 (1997): 756–758.
100. Ebbesen, Thomas W., and Pulickel M. Ajayan. “Large-scale synthesis of carbon nanotubes.” *Nature* 358, no. 6383 (1992): 220–222.
101. Iijima, Sumio. “Helical microtubules of graphitic carbon.” *Nature* 354, no. 6348 (1991): 56–58.
102. Iijima, Sumio, and Toshinari Ichihashi. “Single-shell carbon nanotubes of 1-nm diameter.” *Nature* 363, no. 6430 (1993): 603–605.
103. Sugai, Toshiki, Hiromichi Yoshida, Takashi Shimada, Toshiya Okazaki, Hisanori Shinohara, and Shunji Bandow. “New synthesis of high-quality double-walled carbon nanotubes by high-temperature pulsed arc discharge.” *Nano Letters* 3, no. 6 (2003): 769–773.
104. Guo, Ting, Pavel Nikolaev, Andreas Thess, Daniel T. Colbert, and Richard E. Smalley. “Catalytic growth of single-walled manotubes by laser vaporization.” *Chemical Physics Letters* 243, no. 1–2 (1995): 49–54.
105. Steiner III, Stephen A., Theodore F. Baumann, Bernhard C. Bayer, Raoul Blume, Marcus A. Worsley, Warren J. MoberlyChan, Elisabeth L. Shaw, et al. “Nanoscale zirconia as a nonmetallic catalyst for graphitization of carbon and growth of single-and multiwall carbon nanotubes.” *Journal of the American Chemical Society* 131, no. 34 (2009): 12144–12154.
106. Liu, Zhongfan, Liying Jiao, Yagang Yao, Xiaojun Xian, and Jin Zhang. “Aligned, ultralong single-walled carbon nanotubes: From synthesis, sorting, to electronic devices.” *Advanced Materials* 22, no. 21 (2010): 2285–2310.
107. Lim, SanHua, Zhiqiang Luo, ZeXiang Shen, and Jianyi Lin. “Plasma-assisted synthesis of carbon nanotubes.” *Nanoscale Research Letters* 5, no. 9 (2010): 1377–1386.
108. Dinh, Le N.M., Bich Ngoc Tran, Vipul Agarwal, and Per B. Zetterlund. “Synthesis of Highly Stretchable and Electrically Conductive Multiwalled Carbon Nanotube/Polymer Nanocomposite Films.” *ACS Applied Polymer Materials* 4, no. 3 (2022): 1867–1877.
109. Hu, Shengliang, Fei Tian, Peikang Bai, Shirui Cao, Jing Sun, and Jing Yang. “Synthesis and luminescence of nanodiamonds from carbon black.” *Materials Science and Engineering: B* 157, no. 1–3 (2009): 11–14.
110. Azami, Takeshi, Daisuke Kasuya, Ryota Yuge, Masako Yudasaka, Sumio Iijima, Tsutomu Yoshitake, and Yoshimi Kubo. “Large-scale production of single-wall carbon nanohorns with high purity.” *The Journal of Physical Chemistry C* 112, no. 5 (2008): 1330–1334.
111. Azami, Takeshi, Daisuke Kasuya, Tsutomu Yoshitake, Yoshimi Kubo, Masako Yudasaka, and Toshinari Ichihashi. “Production of small single-wall carbon nanohorns by CO2 laser ablation of graphite in Ne-gas atmosphere.” *Carbon (New York)* 45, no. 6 (2007): 1364–1367.
112. Jorio, Ado, and Gene Dresselhaus. *Carbon nanotubes: Advanced topics in the synthesis, structure, properties and applications*. Edited by Mildred S. Dresselhaus. Vol. 111. Berlin: Springer, 2008.
113. Casteignau, Fanny, Taki Aissou, Charlotte Allard, Christian Ricolleau, Jocelyn Veilleux, Richard Martel, and Nadi Braidy. “Synthesis of Carbon Nanohorns by Inductively Coupled Plasma.” *Plasma Chemistry and Plasma Processing* 42, no. 3 (2022): 465–481.

114. Hernandez, Yenny, Valeria Nicolosi, Mustafa Lotya, Fiona M. Blighe, Zhenyu Sun, Sukanta De, Ignatius T. McGovern, et al. "High-yield production of graphene by liquid-phase exfoliation of graphite." *Nature Nanotechnology* 3, no. 9 (2008): 563–568.
115. De, Sukanta, Paul J. King, Mustafa Lotya, Arlene O'Neill, Evelyn M. Doherty, Yenny Hernandez, Georg S. Duesberg, and Jonathan N. Coleman. "Flexible, transparent, conducting films of randomly stacked graphene from surfactant-stabilized, oxide-free graphene dispersions." *Small* 6, no. 3 (2010): 458–464.
116. Hao, Rui, Wen Qian, Luhui Zhang, and Yanglong Hou. "Aqueous dispersions of TCNQ-anion-stabilized graphene sheets." *Chemical Communications* 48 (2008): 6576–6578.
117. Lotya, Mustafa, Yenny Hernandez, Paul J. King, Ronan J. Smith, Valeria Nicolosi, Lisa S. Karlsson, Fiona M. Blighe, et al. "Liquid phase production of graphene by exfoliation of graphite in surfactant/water solutions." *Journal of the American Chemical Society* 131, no. 10 (2009): 3611–3620.
118. Lotya, Mustafa, Paul J. King, Umar Khan, Sukanta De, and Jonathan N. Coleman. "High-concentration, surfactant-stabilized graphene dispersions." *ACS Nano* 4, no. 6 (2010): 3155–3162.
119. Cai, Minzhen, Daniel Thorpe, Douglas H. Adamson, and Hannes C. Schniepp. "Methods of graphite exfoliation." *Journal of Materials Chemistry* 22, no. 48 (2012): 24992–25002.
120. Brodie, B.J.A.C.P. "Note sur un nouveau procédé pour la purification et la désagrégation du graphite." *Annales de Chimie et de Physique* 45 (1855): 351–353.
121. Staudenmaier, L. "Verfahren zur darstellung der graphitsäure." *Berichte der deutschen chemischen Gesellschaft* 31, no. 2 (1898): 1481–1487.
122. Hummers Jr, William S., and Richard E. Offeman. "Preparation of graphitic oxide." *Journal of the American Chemical Society* 80, no. 6 (1958): 1339- 1339.
123. Jang, J.W., S. Cho, G-H. Moon, K. Ihm, J.Y. Kim, D.H. Youn, S. Lee, Y. H. Lee, W. Choi, K.-H. Lee, and J. S. Lee "Photocatalytic Synthesis of Pure and Water-Dispersible Graphene Monosheets." *Chemistry–A European Journal* 18 (2012): 2762–2767.
124. Eda, Goki, and Manish Chhowalla. "Chemically derived graphene oxide: Towards large-area thin-film electronics and optoelectronics." *Advanced Materials* 22, no. 22 (2010): 2392–2415.
125. Zhang, Qifeng, Christopher S. Dandeneau, Xiaoyuan Zhou, and Guozhong Cao. "ZnO nanostructures for dye-sensitized solar cells." *Advanced Materials* 21, no. 41 (2009): 4087–4108.
126. Choi, Soo Ho, Seok Joon Yun, Yo Seob Won, Chang Seok Oh, Soo Min Kim, Ki Kang Kim, and Young Hee Lee. "Large-scale synthesis of graphene and other 2D materials towards industrialization." *Nature Communications* 13, no. 1 (2022): 1–5.
127. Abdelkhalek, Amr, Mona Abd El-Latif, Hesham Ibrahim, Hesham Hamad, and Marwa Showman. "Controlled synthesis of graphene oxide/silica hybrid nanocomposites for removal of aromatic pollutants in water." *Scientific Reports* 12, no. 1 (2022): 1–22.
128. Herron, Christopher R., Karl S. Coleman, Rebecca S. Edwards, and Budhika G. Mendis. "Simple and scalable route for the 'bottom-up'synthesis of few-layer graphene platelets and thin films." *Journal of Materials Chemistry* 21, no. 10 (2011): 3378–3383.
129. Liu, Weihua, Tao Dang, Zhihao Xiao, Xin Li, Changchun Zhu, and Xiaoli Wang. "Carbon nanosheets with catalyst-induced wrinkles formed by plasma-enhanced chemical-vapor deposition." *Carbon* 49, no. 3 (2011): 884–889.
130. Wang, Jianjun, Mingyao Zhu, Ron A. Outlaw, Xin Zhao, Dennis M. Manos, and Brian C. Holloway. "Synthesis of carbon nanosheets by inductively coupled radio-frequency plasma enhanced chemical vapor deposition." *Carbon* 42, no. 14 (2004): 2867–2872.
131. Kuang, Qin, Su-Yuan Xie, Zhi-Yuan Jiang, Xian-Hua Zhang, Zhao-Xiong Xie, Rong-Bin Huang, and Lan-Sun Zheng. "Low temperature solvothermal synthesis of crumpled carbon nanosheets." *Carbon* 42, no. 8-9 (2004): 1737–1741.
132. Che, Xiaogang, Juan Yang, Siyu Liu, Man Wang, Songjie He, and Jieshan Qiu. "Multilayer-dense porous carbon nanosheets with high volumetric capacitance for supercapacitors." *Industrial & Engineering Chemistry Research* 61, no. 25 (2022): 8908–8917.
133. Wang, Xiaoxu, Jinhui Zhao, Jingya Zhang, Yingqiang Zhao, Peng Zhao, Lei Ni, Qinxing Xie, and Jianqiang Meng. "Ball-milled silicon with amorphous Al2O3/C hybrid coating embedded in graphene/graphite nanosheets with a boosted lithium storage capability." *Langmuir* 38, no. 28 (2022): 8555–8563.
134. Jiao, Liying, Xinran Wang, Georgi Diankov, Hailiang Wang, and Hongjie Dai. "Facile synthesis of high-quality graphene nanoribbons." *Nature Nanotechnology* 5, no. 5 (2010): 321–325.
135. Terrones, Mauricio, Andrés R. Botello-Méndez, Jessica Campos-Delgado, Florentino López-Urías, Yadira I. Vega-Cantú, Fernando J. Rodríguez-Macías, Ana Laura Elías, et al. "Graphene and graphite

nanoribbons: Morphology, properties, synthesis, defects and applications." *Nano Today* 5, no. 4 (2010): 351–372.
136. Li, Xiaolin, Xinran Wang, Li Zhang, Sangwon Lee, and Hongjie Dai. "Chemically derived, ultrasmooth graphene nanoribbon semiconductors." *Science* 319, no. 5867 (2008): 1229–1232.
137. Lu, J., J.X. Yang, J.Z. Wang, A.L. Lim, S. Wang, and K.P. Loh. "One-pot synthesis of fluorescent carbon nanoribbons, nanoparticles, and graphene by the exfoliation of graphite in ionic liquids." *American Chemical Society Nano* 3, no. 8 (2009): 2367–2375.
138. Dobner, Christoph, Gang Li, Mamun Sarker, Alexander Sinitskii, and Axel Enders. "Diffusion-controlled on-surface synthesis of graphene nanoribbon heterojunctions." *RSC Advances* 12, no. 11 (2022): 6615–6618.
139. Way, Austin J., Robert M. Jacobberger, Nathan P. Guisinger, Vivek Saraswat, Xiaoqi Zheng, Anjali Suresh, Jonathan H. Dwyer, Padma Gopalan, and Michael S. Arnold. "Graphene nanoribbons initiated from molecularly derived seeds." *Nature Communications* 13, no. 1 (2022): 1–10.
140. Bonifazi, Davide, Olivier Enger, and François Diederich. "Supramolecular [60] fullerene chemistry on surfaces." *Chemical Society Reviews* 36, no. 2 (2007): 390–414.
141. Imahori, Hiroshi. "Giant multiporphyrin arrays as artificial light-harvesting antennas." *The Journal of Physical Chemistry B* 108, no. 20 (2004): 6130–6143.
142. Guldi, Dirk M., G.M. Aminur Rahman, Vito Sgobba, and Christian Ehli. "Multifunctional molecular carbon materials—From fullerenes to carbon nanotubes." *Chemical Society Reviews* 35, no. 5 (2006): 471–487.
143. Li, Gang, Vishal Shrotriya, Jinsong Huang, Yan Yao, Tom Moriarty, Keith Emery, and Yang Yang. "High-efficiency solution processable polymer photovoltaic cells by self-organization of polymer blends." *Nature Materials* 4, no. 11 (2005): 864–868.
144. Kim, Jin Young, Kwanghee Lee, Nelson E. Coates, Daniel Moses, Thuc-Quyen Nguyen, Mark Dante, and Alan J. Heeger. "Efficient tandem polymer solar cells fabricated by all-solution processing." *Science* 317, no. 5835 (2007): 222–225.
145. Günes, Serap, Helmut Neugebauer, and Niyazi Serdar Sariciftci. "Conjugated polymer-based organic solar cells." *Chemical Reviews* 107, no. 4 (2007): 1324–1338.
146. Kim, Youngkyoo, Steffan Cook, Sachetan M. Tuladhar, Stelios A. Choulis, Jenny Nelson, James R. Durrant, Donal D.C. Bradley, et al. "A strong regioregularity effect in self-organizing conjugated polymer films and high-efficiency polythiophene: Fullerene solar cells." In *Materials for sustainable energy: A collection of peer-reviewed research and review articles from Nature Publishing Group*, pp. 63–69. Nature Publishing Group, 2011.
147. Luo, P.G., F. Yang, S.T. Yang, S.K. Sonkar, L. Yang, J.J. Broglie, Y. Liu, and Y.P. Sun. "Carbon-based quantum dots for fluorescence imaging of cells and tissues." *RSC Advances* 4 (2014): 10791–10807.
148. Baidakova, Marina. "New prospects and frontiers of nanodiamond clusters." *Journal of Physics D: Applied Physics* 40, no. 20 (2007): 6300.
149. Mochalin, Vadym, Olga Shenderova, Dean Ho, and Yury Gogotsi. "The properties and applications of nanodiamonds." *Nano-Enabled Medical Applications* (2020): 313–350.
150. Huang, J.Y., S. Chen, Z.Q. Wang, K. Kempa, Y.M. Wang, S.H. Jo, G. Chen, M.S. Dresselhaus, and Z.F. Ren. "Superplastic carbon nanotubes." *Nature* 439, no. 7074 (2006): 281- 281.
151. Ding, Feng, Kun Jiao, Yu Lin, and Boris I. Yakobson. "How evaporating carbon nanotubes retain their perfection?." *Nano Letters* 7, no. 3 (2007): 681–684.
152. Lee, Changgu, Xiaoding Wei, Jeffrey W. Kysar, and James Hone. "Measurement of the elastic properties and intrinsic strength of monolayer graphene." *Science* 321, no. 5887 (2008): 385–388.
153. Zhu, Yanwu, Shanthi Murali, Weiwei Cai, Xuesong Li, Ji Won Suk, Jeffrey R. Potts, and Rodney S. Ruoff. "Graphene and graphene oxide: Synthesis, properties, and applications." *Advanced Materials* 22, no. 35 (2010): 3906–3924.
154. Rao, C.E.E., A.E. Sood, K.E. Subrahmanyam, and A. Govindaraj. "Graphene: The new two-dimensional nanomaterial." *Angewandte Chemie International Edition* 48, no. 42 (2009): 7752–7777.
155. Finkenstadt, Daniel, G. Pennington, and M.J. Mehl. "From graphene to graphite: A general tight-binding approach for nanoribbon carrier transport." *Physical Review B* 76, no. 12 (2007): 121405.
156. Owens, F.J. "Electronic and magnetic properties of graphene nanoribbons." *Molecular Physics* 104, no. 19 (2006): 3107–3109.

157. Yang, Li, Cheol-Hwan Park, Young-Woo Son, Marvin L. Cohen, and Steven G. Louie. "Quasiparticle energies and band gaps in graphene nanoribbons." *Physical Review Letters* 99, no. 18 (2007): 186801.
158. Makarova, T.L. "Magnetic properties of carbon structures." *Semiconductors* 38, no. 6 (2004): 615–638.
159. Bandow, Shunji. "Magnetic properties of nested carbon nanostructures studied by electron spin resonance and magnetic susceptibility measurements." *Journal of Applied Physics* 80, no. 2 (1996): 1020–1027.
160. Bandow, S., F. Kokai, K. Takahashi, M. Yudasaka, and S. Iijima. "Unique magnetism observed in single-wall carbon nanohorns." *Applied Physics A* 73, no. 3 (2001): 281–285.

5 Carbon Allotropes

Covalent and Noncovalent Functionalization and Characterization

Abhinay Thakur
Department of Chemistry, School of Chemical Engineering and Physical Sciences, Lovely Professional University, Phagwara, Punjab, India

Ashish Kumar
NCE, Bihar Engineering University, Department of Science and Technology, Government of Bihar, India

CONTENTS

5.1 INTRODUCTION

Owing to the capability of carbon to establish stable bonds with the other carbon atoms, nitrogen, oxygen, sulfur, and numerous other elements in Mendeleev's Periodic Table, carbon, the sixth element in the periodic table signified by the letter "C" and the truthful component of life, offers the chemical foundation for existence on Earth. Among the most prevalent substances on earth, carbon is present nearly all over (González-García, 2015; Islam et al., 2022; Kamedulski et al., 2019; Majdoub et al., 2020). It is the 15th most frequent element in the crust of the planet and the fourth most frequent element overall. All biology on Earth, from enzymes to mature trees, includes numerous types of carbonic molecules. Robust individual and multiple carbon–carbon covalent

DOI: 10.1201/9781003323976-6

bonds are also accountable for the development of a variety of carbon-based inorganic substances. Catenation is the mechanism through which an element may combine with self to create lengthy chains. There have only ever been two recognized ordinary carbon allotropes: diamond, which is bright, translucent, and exceedingly rigid, and graphite, which is black, smooth, and conductive. Novel synthesized carbon allotropes like fullerenes, carbon nanotubes, and graphene have only recently been identified. Their exceptional qualities and existing and expected uses attest to their special research and technical significance. In contrast, numerous more carbon structures that have been discovered or are currently being expected have been revealed recently (Alshehri et al., 2016; Backes et al., 2011; Bhawani et al., 2018; Marcia et al., 2017; Menaa et al., 2021). There are numerous allotropes of carbon, that could be categorized based on the type of chemical bond involved in hybridization (sp, sp^2, or sp^3). These include 0D sp^2 fullerenes, the 2D sp^2 honeycomb geometric of graphene, or 3D sp^3 crystals such as lonsdaleite, diamond, and C8. Examples of the latter include alkanes and diamonds. The kind of hybridization sp, sp^2, or sp^3 can be effectively linked to the origins of these characteristics (He et al., 2013; Saifuddin et al., 2013; Veeman et al., 2021). Although sp^3 possesses insulating qualities, the first two retain the capacity to create electrically conductive bonds. Mechanical and electronic characteristics change substantially between each allotrope. For example, graphene exhibits the typical electrical properties of a semimetal, including a straight band distribution and extremely rapid electron motility. Diamond, on the other hand, is among the recognized toughest natural substances and a wide-band-gap insulator. The characteristics of the diamond, sp^3-hybridized carbon allotrope, and sp^2-hybridized carbon allotropes, fullerenes, graphite, graphene, and nanotubes, are well known. On the other side, a carbon allotrope made of sp-hybridized carbon atoms (carbyne) is still uncommon and poses a difficult synthetic task. High surface area carbon materials mainly include activated carbon, porous carbon, carbide-derived carbon, carbon nano onions, carbon aerogels, carbon nanotubes, carbon shells, graphene, and graphene quantum dots (Thakur et al., 2022c; Thakur and Kumar, 2022, 2021). There are many other classifications of carbon allotropes. As previously discussed, carbon (C) atoms can arrange themselves into a variety of configurations, as seen in Figure 5.1. Stacks of 2D single layers known as graphite crystal is created when C atoms are placed in a honeycomb layout. The graphene atomic crystal is made up of a solitary sheet of graphite. The Nobel laureates Novoselov and Geim extracted graphene nanostructures (GNs) for the first time in 2010. Carbon nanotubes (CNT) are another carbon-based structure. It was found by Iijima in 1991,

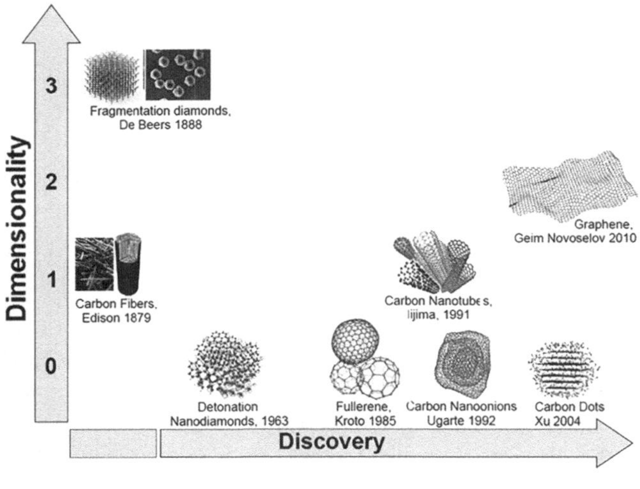

FIGURE 5.1 An arrangement of carbon nanostructures based on dimensions and time since invention. Reprinted with permission from Speranza (2021). Copyright (2021) MDPI. Distributed under the Creative Common attribution-based License CCBY 4.0.

and it can be considered as single-layer graphene stretched along an axis parallel to the crystallographic orientations of graphene. Carbon fibers (CFs) and CNTs are both 1D systems. But CFs are chaotic, entangled formations with a 2D long-range organization of C atoms arranged in planar hexagonal matrices; in contrast, CFs only exhibit a short order in the plane orthogonal to such surfaces owing to analogous plane layering.

The label "carbon nanotropes" was also given to the nanoscale carbon allotropes; however, it is currently essentially unheard of. Fascinating and unexpected carbon structures like protomene, which has a hexagonal crystal shape and a completely stretched primitive cell containing 48 atoms, or a suggested sequence of carbon allotropes named novamene are introduced in reports about novel DFT-predicted carbon allotropes regularly (Kumar and Thakur, 2020; Thakur et al., 2022a, 2022b, 2022d, 2022e; Thakur and Kumar, 2021). To increase the dispersal and interfacial adhesion of thermally rGO nanoplatelets using an epoxy resin, Naebe et al. (2014) covalently functionalized the nanoplatelets through the Bingel reactions. Microscopical, thermodynamic, and spectral methods were used to characterize functionalized graphene. When contrasted to GO, functionalized graphene had much improved thermal durability, according to thermal studies. Only 0.1 wt.% of functionalized graphene increased the tensile stiffness and retention modulus of an epoxy resin by 22% and 18%, respectively. According to microscopic studies, the enhanced mechanical characteristics of nanocomposites were caused by the homogeneous distribution of functionalized graphene and the robust interfacial connection between epoxy resin and customized graphene. In this research, organic flake graphite was oxidized in a mixture of potassium chlorate, HNO_3 and H_2SO_4 for 96 hours to prepare and chemically functionalize graphene. A convection oven was used to dry graphene oxide (GO) for 24 hours at 80°C. In a quartz tube, vacuum-dried GO was put before being washed with Ar for 10 minutes. To acquire thermally reduced graphene (TRG), the quartz tube was then swiftly placed into a Lindberg tube furnace that had been warmed to 1050°C. As per the Bingel process described by Coleman et al., TRG was functionalized. Employing sonication, a TRG dispersion in dried dichlorobenzene (DCB) was created. 1,8-diazabicyclo[5.4.0] undecene (DBU) and diethylbromomalonate were introduced and agitated for 15 hours in an ambient environment. The solution was strained, and then rinsed with DCB and ethanol. The end product was then extracted after being disseminated and agitated in a solution of methanol, NaOH, and THF at ambient temperature. The material was dissolved in HCl media at ambient temperature after the supernatant liquid was removed. The solution was agitated, and distilled H_2O was used to rinse it several rounds. The functionalized graphene referred to as FG henceforth was subsequently processed in the oven for 5 hours at 60°C. The synthesis process is shown in Figure 5.2.

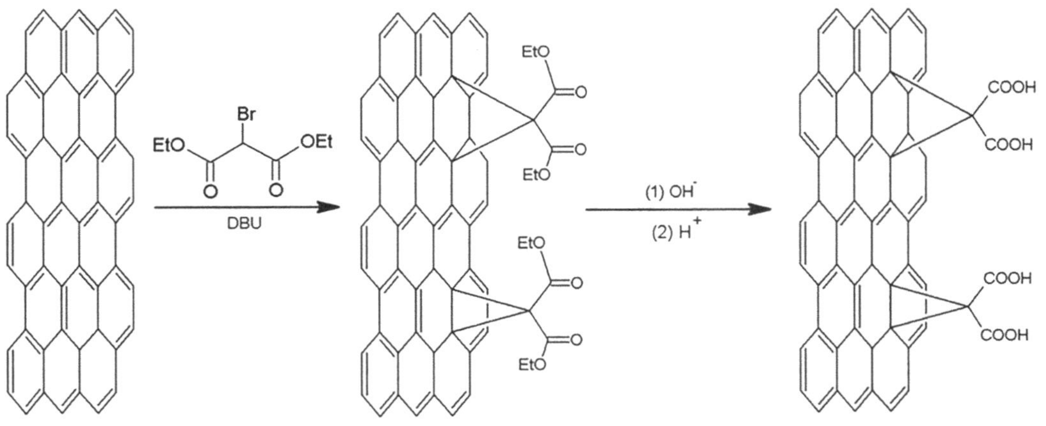

FIGURE 5.2 Generation of functionalized graphene using the Bingel pathway. Reprinted with permission from Naebe et al. (2014). Copyright (2014) Scientific reports. Distributed under the Creative Common attribution-based License CCBY 4.0.

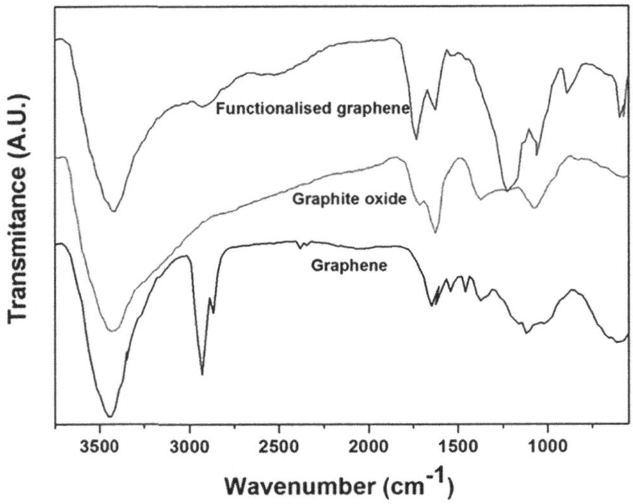

FIGURE 5.3 FTIR spectrum of graphene, thermally rGO, and functionalized graphene. Reprinted with permission from Naebe et al. (2014). Copyright (2014) Scientific reports. Distributed under the Creative Common attribution-based License CCBY 4.0.

The peaks at 1711 and 3432 cm^{-1} in the FT-IR spectra of the GO specimen (Figure 5.3) are attributable to the C=O and –OH bands, correspondingly. The C5O band vanishes after the reduction of GO to graphene, whereas bands at 2865 and 2928 cm^{-1} emerge to indicate the C–H stretch oscillations of the methylene group. Functionalized graphene exhibits a peak at 1731 cm^{-1}, which is a distinctive band for the COOH group's C=O stretch. The broad band at 3412 cm^{-1} serves as additional confirmation of the existence of a carboxylic functional group.

Similarly, using computational modeling, empirical NMR and IR spectrometry, and compounds of the secondary amine piperidine and C70, Lo et al. (2021) were able to demonstrate the presence of the covalent/dative complex and gain further understanding of its structural specificity. The creation of dative bonds at the area with six-membered rings was explained by film out-of-plane variations that transpired in both orientations of the pyrene fragments. The findings show that carbon units involved in both five- and six-member rings have firmer bonds than carbon units in the region where only six-member rings exist. According to the computed free energy, the sole complexes that were generated with carbon units on the rugby-ball structure's poles were thermally viable. This research revealed previously unrecognized details about the interactions of C70 with secondary amines that could be used to synthesize fullerene variants for (electro)catalysis, spintronics, and energy storage.

5.2 FUNCTIONALIZATION OF CARBON ALLOTROPES

The only identified members of the carbon family before the 1980s were diamond and graphite. With the finding of molecular carbon allotropes like CNT, fullerenes, and most notably, 2-D graphene, this has completely transformed. The research world has thus become increasingly interested in researching many topics, especially the surface alteration of carbon allotropes. Carbon allotrope functionalization and diffusion are essential for their applicability. Chemically functionalized graphene, for instance, could be subsequently treated using solvent-assisted methods including spin-coating, layer-by-layer assembly, and purification (Díez-Pascual, 2021; Eigler and Hirsch, 2014; He et al., 2013; Saifuddin et al., 2013; Veeman et al., 2021). The intrinsic features of graphene are preserved, and the aggregation of thin sheet graphene after reduction in the solvent is also prevented by suitable chemical functionalization of graphene. A common initial point for the production of producible graphene is GO. The enhanced Hummers approach is extremely successful in producing GO using natural graphite among the various ways available. The interfaces of GO sheets are extensively oxidized and contain moieties such as epoxide, hydroxyl, diol, ketone, and carboxyl, which could drastically change the van der Waals interactions and result in a variety of solubility in water and organic solvents. GO sheets are highly hydrophilic and can easily expand and scatter in water due to the extended carboxyl and carbonyl groups that are present at the edges of the layers. Academia has developed a variety of prototype GO complexes depending on these oxygen capabilities. The reduction of GO is accomplished thermally, chemically, or photochemically to produce graphene. Furthermore, the quick, irrevocable agglomeration of graphene sheets causes the deposition of graphite granules when GO dispersal is reduced in the absence of a stabilizer. Because of this, GO sheets' surfaces are typically modified by covalent alterations or noncovalent functionalizations before reduction. A common process results in persistent suspensions of functionalized graphene films in organic solvents after reductions of alkylamine-modified GO (Torres, 2011; Vasilescu et al., 2018). By creating sulfonate or carboxylic units on the graphene's basal planes, it is possible to create water-soluble layers of graphene. The direct synthesis of dissolvable graphene using natural graphite has also been attempted. Electrochemically synthesizing graphene with ionic liquid modifications using natural graphite. For the synthesis of steady dispersal of non-functionalized graphene having high yields, numerous more techniques have also been described. For instance, Figure 5.4 shows the functionalization of carbon-based nanomaterials using photochromic molecules.

In an experiment, for the precise detection of dopamine (DA) in biological and pharmaceutical specimens, Karuppiah et al. (2016) presented a noncovalent functionalization centered on a copper tetraphenylporphyrin/chemically rGO (Cu-TPP/CRGO) nanocomposite. Cu-TPP's noncovalent stacking engagement with the graphene sheets resulted in a uniform, electron-rich state. Through the use of cyclic voltammetry and differential pulse voltammetry, the electrocatalytic efficiency of DA was assessed. With escalating DA content in the region of 2–200 μM, the oxidation peak current (Ipa) of DA gradually increased. With a remarkable sensibility of 2.46 $\mu A\ \mu M^{-1} cm^{-2}$, the limit of detection was predicted to be 0.76 μM. The suggested DA sensor's viability was assessed using DA hydrochloride injection, saliva, and human urine, and the findings for the detection of DA were acceptable. The improved Cu-TPP/CRGO nanocomposite electrode also demonstrated outstanding durability, reproducibility, and consistency when used to detect DA.

The noncovalent functionalization method was used by Qu et al. (2013) to construct a novel graphene-fullerene composite (rGO-pyrene-PCBM), wherein [6,6]-phenyl-C61-butyric acid methyl ester (PCBM) was linked to rGO. A simple esterification process was performed to create the pyrene-PCBM compound, and pyrene was then utilized as an anchoring scaffold to unite the elements of PCBM and rGO. SEM, Raman, and AFM investigations of the rGO-pyrene-PCBM geometric structure indicate that the C60 component was separated from the graphene layer and attached to it by the pyrene anchorage. rGO-pyrene-PCBM was ultimately employed effectively as an electron extraction film for P3HT: PCBM heterojunction polymer solar cell (BHJ-PSC) devices, providing a PCE of 3.89%. However, the comparison methods using either rGO or pyrene-PCBM as the electron extraction surface displayed significantly reduced PCE, demonstrating the significance of composite creation involving pyrene-PCBM and rGO constituents for its electron extraction capability.

5.2.1 Covalent Versus Noncovalent Approaches

Wet-chemical approaches to the noncovalent and covalent functionalization of the baseline layer in carbon allotropes have received a significant amount of attention. This is largely intended to get around problems with carbon allotropes' poor tunability, restricted durability, low miscibility, etc. in almost any solvent. A primary pathway is that carbon allotropes are exfoliated or functionalized by a variety of oxidation processes that provide substrate, followed by chemical alteration and,

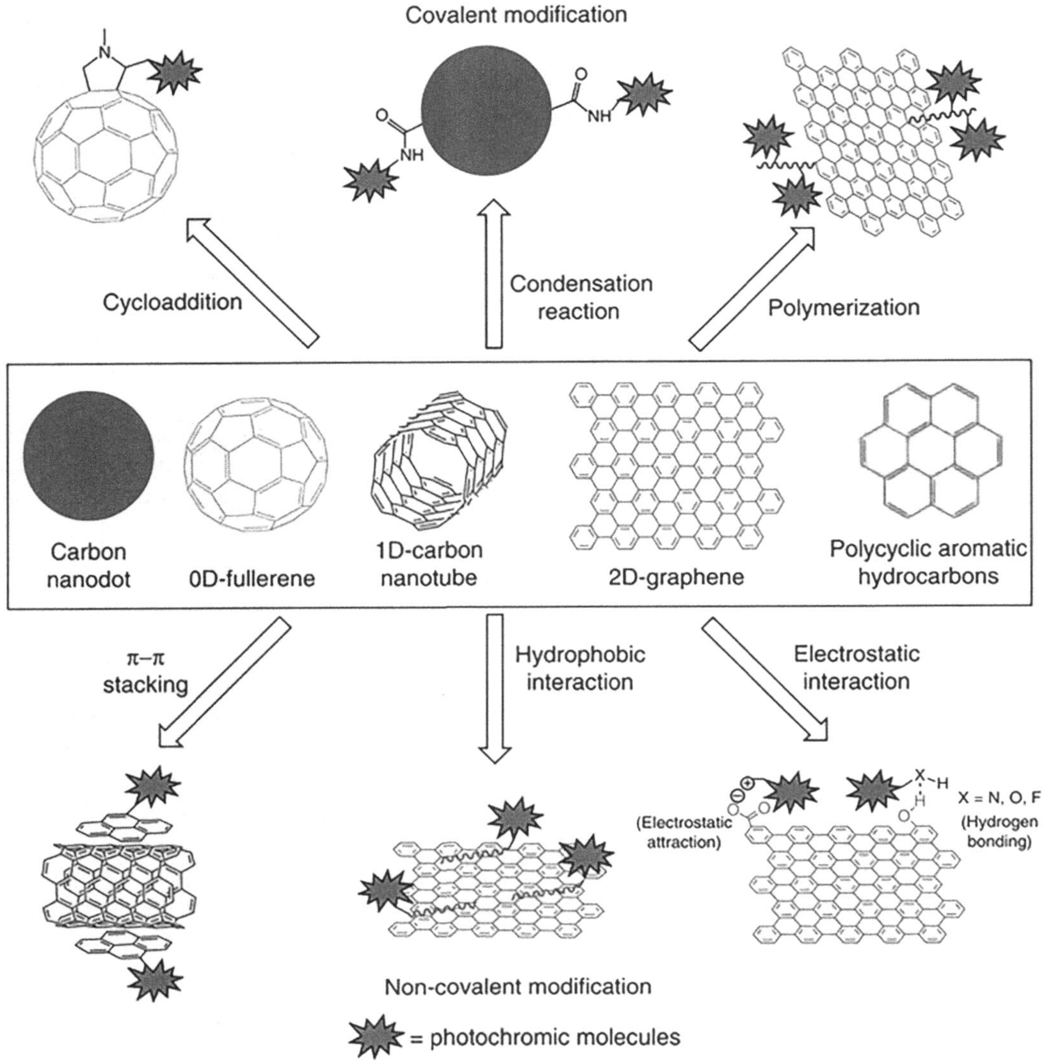

FIGURE 5.4 Functionalization of carbon-based nanomaterials using photochromic molecules. Reprinted with permission from Zhang et al. (2016). Copyright (2016) Nature. Distributed under the Creative Common attribution-based License CCBY 4.0.

finally, reduction to produce single-layered, functionalized carbon allotropes (Han et al., 2019; Zhang et al., 2020). The alteration of the intrinsic electronic characteristics of carbon allotropes when they are converted into functionalized carbon allotropes is one of the main drawbacks of this process. Nevertheless, although functionalized carbon allotropes are highly stable and relatively dispersible in water and organic solvents, they are frequently used as precursor materials to attach functional groups. A variety of noncovalent interactions, depending, among many others, on π-π and electrostatic attractions, contribute significantly to the chemical transformation of functionalized carbon allotropes. By electron transfer processes among layers of functionalized carbon allotropes and alkali metals and/or alloys in the initial state, a very unique route is established. After that, the appropriately produced cationic entities are intercalated into the reduced carbon allotropes to produce compounds with fascinating properties that are functionalized carbon allotrope complexation substances. To this objective, covalently functionalized graphene of very excellent quality is produced via electrophilic addendum using diazonium salts and alkyl halides (Alwan et al., 2021). In essence, targeted defunctionalization during the tempering of covalently functionalized graphene helps to restore the fundamental sp^2-hybridized carbon matrix of almost flawless graphene. The liquid-phase exfoliation of graphite using moderate sonication in solvents like N-methyl pyrrolidone (NMP) has garnered a lot of interest in comparison to the scenario of the above-mentioned method. In reality, it has developed into a flexible method for making persistent graphene suspensions. It is noteworthy that the produced graphene substance, which is existent but only has a couple of layers, is easily used in subsequent chemical modifications or functionalization procedures. Particularly, solvent-mediated dispersion of graphene sheet was exposed to radical, cycloaddition, and nitrene inclusion to show the covalent bonding of unmodified graphene produced by liquid-phase exfoliation of graphite. Furthermore, it is important to emphasize the noncovalent functionalization pathway using functional groups, such as surfactants. Simultaneous functionalization of graphite is made possible as well as the exfoliation that results from this. Ultimately, this method paves the way for hybrids that have even more intriguing properties than carbon allotropes. Controlling the doping concentration in graphene to alter the conductivity and the ensuing charge carrier mode transit is a prominent instance in the view of the recent Flow (Maślana et al., 2020; O'Bannon et al., 2020). Specifically, a solution approach to p-dope graphene has utilized an electron-accepting TCNQ.

In research, using a one-pot reductive diazotization method, Dasler et al. (2017) were able to effectively create graphene-porphyrin nanohybrid substances with a precise covalent bond across the functional porphyrin component and graphene carbon system. THF was used to dissolve a graphite potassium intercalation complex (KC8) before several purified porphyrin-diazonium salts were introduced. Raman and spectrometry were used to directly attach the functional hybrid substance and characterize it with great precision. As a result of the porphyrin's perpendicular alignment and single σ-bond linkage to the graphene layers in the end compound, the functional moiety was permitted to rotate. Raman and fluorescence analyses were used to thoroughly describe this extremely functional graphene–porphyrin composite complex. Additionally, LDI-ToF-MS was presented as a flexible and incredibly sensitive technique for the investigation of derivatives of covalently functionalized graphene. Combining the methodologies used made it possible to make inferences about the engagement and interaction of functional units with graphene. This sort of covalently connected hybrid composite can be used in numerous activities in a variety of fields, including corrosion mitigation, solar conversion devices, and electrical gadgets, taking into account the specific features of both porphyrin and graphene.

Similarly, by solution mixing using a solvent of hexafluoro-2-propanol, Cho et al. (2017) created a variety of polyketone (PK) nanocomposite sheets utilizing different contents of noncovalently functionalized graphene nanoplatelets with 1-aminopyrene (GNP/APy). The carbonyl substituents of the PK matrices and amine functional group of APy/GNP efficiently elicited specific contact, including hydrogen bonding, by the use of a simple approach for generating GNP/APy. In-depth research on the combination of the matrix and filler was accomplished using comparisons of GNP and GNP/Py as relevant sources. Additionally, the 3D nondestructive X-ray microcomputed tomography analysis of the scattering state of the functionalized GNP (f-GNPs; GNP/Py and GNP/APy) in the PK lattice revealed that the enhanced dispersion state of those admixtures results in notable enhancements in the water vapor transmission rate (WVTR). At a filler dosage of 1 wt.%, the PK/GNP/APy nanocomposite film's improved WVTR results in a barrier efficacy that was around two times greater than that of the PK/GNP nanocomposite sheet and a WVTR decrease of about 92% when contrasted to the condition of pure PK films. They anticipate that this simple technique of graphene functionalization would be applied extensively to increase the possibilities of graphene films for barrier applications such as corrosion mitigation of metals and alloys by improving graphene dissolvability and interface engagement with the polymer matrices.

5.2.2 Graphene

Oxygenated functional moieties, the most common of which are carbonyl, carboxyl, epoxy, and hydroxyl, constitute GO. Graphene has polar functional groups, which makes it extremely hydrophilic. Moreover, throughout the rGO manufacturing procedure, the oxygen functions in GO nanosheets are eliminated, which causes them to lack their solubility in water and subsequently go through irreversible agglomeration. Owing to significant π-π stacking or van der Waals pressures, graphene derivatives also readily agglomerate in solvents, creating considerable difficulties when preparing for further applications (Hussain et al., 2020; Serhan et al., 2019). To alleviate this problem, several functionalization techniques are used to enhance graphene dissolvability in polar solvents like water. Enhancing graphene miscibility in diverse solvents is the end objective of functionalization, which is an important stage in creating graphene-based nanocomposites for coatings and adhesives. Functionalized graphene (FGs) is a suitable substrate for catalytic activities because the functional units in FGs enhance catalyst intercalation and adsorption. The interface involving the polymer matrices and nano-reinforcing substance is improved by the functionalization of graphene, in addition to the latter's solubility. Charge transfer could be enhanced through the employment of oxygen-containing molecules in functionalization, which had the enormous possibility for monitoring and sensing gas and organic molecules. As a result, the investigation of graphene functionalization has substantially increased the applicability potential for graphene. Two methods have been used to functionalize graphene. They are graphene modifications that are covalent and noncovalent (Kharisov and Kharissova, 2019).

When an appropriate functional unit establishes a covalent bond to the sp^2 carbon network of graphene, particularly at the margins or underlying planes, covalent functionalization takes place. The aromatic nature of graphene could be modified by adding moiety via a covalent method, which can alter its electrical characteristics, enhancing stability and solubility and widening the bandgap. The development of covalent connections among the oxygen units of GO and functional groups, including biomolecules, organic molecules, metal nanoparticles, polymers, and others, has led to the development of a variety of covalent functionalization techniques. With organic compounds, including carboxylic acid groups, epoxide, amino groups, and hydroxyl groups, graphene could be covalently functionalized effectively. To increase dissolvability, for instance, the carboxylic acid units in the GO edges could be swapped out for amine groups like ethanolamine and ethylenediamine. Moreover, the aromatic basal planes of GO could be effectively functionalized with extremely reactive precursors including aryl diazonium salts and nitriles. This process may change the basal plane's sp^2 hybridization to sp^3 after functionalization. In comparison to non-functionalized GO, GO functional groups considerably increased dispersibility in dimethyl sulfoxide, water, and N, N-dimethylformamide (DMF), uniformly. The GO could be covalently functionalized using functional groups like the hydroxyl, carboxyl, and epoxy groups. Similar to this, polymers' amine groups and the carboxylic acid group of GO may undergo an amidation reaction. Following the interaction of ethyl(dimethylaminopropyl) carbodiimide/N-hydroxy succinimide EDC/NHS with the carboxylic acid group, PEI functionalization was accomplished on poly(acrylic acid) enhanced GO. The GO-PEI nanocomposites' in vitro proliferating and localized adhesions in hMSCs were enhanced by this covalent functionalization. PEI could also be electrostatically attached to the GO interface owing to its extensive cationic charge density (as described above), making it a flexible substrate for additional functionalization based on amine chemistry.

The binding of functional units to graphene is possible via noncovalent functionalization through π-interactions while affecting the material's electrical architecture. In this instance, the functionalities are not covalently bonded to graphene; rather, they are commonly adsorbed on the interface of graphene by relatively mild associations, including via anion-π, cation-π, H-π, π-π, and surfactant interactions. For instance, the negatively charged graphene electron cloud interacts in a way called " π-π stacking" with external molecules that have orbitals. According to this, metal cations like Cu^+, Na^+, and Ag^+ engage with the electron cloud of graphene through cation interactions. Although noncovalent functionalization won't alter the intrinsic electrical characteristics, it has been demonstrated that the noncovalent functionalization of graphene will increase its water solubility. To utilize them as supercapacitor electrode components, organic compounds with electrochemically active moieties have subsequently been inserted via noncovalent interaction into graphene layers. Currently, it has been discovered how to create noncovalently functionalized graphene sheets, that are largely employed in corrosion mitigation, solar cells, biosensors, and electrocatalysis. These aromatic dyes include methylene blue/green/congo red. This procedure uses forces like π-π stacking, hydrophobic interactions or electrostatic binding among the active compounds and graphene-based substances. The polar portions of surfactants engage with the hydrophobic attractions to integrate the aliphatic portions of surfactants over onto graphene, improving the durability of surfactant-functionalized rGO in polar solvents. Although GO layers are polarized, a positively charged biomolecule could be electrostatically bound over the top of the GO surface, which is another method for functionalizing these materials. The GO surface could also be adaptably modified using the same method using polymers and targeted ligands. According to a review by Li and Papadakis (2021), the addition of (poly)saccharides to graphene enables the resultant GBMs to be even more biocompatible and may pave the way for new click chemistry-based biological, anticorrosion, and biomedical implications. Diels Alder operations or other kinds of cycloadditions could also be used to directly functionalize graphene by click chemistry (see Figure 5.5).

To exfoliate graphite onto graphene layers and noncovalently functionalize a graphene surface using poly-N-vinyl-2-pyrrolidone (PNVP), Perumal et al. (2021) employed a one-step in situ polymerization technique. The resulting composite of poly-N-vinyl pyrrolidone (GPNVP) and graphene underwent extensive characterization. Calculating the proportion of PNVP on the graphene layer in the GPNVP composite required thermogravimetric investigation. Numerous investigations have supported the effective creation of the GPNVP composite and functionalization of the graphene sheets. The usual three-electrode method was used to conduct the EIS research of the GPNVP composite and the CV assessment at various scan rates (5–500 mV s^{-1}). With its capacitance feature, the GPNVP composite offers outstanding rate capabilities. The functionalization of graphene sheets and one-pot exfoliation with the heterocyclic polymer PNVP was demonstrated in this research; the resultant GPNVP composite would be an excellent alternative for a variety of electrode materials and anticorrosion materials. Figure 5.6 depicts the engagement of graphene substrates and PNVP to form a composite called GPNVP.

The X-ray diffraction (XRD) characteristics of the GP and GPNVP composite are shown in Figure 5.7a. Figure 5.7a makes clear the three main GP distinctive peaks, (101) at 44.6, (002) at 26.65, and (004) at 54.79. Three peaks may be seen on the GPNVP composite: (101) at 44.58, (002) at 26.54, and (004) at 54.64. In comparison to GP, the GPNVP composite exhibits a substantial reduction in levels, expanded dispersion spikes, and a minor change in the peak locations. These variations and alterations in peak locations point to the functionalization of the graphene sheets using PNVP as well as the exfoliating of graphene sheets in the GPNVP composites. Exfoliation was additionally demonstrated by the GPNVP composite's predicted interlayer separation of 0.336 nm, which was greater than GP (0.334 nm). The small enhancement in the interfacial grating of the GPNVP composite shows that the full or partial exfoliating of graphite into graphene sheets was aided by the in situ polymerization of NVP monomers onto graphene interfaces. The thermogravimetric analysis (TGA) profiles of the GP and GPNVP composite are displayed in Figure 5.7b. For GP, a minor weight reduction can be seen above 100°C, which would be due to moisture/water evaporating. Following this, no more significant weight reduction was seen, and a minor initial mass was still present. The GPNVP composite exhibits a weight reduction beyond 100°C, which corresponds to water molecules that have been adsorbed. PNVP degrades at temperatures between 350°C and 420°C, losing 84% of its weight along the way. The full breakdown of PVP occurs in the last stage, around 420°C–570°C, with an 83% weight reduction

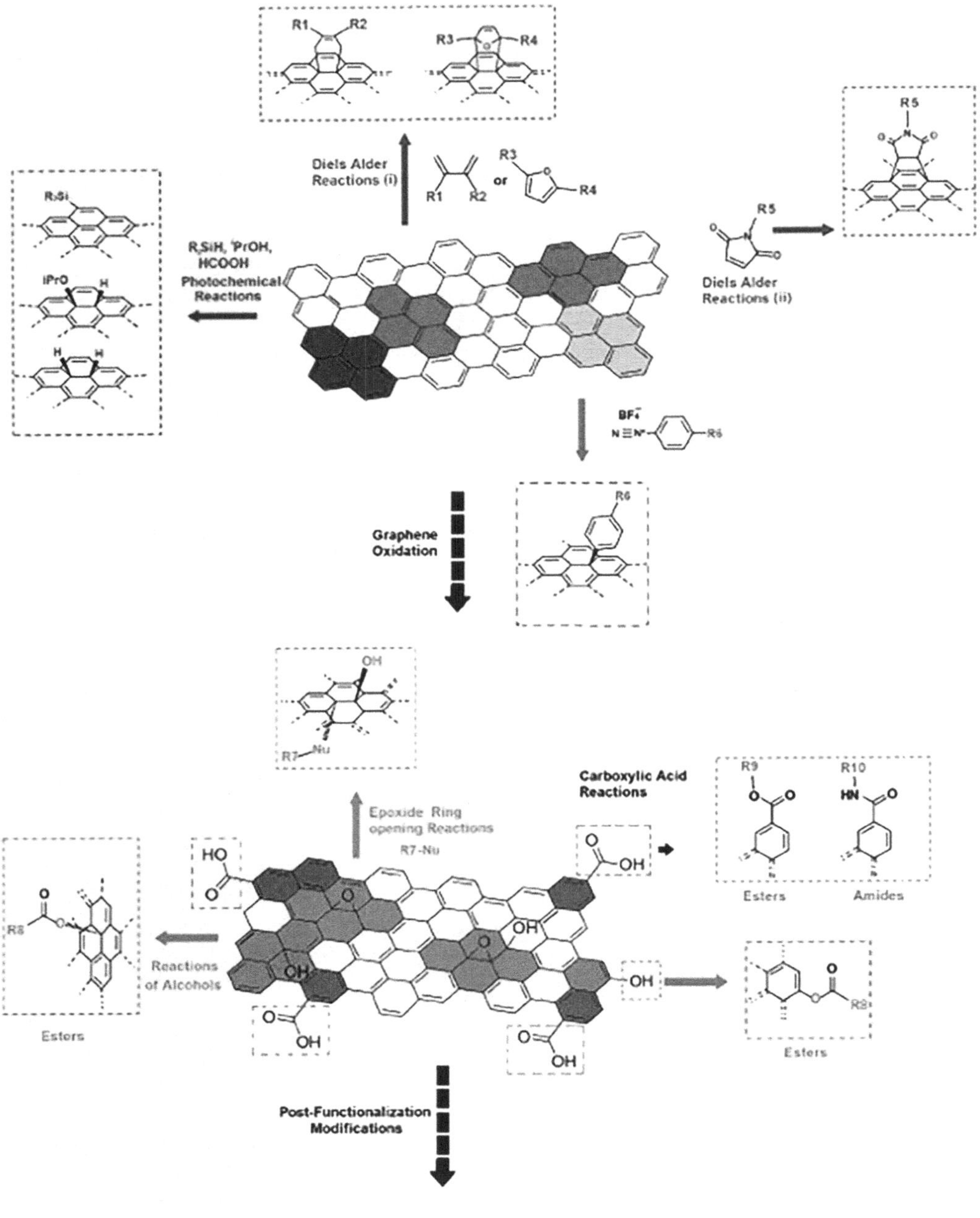

FIGURE 5.5 Different graphene and GO functionalization processes that are pertinent to and enhance each other's activity in click chemistry. Reprinted with permission from Li and Papadakis (2021). Copyright (2021) MDPI. Distributed under the Creative Common attribution-based License CCBY 4.0.

and about 80% carbon residual remaining. The proportion of PNVP functionalized onto graphene substrates in the GPNVP composite was estimated from this TGA analysis to be around 15%.

Through the use of carbon disulfide (CS_2) functionalization for Pb(II) adsorption, Lian et al. (2020) demonstrated the impact of the surface reactivity of GO. CS_2 functionalized GO (GOCS) was effectively prepared, and precise methods were then used to examine Pb(II) adsorption over GOCS. The highest adsorption capability of Pb(II) on GOCS was found to be 31% higher than that

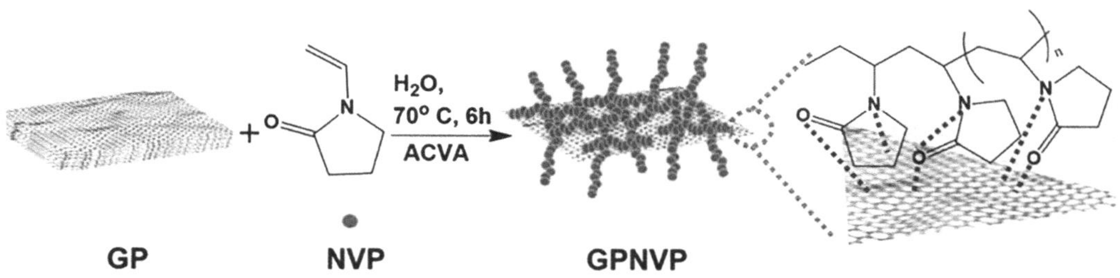

FIGURE 5.6 In situ polymerization used to generate graphene with a poly-N-vinyl-2-pyrrolidone functionalized interface. Reprinted with permission from Perumal et al. (2021). Copyright (2021) MDPI. Distributed under the Creative Common attribution-based License CCBY 4.0.

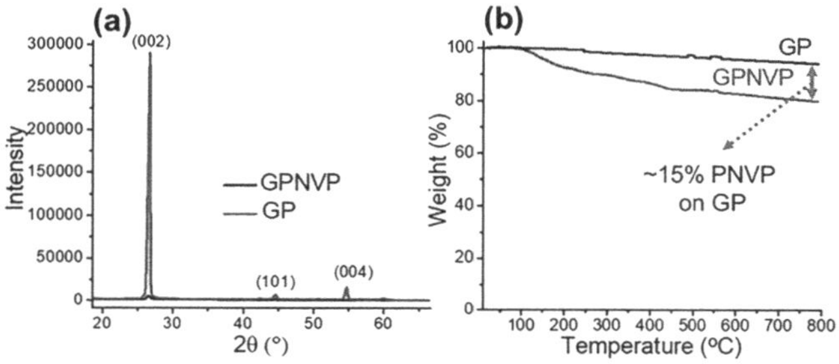

FIGURE 5.7 (a) GP and GPNVP composite XRD spectra and (b) TGA curves in an N_2 environment. Reprinted with permission from Perumal et al. (2021). Copyright (2021) MDPI. Distributed under the Creative Common attribution-based License CCBY 4.0.

of GO, expanding from 292.8 to 383.4 mg g^{-1}, according to the outcomes, which showed that the innovative sulfur-containing functional factions added to GOCS substantially augmented Pb(II) adsorption capability on GOCS than those of GO. At the same preliminary dosage of 150 mg L^{-1} and an ideal pH of 5.7, the equilibrium adsorption capability of GOCS was 280.2 mg g^{-1}, an increase of 83.2% over those of 152.97 mg g^{-1} for GO. The outcomes of the adsorption studies also demonstrated a remarkable agreement with the Langmuir and pseudo-second-order frameworks, which represent chemical and monolayer adsorption, correspondingly. The coordinated, electrostatic forces, cation-π interrelations, and Lewis acid–base associations were suggested as the mechanisms for Pb(II) adsorption on GOCS. Following five regeneration processes, the rejuvenation investigation revealed that GOCS exhibited a respectable degree of reusability for Pb(II) adsorption, with an adsorption capacity of 208.92 mg g^{-1}. In conclusion, it has been demonstrated that GOCS is an innovative, practical, and presumably affordable adsorbent for the highly effective extraction of Pb(II) from aqueous systems.

5.2.3 Carbon Nanotubes

Carbon nanotubes (CNTs) are sp^2 hybridized massive tubular structures with a hexagonal lattice of carbon atoms (C-C spacing of about 1.4 Å). One or more sheets of graphene layers constitute the CNTs' sidewall. Single-walled carbon nanotubes (SWCNTs) are generated from single-rolled graphene layers, whereas multi-walled carbon nanotubes (MWCNTs) are generated by wrapping multiple graphene layers (Malig et al., 2013; Ling et al., 2013; Lo et al., 2021). In a hemispherical configuration of carbon systems termed as the fullerenes deformed up by the graphene layer, MWCNTs and SWCNTs are sealed at both extremities of the ducts. Whereas SWCNTs only possess a dimension in the region of 0.6–2.4 nm, MWCNTs possess interior graphene sheet distinctions that are on mean 0.34 nm, resulting in single tubes with wider outside diameters of 2.5–100 nm. In contrast to MWCNTs, which contain minor morphological flaws and exhibit a slightly unsteady nanostructure, SWCNTs possess a well-defined wall architecture. CNTs are excellent prospects for the distribution of a variety of chemicals due to their compact structure, excellent surface region to volume fraction, and capacity to hold biomolecules.

Due to their inherent hydrophobicity, which makes them extremely challenging to scatter in the aqueous solution, carbon nanotubes (CNTs) underwent functionalization. Consequently, to increase their mobility and assimilation with the biological environments, their interfaces are typically functionalized either covalently or noncovalently containing hydrophilic functional units or amphoteric macromolecules like lipids, polysaccharides, and surfactants. Functionalization via covalent bonds involves direct side-wall functionalization or defective group functionalization, which includes transforming or rehybridizing sp^2 carbon into sp^3 geometries and ultimately generating covalency to invading molecules, which are two methods for achieving covalent changes. Cycloaddition of azomethine ylides, Halogenation, or the inclusion of radicals can all be used to do this. The surface alterations of the defect component are completed by creating flaws by oxidation to provide carboxylic acid functions, which gives it the ability to bind other targeting units like proteins, peptides or antibodies using esterification or amidation. Multifunction CNTs, which could be employed for medication delivery or anticorrosive materials, are made via triple and double covalent functionalization. The amidation/esterification, 1,3-dipolar cycloaddition and cyclopropanation, amidation, amidation,double arylation and cyclopropanation, double 1,3-dipolar cycloaddition, etc. are all methods for achieving double covalent functionalization. By simultaneously functionalizing with several aryl diazonium salts, triple functionalization could be achieved. In their discussion, Díez-Pascual et al. (2021) concentrated on the covalent functionalization through "grafting onto" or "grafting through" approaches and gave a quick summary of the various methods for CNT modification utilizing polymers. They claimed that carbon atoms transition between sp^2 to sp^3 hybridization when a covalent alteration takes place. One method involved the extreme-temperature air-based oxidation of CNTs, which caused the end caps to unfold. Carboxyl, ketone, hydroxyl, carbonyl, and epoxy units were created on the CNT interface as a result of oxidation, which enhances the CNTs' engagement with polymers. However, chemical functionalization altered the electrical structure of the carbon nanotubes, causing an increase in sp^3 hybridization, which in turn caused a band gap to expand, lowering the conductivity. As previously mentioned, there are two types of functionalization of CNTs using polymers: noncovalent or covalent attachment of polymerization to nanotubes. In the second technique, CNTs have been irreversibly functionalized through integrating polymers to their rims (sidewall functionalization) or at flaw locations (flaw functionalization), which were also commonly found at the frontiers and could be achieved using either "grafting onto" or "grafting through" methodologies. The harmful consequences of the covalent alterations of CNTs were avoided by using noncovalent functionalization techniques. Wrapping or adsorption of surfactants, biopolymers, and polymers over the tubular interface via van der Waals effects and π-π stacking may result in noncovalent functionalization. For instance, when immersed in a liquid of the polymer, poly(vinylpyrrolidone), polyethylene glycol (PEG), tetraalky-lammonium, poly (meta phenylene vinylene), etc. are used to construct a casing surrounding the SWCNTs. Surfactants that adsorb on the interface of CNTs via interfaces, including sodium dodecylbenzene sulfonate (SDBS), sodium dodecyl sulfate, Triton X-100, etc., assisted to increase dissolvability, dispersion, and penetration across the system.

To create MWCNTs containing hydroxyl groups, Ling et al. (2013) subjected MWCNTs to four moist chemical oxidants (Na_3BO_3, Na_2ZnO_2, $NaAlO_2$, and NaOH). The findings show that modified MWCNTs have hydroxyl groups on their surfaces whilst yet maintaining their structural integrity. According to their dispersal patterns, modified MWCNTs may disseminate in chloroform more easily than untreated MWCNTs. The percentage of hydroxyl groups in the four MWCNTs processed with $NaAlO_2$ was the maximum. By covalently bonding carboxylic acid-functionalized

MWNTs by the diamino-functionalized beta-cyclodextrin via amide interconnections utilizing a water-soluble condensation reagent at ambient temperature, Bhoi et al. (2013) were able to immobilize beta-cyclodextrins upon the substratum of MWCNTs in an aqueous solution. The final substance, known as beta-cyclodextrin-modified MWCNTs, was very soluble in water. By using TGA, FTIR, EDS, and TEM, the covalent surface functionalization of MWNTs by beta-cyclodextrin was identified. According to the TGA studies, MWNTs have about 70% beta-cyclodextrin bonded to their surface. Additionally, the fluorometric assessment for the sorption process of rhodamine 6G dye recommended that beta-cyclodextrin-dye incorporation conjugate was formed previous to the dyes adhering to MWNTs and that consistent MWNT distribution following surface immobilization enables outstanding fluorescence quenching rather than the immaculate and oxidized MWNTs.

To create a novel nanocomposite electrode for high-performance supercapacitors, Sathish Kumar et al. (2021) deliberately combined ferrous tungstate with functionalized MWCNTs (FeO4W/ f-MWCNTs). The intrinsic capacitance of this supercapacitor electrode was as large as 875 Fg $^{-1}$. The cycle testing revealed that following a massive 10,000 cycles in a three-electrode setup, capacitance stability was 89.17%. Additionally, the FeO4W/f-MWCNT-based symmetrical supercapacitor device produced a significant specific capacity with excellent capacitance persistence. The existence of functionalities in the composites was confirmed by analytical methods like UV and FTIR spectroscopy, whereas TGA evaluation of the nanocomposite reveals superior thermal stability than its competitors. The XRD methodology was used to determine the crystallinity of the composite. With the use of TEM and a few area electron diffraction techniques, the interplanar spacing (dhkl) in the nanomaterial was calculated.

5.2.4 Fullerenes

Fullerenes, C60, and buckminsterfullerene are examples of endohedral fullerenes, often referred to as buckyballs or buckyball clusters. These molecules each contain smaller than 300 carbon atoms. To assist cellular scapegoating, fullerenes have a variety of functional sites that allow the binding of chemical units of targeting ligands in 3D orientations. Additionally, it is feasible to change the properties of their allotropes to enhance their therapeutic effects, thickness, wettability, and colloidal durability in a biological context. Corannulene, naphthalene, or further polycyclic compounds are examples of polycyclic aromatic hydrocarbons that might be pyrolyzed to produce fullerenes. With the breakage of hydrogen bonds and the generation of mostly C60 and C70, these compounds disintegrate at extreme temps (about 1000°C) in the vicinity of an inert gas (argon) (Karuppiah et al., 2016; Kuila et al., 2012). Fullerenes can be easily made soluble in both water and organic solvents by surface functionalization. Two alternative chemical processes can be used to modify the fullerene surface: (i) complexation using a solubilizing reagent to substantially cover the hydrophobic fullerene outer layer; and (ii) covalent functionalization of the fullerene substratum. Employing potent acids at extreme temps, primarily hydroxyl groups, oxygen-based functional groups may be transplanted onto the surface of fullerenes. At a temperature of 85°C–115°C, C60 and C70 fullerenes were combined with an aqueous solution containing nitric and sulfuric acids. The fullerenes' interfaces were oxidized by the acid, which led to a mean grafting of 15 hydroxyl groups per fullerene. If tetrabutylammonium hydroxide is present and functioning as a catalyst, introducing strongly basic NaOH to C60 benzene mixtures could also result in the formation of OH-functionalized fullerenes (fullerenols). There are more attached hydroxyl groups as a consequence of the reaction's better efficiency than that of stronger acids.

To address the issues of poor water dissolution and weak dispersion in a variety of organic solvents, fullerenes are functionalized. Covalent functionalization and the application of solubilizing agents to substantially conceal the interface of the fullerenes could be used to modify their interface. The fullerene framework has double bonds, which allows for significant modification by desired functional groups. Fullerenes are made of carbon molecules that have irregular sizes and resemble hollow spheres or tubes. They are made entirely of carbon and come in a variety of forms that resemble hollow spheres, ellipsoids, or tubes. Their strong apolar nature allows them to form lipid-like systems, which could readily pass through biological membranes. In their research, Maleki et al. (2020) created their smart drug carriers and Ravi used molecular dynamics (MD) simulations to examine how the PAX and DOX loaded and released onto the pH-sensitive functionalized Fullerene carrier. They also investigated how the chitosan polymer affected the Fullerene carrier's functionalization. Additionally, to enhance the loading and discharging characteristics of the PAX and DOX, the fullerene carrier was designed and synthesized with carboxyl groups. The findings demonstrated that the DOX is effectively absorbed onto fullerene, which has been functionalized using a carboxyl group, and that its distribution into malignant tissues was under regulation. In contrast to PAX, the functionalized fullerene could be a suitable carrier for DOX, as determined by the Van der Waals and electrostatic interactions. By extending the period at an acidic pH, the Van der Waals energy approaches zero, which promotes the discharge of the PAX in malignant cells. The trimethyl chitosan (TMC) polymer might enhance the Van der Waals affinities of the fullerene and PAX. At acidic and neutral pHs, correspondingly, the carboxyl group that was used as a functionalizing reagent might also enhance the number of hydrogen bonds for the DOX and PAX. The addition of TMC to the carrier also resulted in a notable increase in the number of hydrogen bonds across the fullerene and PAX at neutral pH. At acidic pH, a greater reduction in gyration radius was seen for the DOX, indicating that the DOX containing TMC-fullerene was the more robust carrier. So, a potential formulation for smart cancer therapy was implemented in the form of this smart nanomedicine platform.

To fully regulate the kinetics and axial regioselectivity of Bingel-Hirsch cyclopropanation processes of a contained C60 guest, Fuertes-Espinosa et al. (2020) employed a nanocapsule as a supramolecular tetragonal prismatic cover. This means that equatorial bis-, tris-, and tetrakis-C60 homo-adducts can only be produced gradually. For the first occasion, hetero-Th-hexakis-adducts or isomer-pure equatorial hetero-tetrakis-adducts are produced in a single pot. To fully regulate the functionalization of encased C60 in respect of unique axial regioselectivity and sequential synthesis of the bis-, tris-, and tetrakis-C60 homo-adducts, the supramolecular mask's four cross-shaped open windows were essential. Additionally, for the first moment, hetero-Th-hexakis-adducts or isomer-pure equatorial hetero-tetrakis-adducts were produced in a single pot. By utilizing catalytic and anion metathesis quantities of the nanocapsule, a synthesized relevant biphasic procedure was constructed to produce precise and precise production of solely tetrakis-adducts, upgrading this method into a catalytic regime. The supramolecular mask technique described here offered a substantial enlargement of the regioselective technique in contrast to prior instances of functionalization of C60 in bowl-shaped voids restricted to DA mono-adducts. Astounding pure-isomer poly-functionalized C60-based compounds were produced in this research in fact, and they are candidates for evaluation in corrosion, solar cell devices and biological activities.

5.2.5 Nanodiamonds

The freshest carbon allotropes to be discovered are nanodiamonds (NDs), which have a significant surface region and a –single-crystalline with a diameter of 2–8 nm(Karuppiah et al., 2016; Kuila et al., 2012; Sakellariou, 2015; Walters et al., 2020). The implosion method and chemical vapor deposition techniques are the main ways that NDs are produced. They are nanocrystals containing tetrahedral carbon atoms in a 3D cubic trellis that have the electrical characteristics of a diamond and are covered in a casing of graphite that resembles an onion. Their surface consists of functional groups, typically including oxygen, ketone, hydroxyl, carboxylic acid, lactone, and ether, that were produced on the ND interface following purification as a consequence of the detonation procedure. Fluorescent NDs (FNDs) comprise NDs containing nitrogen voids in the center that display photoluminescence properties and have developed a variety of extremely intriguing applications in bioimaging. The structure and surface characteristics of NDs, among other physicochemical characteristics, play a major role in determining their fluorescence qualities. The aggregation of NDs dictates their size after application because single crystals of them are extremely uncommon in solution. However, because of the functional groups, such aggregations are frequently hydrophilic and may still create a persistent aqueous colloidal suspension, which is essential for their biological usefulness. Owing to this characteristic, NDs have also been considered drug transporters. Covalent and noncovalent functionalization techniques are used to bind or attach a variety of moieties to the NDs' interface. The specific surface chemistry of ND is difficult to describe and regulate due to the variety of the ND surface caused by the manufacturing process. The covering of ND to casings of other materials that could harbour molecules in many elevated quantities than what could be adsorbed onto the native ND substratum and otherwise enhance their biomedical potential application is yet another frequently used method to take advantage of the innate photoluminescent characteristics of ND while expanding their drug-carrying ability. The amount of drug elimination from the liquid, maximum monolayer capability, and binding strength are all factors that affect how well a drug binds to native nanodiamonds. The ND functionalization techniques involve modifications to the diamond surface chemistry, which is largely established. H_2O_2 and $FeSO_4$ (Fenton), which rapidly assault the non-diamond stages and endow the surfaces with OH groups, can likewise be used to dissolve the graphitic shell. Additionally, mechanical techniques like grinding and ultrasounds in water that trigger radical processes and the -(OH) terminating can be used to generate diamond surface hydroxylation. When the graphitic/amorphous covering on the diamond core is removed using highly oxidizing chemicals, including singlet oxygen in NaOH, strong acids, strong ozone, and air processed in the context of a catalyst, the ND surface can get oxidized. In the manufacturing of composite materials and drug delivery, this chemical functionalization has been extensively utilized. Noncovalent functionalization is a simple technique, but because the adsorption mechanism cannot be controlled, it has low repeatability. To fuse further molecules, the covalent functionalization uses hydroxyl and carboxylic groups. Examples include the reaction between amines and carboxylic groups following activation using thionyl chloride. The PEGylated NDs were made using this procedure. We note that PEG-chains could be finished with azido, amino, and thiol groups, which are often employed to covalently attach several other biomolecules. Derivatization of the hydroxyl groups through esterification produced through extreme temperature -OH carboxylation succinic anhydride under extreme temps is another commonly used functionalization. NDs can also be derivatized using siloxanes as an alternate method. Furthermore, this type of functionalization only provides a limited amount of flexibility over the density of the siloxane coating on the ND core, which also depends on the employment of the siloxane precursor. Siloxane functionalization has been used to covalently bind dyes and receptors and deliver drugs. Other forms of functionalization involve reactions between hydroxylated NDs and alkyl chlorides, functionalizing H-terminated NDs using alkenes, and using diazonium salts. Ultimately, it is possible to convert carboxyl groups on the ND surface into azido groups that are prepared for additional click chemical reactions.

To increase the colloidal dispersion and oral absorption of the formulation, Cheng et al. (2018) created a D-tocopherol polyethylene glycol 1000 succinate (TPGS) adorned nanodiamond (ND) combination loaded water-insoluble curcumin (ND/CUR/TPGS). Physically loading CUR into ND clusters was followed by coating TPGS over the ND/CUR compound to create an amorphous nanostructure on the ND substrate's interparticle nanocage. The response surface approach was used to optimize the composition of the nanocomplexes, and the ideal ND/CUR/TPGS demonstrated tiny particle size (196.32 nm), excellent drug loading efficacy (81.59%), and core configuration. An examination of the nanocomplexes' in vitro discharge behavior showed that they had persistent discharge behavior. When contrasted to CUR suspensions, the absorptive content of ND/CUR/TPGS was much higher in the entire intestinal tract, and the absorption was regulated by several transcytosis pathways. Additionally, the pharmacokinetic tests showed that, in comparison to CUR solutions, ND/CUR/TPGS had considerably greater Cmax (4.50-fold), higher AUC0-t

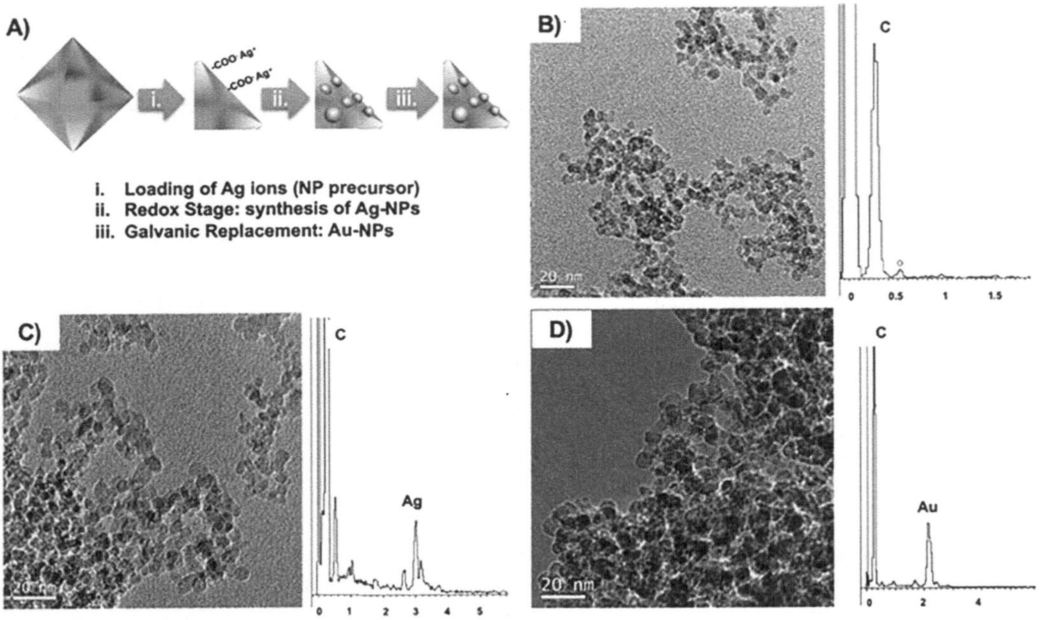

FIGURE 5.8 (a) A schematic of the functionalization procedure, (b) nanodiamonds (NDs) TEM scans and EDS, (c) Ag nanodiamonds TEM scans and EDS, (d) Au nanodiamonds TEM images and EDS. Reprinted with permission from Montes et al. (2020). Copyright (2020) MDPI. Distributed under the Creative Common attribution-based License CCBY 4.0.

(10.67-fold), and prolonged MRT0-t (3.07-fold). Consequently, ND/CUR/TPGS offered excellent promise for the oral distribution of insoluble and inadequately permeable medicines.

Montes et al. (2020) created viable (bio)sensing platforms and Ravi assessed the suitability of NDs enhanced using silver (Ag@NDs) and gold (Au@NDs) nanoparticles. To enhance the electroanalytical capabilities of the created (bio)sensors, a thorough geometrical and electrochemical evaluation as a function of the synthesized nanocomposite composition was carried out. The best ratio in the current study was 3.5% for Au@NDs and 1% for Ag@NDs on the nanocomposite matrices. Utilizing a (bio)sensor constructed on graphite-epoxy-Ag@NDs, positive outcomes were attained in the assessment of the ideal combination of glucose and hydrogen peroxide as a model analyte (17:82:1). The design for the ND modification is shown in Figure 5.8, along with TEM pictures and the accompanying EDS for all three systems (NDs, Ag@NDs, and Au@NDs). In the ND dimension, a strong monodispersity could be detected. Additionally, the mean size of the NDs was <4 nm (3.5 ± 0.3), and it expanded when Ag and Au NPs were added, attaining 4.5 nm (Ag@NDs) and 6.7 nm (Au@NDs), respectively. Additionally, the Ag@NDs and Au@NDs systems' EDS tests verified the existence of the Ag and Au NPs bonded to the NDs.

5.2.6 Graphite

A crystalline allotrope of carbon, graphite is a type of coal and a naturally occurring elemental mineral. Extremely coarse flake graphite, also known as amorphous, lump graphite, is a type of natural graphite that can be found as microscopic, crystalline flakes (Mountrichas et al., 2009). The physical characteristics of graphite's alpha (hexagonal) and beta (rhombohedral) forms are remarkably identical. Although this allotrope has been a traditional form of carbon for two centuries, researchers are still very interested in it. Under normal settings, graphite is the most durable form of carbon. Graphite has a planar, sp^2 hybridized structure of graphene layers.

D46h (P63/mmc) hexagonal lattice describes the crystalline structure. Each layer's carbon atoms are organized in a honeycomb crystal lattice with a 0.142 nm spacing between them, and there is 0.335 nm between each plane. Only three of the four possible bonding spots are achieved, resulting in the covalent bonding of the atoms in the plane. Graphite is electrically conductive because the fourth electron is unimpeded in its ability to move along the plane. Electricity is typically conducted along layers rather than perpendicular to them, that is, inside the plane of the layers. The weaker van der Waals connections used for bonding among the layers enable for easy separation or sliding of the layers of graphite past one another.

Kamedulski et al. (2019) reported findings on the carbonization of particular precursors in the context of a rigid template, resulting in the 3D functionalization of graphite-originated flakes to graphene. As a rigid framework, in situ deposited $CaCO_3$ nanopowder or Na_2CO_3 nanocrystals were utilized. Conventional graphite was exfoliated using wet chemistry to produce graphene flakes. The non-specific binder and hard template were combined with the flakes beforehand, and they were carbonized at temperatures between 700°C and 800°C while being blown with nitrogen. Although the surface area of raw graphite was reported to be 25 $m^2\ g^{-1}$, the inclusion of a template enabled increases in the surface region of 333 $m^2\ g^{-1}$ for the $CaCO_3$ template and 287 $m^2\ g^{-1}$ for the Na_2CO_3 template. The produced 3D-graphene materials were characterized using a variety of experimental techniques, including combustion elemental characterization, TEM, SEM, XPS, Raman spectroscopy, and N_2 adsorption at moderate temperatures. It was shown how the surface area of the 3D structured graphene was affected by the carbonization temperature and the insertion of a template. The deglomeration of graphene flakes into few (FLG) and double (DLG) layered graphene was accomplished using the wet-chemistry approach.

To efficiently convert extended graphite to FFG, Bu et al. (2022) thoroughly examined the use of gallnut extract (GNE) throughout the high-shear exfoliation procedure. Eventually, several variables, including GNE content, graphite content, exfoliation duration, and high-shear mixer angular velocity, were tuned for FFG manufacturing. Utilizing Raman spectra and TEM studies, the produced FFG was examined for surface functioning and shape. Energy-dispersive X-ray spectra also supported the conjugated FFG with Ag. Using the agar well dispersion technique, the Ag-FFG composite displayed antibacterial efficacy towards both Gram-negative and Gram-positive bacteria. This research offered a productive, affordable, and environmentally friendly way to produce FFG and Ag-FFG for biomedical and other crucial uses such as corrosion mitigation of several metals and alloys. The schematic representation for the FFG production procedure is shown in Figure 5.9.

5.2.7 Nanofiber

Carbon nanofibers (CNFs) are incredibly tiny, cylinder-shaped structures made of graphene layers. The Soviet researchers Lukyanovich and Radushkevich made the first electron microscope investigations of CNFs in 1952. In the *Russian Journal of Physical Chemistry*, they wrote a paper regarding hollow graphitic carbon fibers (50 nm in diameter) (Cho et al., 2017; Dasler et al., 2017). Despite several parallels between CNFs and CNTs, scientists paid far less attention to the first class of nanoobjects. Distinct or equivalent methods are utilized for their production; CNFs and CNTs have various morphologies and comparable nanoscale diameters. As a result, the diameter of CNTs is between 1 and 30 nm, whereas that of CNFs is between 50 and 200 nm. Considering the manufacturing process, CNFs and CNTs can have a diameter of several hundred micrometers. CNFs are easier to spread than CNTs because the van der Waals forces are lower in CNFs. When contrasted to CNTs, manufacturing CNFs costs substantially less ($100–$500 per pound). Additionally, because CNFs' graphene edges are visible, they are more reactive than CNTs and may quickly modify their morphology to form chemical bonds with any matrices using heat or chemical functionalization techniques.

Choi et al. (2020) proposed the creation of a cellulose nanofiber membrane with thiol functionalization that successfully adsorbed heavy metal ions. As a result of the deacetylation of

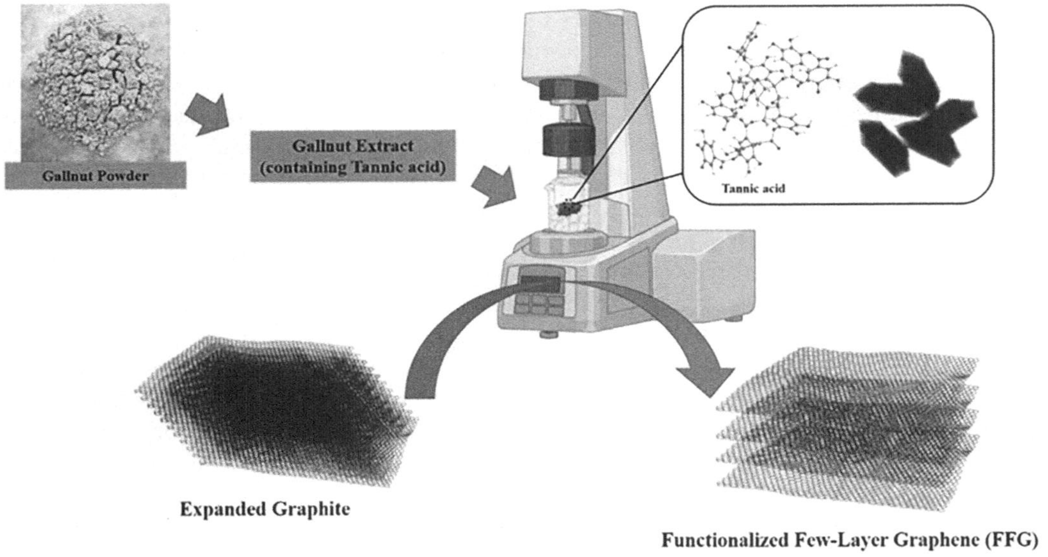

FIGURE 5.9 Illustration of the formation procedure of FFG. Reprinted with permission from Bu et al. (2022). Copyright (2022) MDPI. Distributed under the Creative Common attribution-based License CCBY 4.0.

electrospun cellulose acetate nanofibers and consequent esterification of a thiol precursor molecule, thiol was added to the interface of cellulose nanofibers. Adsorption isotherms were used to evaluate the adsorption process. The Langmuir isotherm provided a good description of adsorption capability as a consequence of adsorbent dosage, indicating that metal ions form a surface monolayer with uniformly dispersed adsorption energy. Cd(II), Cu(II), and Pb(II) ions had the highest adsorption capabilities in the Langmuir isotherm, with values of 45.9, 49.0, and 22.0 mg g^{-1}, correspondingly. The time-dependent sorption capabilities were predicted by a pseudo-second-order kinetic framework, which implies that each doubly charged metal ion undergoes chemisorption when two thiol groups are present on the interface. These findings emphasize the value of surface functioning on nontoxic and ecological cellulose substances to increase their usefulness and promise for water remediation and anticorrosion activities.

Haidyrah et al. (2021) used straightforward manufacturing of functionalized carbon black hybrid composite and carbon nanofibers (f- CNF/CB) to exhibit amperometric identification of the antibacterial medication nitrofurantoin (NFT). The f-CNF/CB composite was made using a straightforward one-pot ultrasonication technique. Many morphological and spectral techniques were used to characterize the f-CNF/CB combination. The glassy carbon electrode (GCE) was used to create the f-CNF/CB composite to expedite the electrode–electrolyte interface's quick electron transmission mechanism. Therefore, in physiological pH conditions, the f-CNF/CB/GCE demonstrated improved electrochemical efficacy over NFT. With a detection limit and sensibility of 16 nM and 15.25 μA μM^1 cm^2, correspondingly, amperometric analysis of NFT employing f-CNF/CB/GCE showed a broad reaction spectrum from 0.05 to 104.66 μM. Additionally, the f-CNF/CB hybrid composite's enhanced electrochemical efficiency might be used to create next-generation anticorrosion materials, sensors and energy storage systems like capacitors and batteries.

5.2.8 Carbon Onion

Other than electron irradiation, many synthesis techniques for carbon nano-onions were established around the end of the twentieth century. Boron trichloride, Acetylene and ammonia precursors could be broken down utilizing hydrogen as a transporter in chemical vapor deposition (CVDs) to form CNOs. As a consequence, quasi-stoichiometric boron-nitride regions are present in the CNOs. In another study, a Ni/Al catalyst assisted CVD to break down methane and create CNOs (Sathish Kumar et al., 2021). Other authors have also used CVD to make metal-containing CNOs. To create the catalysts, a sol-gel comprising Ni and Sn was spin-coated onto a Si substrate, dried, and heated to 600°C. Then, CVD was used to decompose a combination of hydrogen and methane, resulting in the creation of CNOs having Ni_3Sn_2 cores. Ni was used as a catalyst in the production of CNOs and CNTs in counterflow diffusion flames. A base burner utilized a combination of nitrogen and oxygen, whereas an upward burner utilized a combination of 5% ethylene, 15%–45% methane, and the remainder part oxygen. A quasi-stable one-dimensional flame exhibiting operating temperatures between 400°C and 1200°C was produced under low flow rate circumstances. The degradation of methane produced a significant output of CNOs or CNTs by altering the quantities of methane, ethylene and oxygen. Additionally, naphthalene was burned in a combustion flame to yield gram-scale amounts of CNOs. Resonant stimulation of precursor particles and acoustic manipulation of the laminar flame were two techniques used to increase the efficacy of the manufacturing of CNOs as well as alternative precursors such as phenolic resins.

With a prolonged drug distribution characteristic, Mamidi et al. (2020) presented a novel class of pH-controlled composite nanofibers made of polycaprolactone/mercaptophenyl methacrylate functionalized carbon nano-onions (PCL/f-CNOs). SEM and FTIR were used to examine the morphological and structural properties of PCL/f-CNOs nanofibers. According to the morphology findings, FS offered homogenous, bead-free nanofibers with mean diameters ranging from around 215 to 596 nm. Over the course of 15 days, PCL/f-CNOs composite fibers demonstrated pH-responsive emission of DOX; pH 6.5 displayed 87%, and pH 5.0 showed approximately 99%. Drug discharge studies revealed that force-spun PCL/f-CNOs fibers released regulated amounts of DOX as a result of the interactions among f-CNOs and DOX. Due to the addition of f-CNOs, PCL/f-CNOs fibers demonstrated increased tensile endurance (3.16 MPa) in comparison to pure PCL fibers. This demonstrates the degree of f-CNOs' physisorption and colloidal stabilization within the PCL matrices. Additionally, human fibroblast cells were used to test the in-vitro viability of cells, and excellent cellular vitality was found. However, pH-responsive PCL/f-CNOs composite nanofibers with DOX contained in them may have been used in the field of biomedical and anticorrosion research.

To explore the photocatalytic destruction of 2,4-dinitrophenol, Park et al. (2019) focused on the visible-light-induced photocatalytic efficiency of a 3D hybrid composite built on carbon nano-onion (CNO)-functionalized zinc-oxide tetrapods (T-ZnO). A simple one-step procedure was used to effectively create the hybrid CNO-functionalized T-ZnO 3D composite. Through chemical blending, the CNOs were adorned on the surface of T-ZnO after being produced environmentally friendly from flaxseed oil. Such a hybrid composite enables, thorough T-ZnO/CNO interface modification, improvement to visible-light harvesting and contributes to efficient visible-light-induced photocatalysis. The extraordinary charge transmission and dissociation achieved by the 3D geometry of T-ZnO and the combined inherent features of CNOs were associated with the improved photocatalytic efficacy of the T-ZnO-CNO composite. Depending on a bandgap investigation and trapping tests, a hypothesized reaction pathway for the breakdown of DNP was established. Throughout repeated cycling investigations, the photocatalyst also exhibited favorable reusability. The ecological evaluation of the photocatalytic activity was carried out using the propagation of common gramme seeds (Cicer arietinum), and the results demonstrate that the synthetic hybrid 3D composite was minimal in cytotoxicity and was environmentally safe. The data demonstrate the extensive significance of developing multifunctional materials for a variety of cutting-edge applications and demonstrate that the synthetic hybrid 3D composite permits wastewater purification utilizing photocatalytic technology.

5.3 CHARACTERIZATION OF CARBON ALLOTROPES

5.3.1 Raman Spectroscopy

A rapid, easy, non-destructive evaluation method is Raman spectroscopy. It could be performed at ambient pressure and temperature, and most user groups have access to the instruments. The methodology for studying the variations in the characteristics of nanotubes generated under multiple configurations and processes is incredibly accurate. With increasing incident laser intensity, the sample's Raman signal strength rises, allowing for detailed electrical and systemic analysis. Glassy carbon and graphene were used by Nimbalkar et al. (2022) to create a hybrid substance with outstanding electrical and electrolytic characteristics. By (a) the synthesis process, (b) extensive geometrical and textural characterizations, and (c) molecular dynamics modeling, the production of the composites was shown to occur and was verified. Using Raman spectroscopy, chemical bonding and morphologies were studied. For both Gc and GrGC, the anticipated graphitic and disorder peaks were seen at 1340 and 1583 cm^{-1}, correspondingly. The 2D band combines with the adjacent bands in the pure GC to create a slight modulated bump. Owing to the creation of more homogeneous carbon hexagonal rings, the 2D signal for GrGC was significantly more intense than that of GC. Furthermore, the GrGC I2D/IG of 0.27 indicated the existence of several Gr layers. The ID/IG ratio, on the other contrary, showed a comparatively slight distinction between the factions, being 1.03 for GC and 0.99 for GrGC. The elongation of GrGC's peak around 1630 cm^{-1} indicated the existence of carbonyl groups (C–O), whereas the elongation of carboxyl groups (R-COOH) was seen around 1720 and 1740 cm^{-1}. In particular, the GrGC microstructure showed stretching of the alkene (=C–H) peak among 3000 and 3100 cm^{-1}, the hydroxyl (–OH) peak among 2500 and 3300 cm^{-1}, the ether group (C–O–C, C–O) spike among 1300 and 1000 cm^{-1}, and the epoxide (cyclic ethers) spike around 1300 and 1060 cm^{-1}. Additionally, the resonances in the multi-acquire high-loss EELS around 285 and 275 eV provided direct proof of the existence of interfacial dangling bonds and, consequently, functional units in the GrGC hybrid composite.

To study the novel carbon allotrope cyclo[18]carbon, Hussain et al. (2020) employed Raman spectroscopy (C18). In this research, the relationship between C18 and Ag cluster tip was examined regarding their proximity and atomically precise locations. The location of CC bonds was represented in the TERS images. At 2186 and 1760 cm^{-1}, two prominent NRS bands were identified as C–C antisymmetric and symmetric stretching, correspondingly. There were minor weak bands in the range 400–500 cm^{-1} that were attributed to ring delamination, and one weak band at 898 cm^{-1} that was attributed to C-C stretching coupled with bending vibrations. In particular, it is much more appropriate to categorize the vibrational movements as antisymmetric and symmetric for compounds having –C≡C–C≡C– groups. These carbon chains' C–C stretching bands are often relatively feeble, and the Raman spectrum often shows vibrations solely associated with C–C stretching; therefore, they might not be regarded as defining frequencies.

As a follow-up to their earlier research in this area, which reviewed coatings-based formulations for different applications, Tudose et al. (2022) revealed optimized preparations of alcohol-based conductive paints for EMI and anticorrosion activity, that can secure interoperability and lessen the accessibility of computer components. Figure 5.10 displays the Raman spectrum of the first, second, and third layers of the (a), (b), (c), (d), and (e) B1 paints. All specimens' Raman spectrum display three unique Raman maxima for the D band at about 1384 cm^{-1}, the G band at about 1518 cm^{-1}, and the 2D band at about 2680 cm^{-1}, which each represents the quantity of stacked graphitic layers. The proportion IG/I2D, in which the apex intensities are designated as I (D), I(G), and I(2D), correspondingly, for the D, G, and 2D peaks, was another technique to identify the distinctive structural features of the stacked coatings (Figure 5.10). Two techniques were described in the research for determining the ID/IG in carbonic components: estimating the region under the curves using peak fitting for wider peaks or computing the proportion using the linear intensity whenever the peaks are acute. By isolating the intensity distribution with two

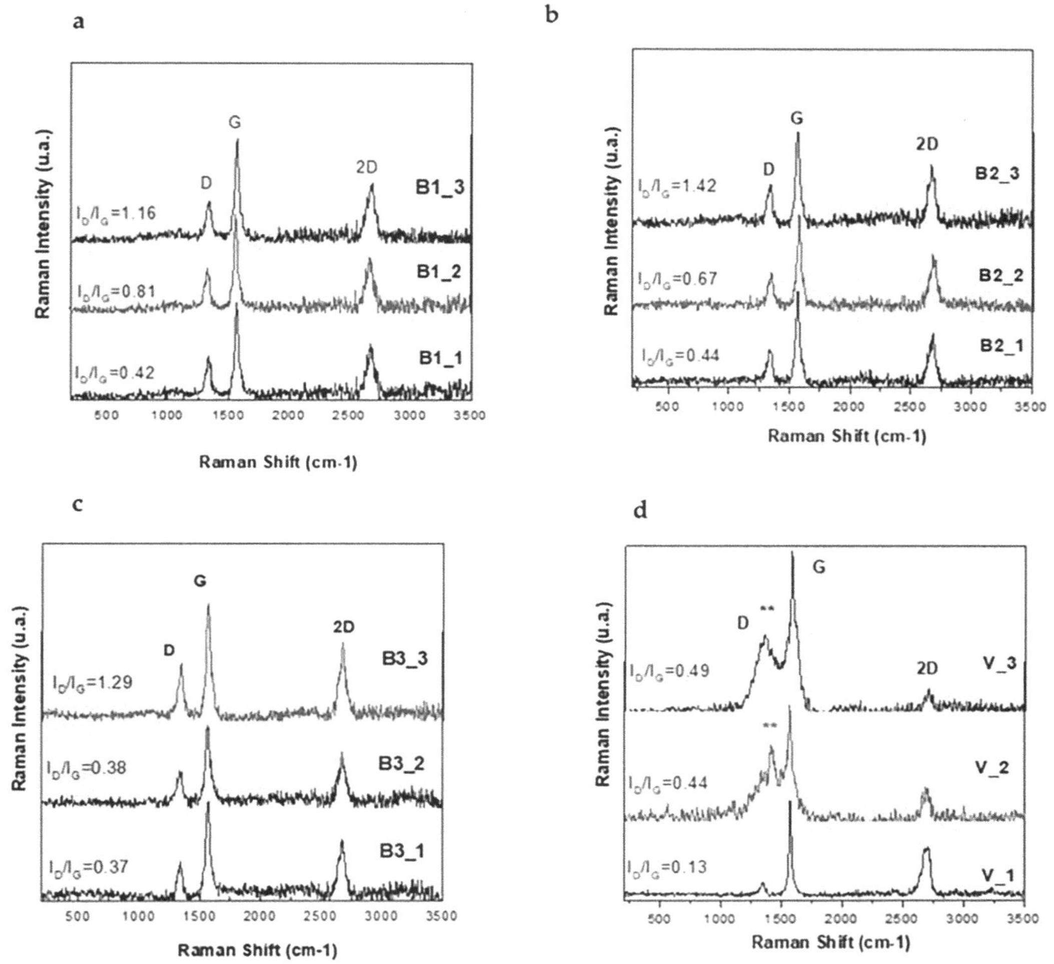

FIGURE 5.10 Raman spectrum of 1, 2, and 3 coatings of (a) B1, (b) B2, (c) B3, and (d) V paints. Reprinted with permission from Tudose et al. (2022). Copyright (2022) MDPI. Distributed under the Creative Common attribution-based License CCBY 4.0.

decimal places for the G-band and D-bands from the analysis of statistical information of the Raman-spectra acquisitions in Witec software, the ratio IG/I2D in this example was determined. The fact that the IG/I2D proportion was greater than 1 in specimens B1–B3 suggests that the graphitic framework was arranged in a stacking-like multilayer configuration in the PVB polymer matrices. For each specimen, the ID/IG ratio notifies us about the structural and domain extent of graphitic material as determined. For moderate and severe flaw densities, ID/IG rises as defect density does. In comparison to the graphitic specimen B1, ID/IG was significantly higher, indicating that a few sp^3 carbon were formed via functionalization using Fe_3O_4 in the B2 specimen and Fe Ore in the B3 specimen. These values only provide a quantitative understanding of how multicomponent composite materials are structured.

5.3.2 Transmission Electron Microscopy (TEM)

By creating a projected representation via the transmission of electrons, TEM is utilized to analyze thin samples. In many aspects, TEM is analogous to the traditional (compound) light microscope.

The diameters of individual CNTs and CNT bundles could be measured using TEM imaging. SWCNT bundles have diameters as small as 4 nm, whereas all the DWCNTs have sizes of at least 2.5 nm. The distance between fringes in various bundles can also be statistically measured using high-magnification TEM images.

$ZnFe_2O_4$ (ZFO)/carbon derivatives (ZFO@CNT, ZFO@GO, and ZFO@Fullerene) nanocomposites were created by Behera et al. (2019) using a one-pot hydrothermal technique and calcination. Under open sunlight luminance, the photocatalytic activities of ZFO@carbon nanocomposites were investigated for the deterioration of hexavalent chromium and antibiotics (Norfloxacin). The findings indicate that adding carbon derivatives to ZFO nanoparticles improves their ability to absorb visible light and their excitation separation effectiveness. ZFO@CNT displays the maximum reactivity across the synthesized composites when compared to other nanocomposites. According to PL and impedance measurements, the maximum activities of ZFO@CNT were caused by the material's excellent charge transport characteristics and minimal photoexcited electron-hole coupling. The created ZFO@CNT nanocomposite additionally displayed the maximum photocurrent density, 2.25 mA cm^{-2}, which was 225 times greater than that of pristine ZFO. ZFO@CNT demonstrated the best photocatalytic performance, with 91.36% disintegration of 50 ppm norfloxacin and 82% decrease of 10 ppm Cr (VI) in 90 and 60 minutes, correspondingly, under solar illumination. The TEM demonstrated that the particulates of pure zinc ferrite are large and irregularly structured, and the majority of the particles are observed to be aggregated. These photos unmistakably show that ZFO particulates are roughly 30–40 nm in size. The HRTEM pictures of ZFO nanoparticles having excellent crystallinity and lattice patterns with interfacial d-spacing values of 0.29 and 0.25 nm corresponding to (2 2 0) and (3 1 1) crystal planes of the spinel ferrite could be observed. Additionally, ZFO@GO TEM images show that the nanoparticles were evenly dispersed across the GO sheet. The ferrite nanoparticles filled the majority of the graphene layer. The HRTEM and SAED pattern demonstrated the same characteristics as those seen for ZFO, namely lattice fringes of 0.29 nm and 0.25 nm, correlating to (2 2 0) and (3 1 1) crystal planes, which describes the preservation of ZFO nanoparticles' crystalline phase despite their being evenly dispersed throughout the GO layer.

Zhang et al. (2020) formed MW carbon nanowires (MWCNWs) having linear carbon chains embedded in MWCNTs (LLCC@MWCNTs) effectively using magnetic field improved hydrogen arc discharge. It has been discovered that adding a static magnetic field to the traditional hydrogen arc discharge approach might increase the spectrum of pretreatment variables for MWCNW specimens while also improving growth efficiency. Integrating the characterization and modeling, it could be said that the magnetic field aids in raising the temperature of the arc plasma, which is crucial to the development of MWCNWs. The study improved MWCNW production and provided a novel, feasible method that will aid future investigation on the novel 1D allotrope.

5.3.3 Atomic Force Microscopy (AFM)

Among the current microscopy approaches, the AFM is the characterization method that is most frequently utilized. The widespread usage of the AFM is attributed to its ability to produce a 3D specimen topographical with atomic scale quickly and for a relatively modest price. The AFM data also includes critical information regarding the structural analysis of the surface. With the help of this technology, we can also determine the width of the bundles and the extent of the nanotubes. In this study, Alwan et al. (2021) synthesized molybdenum and cobalt oxides based on the graphene catalyst CoMo/G and evaluated its performance for the hydro-desulfurization reaction HDS. AFM was used to characterize the catalyst. On the graphene surface, nanosized metal oxide particulates were scattered with some oxide molecule agglomeration propensity. The oxide particulates were distributed with a mean particle size of around 120 nm and roughness of around 18.3 nm (active and promoters portions combined). In coatings/films constituted of polyvinylpyrrolidone (PVP) and oxidized carbon nanohorns (CNHox), Serban et al. (2021) revealed the electrical percolation

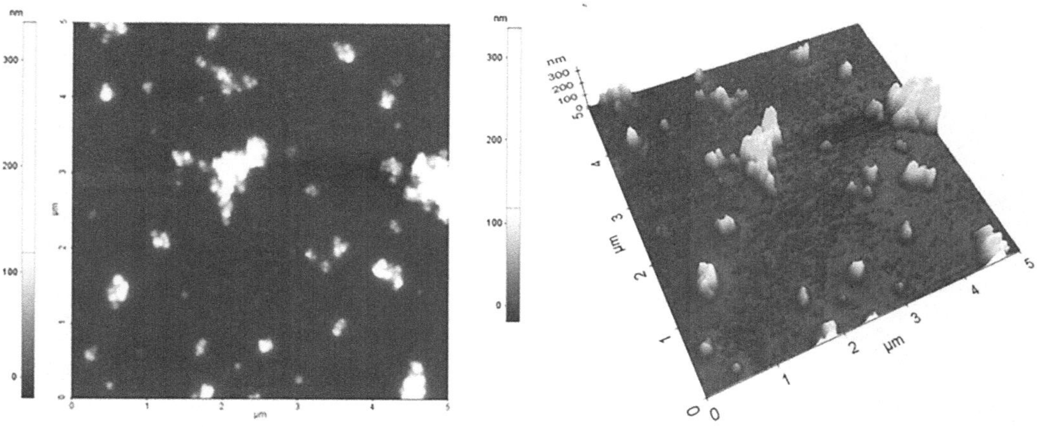

(a) (b)

FIGURE 5.11 AFM scans of 1 wt.% ox-SWCNHs in the PVP matrices (a) upper view and (b) aspect view, respectively. Reprinted with permission from Serban et al. (2021). Copyright (2021) MDPI. Distributed under the Creative Common attribution-based License CCBY 4.0.

threshold. As detecting sheets for resistant relative humidity sensors, they showed how the layers were designed and synthesized. AFM spectrometry was used to look into the shape and composition of the sensor layers. The conductive transducer (IDT) dual-comb arrangements were used to create these nanocomposite thin films to determine the electrical percolation thresholds of CNHox within PVP. The IDTs were manufactured on both a hard Si/SiO_2 substrate having a 10 μm metal digits separation and a foldable polyimide surface having a 100 μm metal digits separation. When conducted on 10 and 100 μm-IDT, the percolation thresholds of CNHox in the PVP matrices were equivalent to 0.05–0.1 wt.% and 3.5 wt.%, correspondingly. The AFM morphology of the thin coatings of the PVP/CNHox nanocomposite is shown in Figures 5.11 and 5.12, exhibiting a mass density of 1 wt.% CNHox (Figure 5.11) and 2 wt.% CNHox (Figure 5.12). The examined region for both cases was 5 × 5 μm^2.

Using nickel oxide-functionalized highly porous graphitic carbon nitride, Maślana et al. (2020) created nitrogen-doped CNT (ex-g-C3N4-NixOy). Ex-g-C3N4-NixOy reduction using ethylene

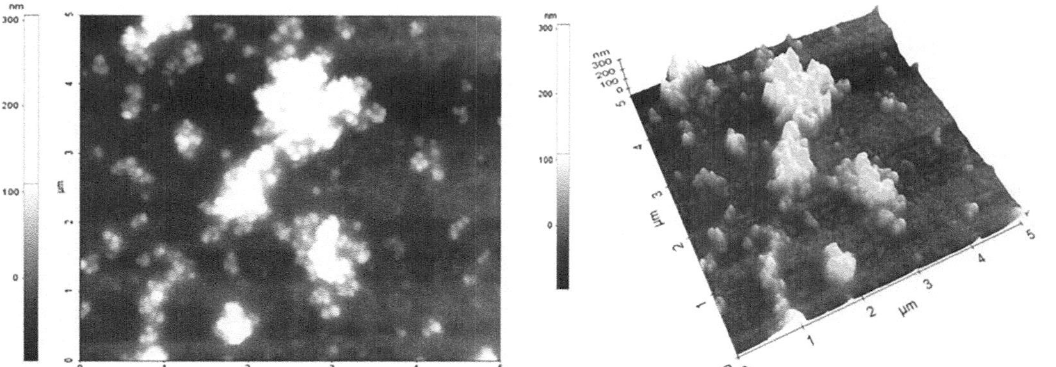

(a) (b)

FIGURE 5.12 AFM scans of the PVP matrices‘ 2 wt.% ox-SWCNHs (a) upper view and (b) aspect view, respectively. Reprinted with permission from Serban et al. (2021). Copyright (2021) MDPI. Distributed under the Creative Common attribution-based License CCBY 4.0.

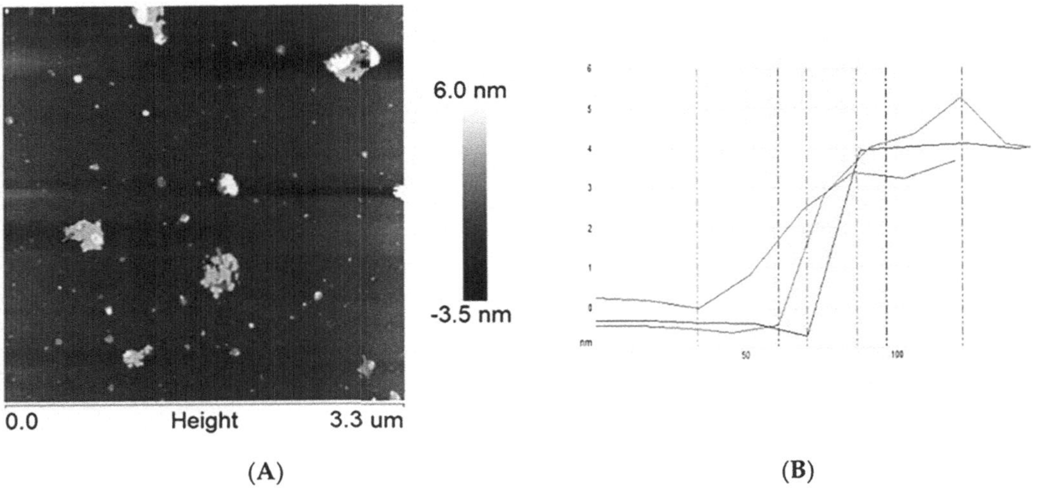

FIGURE 5.13 (a) AFM scans of ex-g-C3N4 and (b) increased profile. Reprinted with permission from Maślana et al. (2020). Copyright (2020) MDPI. Distributed under the Creative Common attribution-based License CCBY 4.0.

and hydrogen-assisted chemical vapor condensation were the two procedures used to create CNT-N at 900°C (CVD). Using AFM, the generated composites were thoroughly characterized. AFM was used to analyze the morphology of the exfoliated g-C3N4 in more detail (Figure 5.13). The acquired findings show that ex-g-C3N4 was around 4–5 nm thick, which indicates that this specimen was made up of 10–13 layers. Bulk graphitic carbon nitride was comprised of between 100 and 170 layers; thus, it might be stated that the exfoliation stage was successful. According to AFM scans, the quantity of ex-g-C3N4 layers was greater than 30 times that of the original material, which was confirmed by TEM.

The outcomes demonstrated the significance of ethylene in the development of tubes. In addition to being a reservoir of carbon in the eventual specimen, it carbon reduced the nickel oxide (II) to metallic nickel, a state that was effective in the formation of nanotubes. The produced N-doped nanotubes could function well in several applications, including anticorrosion, catalysis, supercapacitors, Li-ion batteries, and others.

5.4 CONCLUSION

We covered the covalent and noncovalent functionalization of several carbon allotropes in this chapter, including graphene, CNT, fullerene, nanodiamonds, graphite, and nanofibers, employing a variety of extremely yielding processes like the Bingel reaction. Functional groups could covalently bind to the framework of the carbon allotropes owing to the reaction parameters. In one instance, the graphene that Bingel tweaked exhibited an arbitrarily disorganized form with improved thermal stability. Functionalized carbon allotropes were added, which raised the glass temperature of epoxy resin. This change shows greater interfacial contacts among carbon allotropes and resin (epoxy) as a consequence of the chemical remodeling of thermally reduced carbon allotropes, combined with increased mechanical characteristics. Covalent and noncovalent functionalization are the two main categories into which the functionalization approaches have been categorized. In the former, carbon allotropes and functional groups forms chemical bonds, whereas in the latter, π-π interactions are thought to be the critical component. Covalent functionalization has the drawback of causing deterioration to the initially perfect surface, which could negatively impact the optical and electrical characteristics of carbon allotropes. As usage of carbon allotropes in numerous critical situations remains to develop, research is anticipated to proceed into various chemical techniques to functionalize carbon allotropes (both covalently and noncovalently). The resulting functionalized carbon allotropes exhibit exceptional thermal durability, strong tensile strength, electromagnetic shielding capacity and Young's modulus. While researching novel carbon allotropes that are ideal for particular functionalization, it is essential to consider the type of viscosity, solvent, molecular mass, and dosage. To create functionalized carbon allotropes with adaptable structures, new functions, and superior characteristics, all parameters must be optimized. Therefore, more research on their effects on the human body is needed before any practical deployment.

REFERENCES

Alshehri, R., Ilyas, A. M., Hasan, A., Arnaout, A., Ahmed, F., Memic, A., 2016. Carbon nanotubes in biomedical applications: Factors, mechanisms, and remedies of toxicity. *J. Med. Chem.* 59, 8149–8167. 10.1021/acs.jmedchem.5b01770

Alwan, H. H., Makki, H. F., Al-Hattab, T. A., 2021. Preparation, characterization and activity of CoMo supported on graphene for heavy naphtha hydro-desulfurization reaction. *Iran. J. Catal.* 11, 101–111.

Backes, C., Hauke, F., Hirsch, A., 2011. The potential of perylene bisimide derivatives for the solubilization of carbon nanotubes and graphene. *Adv. Mater.* 23, 2588–2601. 10.1002/adma.201100300

Behera, A., Mansingh, S., Das, K. K., Parida, K., 2019. Synergistic $ZnFe_2O_4$ -carbon allotropes nanocomposite photocatalyst for norfloxacin degradation and Cr (VI) reduction. *J. Colloid Interface Sci.* 544, 96–111. 10.1016/j.jcis.2019.02.056

Bhawani, S. A., Tariq, A., Moheman, A., Ngaini, Z., 2018. *Functionalized Graphene Nanocomposites for Water Treatment, Functionalized Graphene Nanocomposites and Their Derivatives: Synthesis, Processing and Applications*. Elsevier. 10.1016/B978-0-12-814548-7.00005-2

Bhoi, V. I., Imae, T., Ujihara, M., Murthy, C. N., 2013. Surface immobilization of carbon nanotubes by β-cyclodextrins and their inclusion ability. *J. Nanosci. Nanotechnol.* 13, 2604–2612. 10.1166/jnn.2013.7358

Bu, Y., Kushwaha, A., Goswami, L., Kim, B.-S., 2022. Green production of functionalized few-layer graphene–silver nanocomposites using gallnut extract for antibacterial application. *Micromachines* 13, 1232. 10.3390/mi13081232

Cheng, B., Pan, H., Liu, D., Li, D., Li, J., Yu, S., Tan, G., Pan, W., 2018. Functionalization of nanodiamond with vitamin E TPGS to facilitate oral absorption of curcumin. *Int. J. Pharm.* 540, 162–170. 10.1016/j.ijpharm.2018.02.014

Cho, J., Jeon, I., Kim, S. Y., Lim, S., Jho, J. Y., 2017. Improving dispersion and barrier properties of polyketone/graphene nanoplatelet composites via noncovalent functionalization using aminopyrene. *ACS Appl. Mater. Interfaces* 9, 27984–27994. 10.1021/acsami.7b10474

Choi, H. Y., Bae, J. H., Hasegawa, Y., An, S., Kim, I. S., Lee, H., Kim, M., 2020. Thiol-functionalized cellulose nanofiber membranes for the effective adsorption of heavy metal ions in water. *Carbohydr. Polym.* 234, 115881. 10.1016/j.carbpol.2020.115881

Dasler, D., Schäfer, R. A., Minameyer, M. B., Hitzenberger, J. F., Hauke, F., Drewello, T., Hirsch, A., 2017. Direct covalent coupling of porphyrins to graphene. *J. Am. Chem. Soc.* 139, 11760–11765. 10.1021/jacs.7b04122

Díez-Pascual, A. M., 2021. Chemical functionalization of carbon nanotubes with polymers: A brief overview. *Macromol* 1, 64–83. 10.3390/macromol1020006

Eigler, S., Hirsch, A., 2014. Chemistry with graphene and graphene oxide – Challenges for synthetic chemists. *Angew. Chemie – Int. Ed.* 53, 7720–7738. 10.1002/anie.201402780

Fuertes-Espinosa, C., García-Simón, C., Pujals, M., Garcia-Borràs, M., Gómez, L., Parella, T., Juanhuix, J., Imaz, I., Maspoch, D., Costas, M., Ribas, X., 2020. Supramolecular fullerene sponges as catalytic masks for regioselective functionalization of C60. *Chem.* 6, 169–186. 10.1016/j.chempr.2019.10.010

González-García, P., 2015. Surface oxygen functionalities and microstructure of activated carbons derived from lignocellulosic resources. In *Chemical Functionalization of Carbon Nanomaterials: Chemistry and Applications*. Edited by Vijay Kumar Thakur, and Manju Kumari Thakur, 1st Edition. 1101. CRC Press. 10.1201/b18724

Haidyrah, A. S., Sundaresan, P., Venkatesh, K., Ramaraj, S. K., Thirumalraj, B., 2021. Fabrication of functionalized carbon nanofibers/carbon black composite for electrochemical investigation of antibacterial drug nitrofurantoin. *Colloids Surfaces A Physicochem. Eng. Asp.* 627, 127112. 10.1016/j.colsurfa.2021.127112

Han, S., Meng, Q., Araby, S., Liu, T., Demiral, M., 2019. Mechanical and electrical properties of graphene and carbon nanotube reinforced epoxy adhesives: Experimental and numerical analysis. *Compos. Part A Appl. Sci. Manuf.* 120, 116–126. 10.1016/j.compositesa.2019.02.027

He, H., Pham-Huy, L. A., Dramou, P., Xiao, D., Zuo, P., Pham-Huy, C., 2013. Carbon nanotubes: Applications in pharmacy and medicine. *Biomed. Res. Int.* 2013. 10.1155/2013/578290

Hussain, S., Chen, H., Zhang, Z., Zheng, H., 2020. Vibrational spectra and chemical imaging of cyclo[18] carbon by tip enhanced Raman spectroscopy. *Chem. Commun.* 56, 2336–2339. 10.1039/c9cc09130k

Islam, M. S., Shudo, Y., Hayami, S., 2022. Energy conversion and storage in fuel cells and super-capacitors from chemical modifications of carbon allotropes: State-of-art and prospect. *Bull. Chem. Soc. Jpn.* 95, 1–25. 10.1246/bcsj.20210297

Kamedulski, P., Ilnicka, A., Lukaszewicz, J. P., Skorupska, M., 2019. Highly effective three-dimensional functionalization of graphite to graphene by wet chemical exfoliation methods. *Adsorption* 25, 631–638. 10.1007/s10450-019-00067-9

Karuppiah, C., Sakthinathan, S., Chen, S. M., Manibalan, K., Chen, S. M., Huang, S. T., 2016. A non-covalent functionalization of copper tetraphenylporphyrin/chemically reduced graphene oxide nanocomposite for the selective determination of dopamine. *Appl. Organomet. Chem.* 30, 40–46. 10.1002/aoc.3397

Kharisov, B. I., Kharissova, O. V., 2019. *Carbon Allotropes: Metal-Complex Chemistry, Properties and Applications*. Springer. 10.1007/978-3-030-03505-1

Kuila, T., Bose, S., Mishra, A. K., Khanra, P., Kim, N. H., Lee, J. H., 2012. Chemical functionalization of graphene and its applications. *Prog. Mater. Sci.* 57, 1061–1105. 10.1016/j.pmatsci.2012.03.002

Kumar, A., Thakur, A., 2020. Encapsulated nanoparticles in organic polymers for corrosion inhibition. In *Corrosion Protection at the Nanoscale*. Edited by Susai Rajendran, Tuan ANH Nguyen, Saeid Kakooei, Mahdi Yeganeh, Yongxin Li. 345–362, Elsevier. 10.1016/b978-0-12-819359-4.00018-0

Li, H., Papadakis, R., 2021. Click chemistry enabling covalent and non-covalent modifications of graphene with (Poly)saccharides. *Polymers (Basel)*. 13, 1–19. 10.3390/polym13010142

Lian, Q., Ahmad, Z. U., Gang, D. D., Zappi, M. E., Fortela, D. L. B., Hernandez, R., 2020. The effects of carbon disulfide driven functionalization on graphene oxide for enhanced Pb(II) adsorption: Investigation of adsorption mechanism. *Chemosphere* 248, 126078. 10.1016/j.chemosphere.2020.126078

Ling, X. L., Wei, Y. Z., Zou, L. M., Xu, S., 2013. Preparation and characterization of hydroxylated multi-walled carbon nanotubes. *Colloids Surfaces A Physicochem. Eng. Asp.* 421, 9–15. 10.1016/j.colsurfa.2012.12.046

Lo, R., Lamanec, M., Wang, W., Manna, D., Bakandritsos, A., Dračínský, M., Zbořil, R., Nachtigallová, D., Hobza, P., 2021. Structure-directed formation of the dative/covalent bonds in complexes with C70···piperidine. *Phys. Chem. Chem. Phys.* 23, 4365–4375. 10.1039/d0cp06280d

Majdoub, M., Anfar, Z., Amedlous, A., 2020. Emerging chemical functionalization of g-C3N4: Covalent/noncovalent modifications and applications. *ACS Nano* 14(10), 12390–12469. 10.1021/acsnano.0c06116

Maleki, R., Khoshoei, A., Ghasemy, E., Rashidi, A., 2020. Molecular insight into the smart functionalized TMC-Fullerene nanocarrier in the pH-responsive adsorption and release of anti-cancer drugs. *J. Mol. Graph. Model.* 100, 107660. 10.1016/j.jmgm.2020.107660

Malig, J., Jux, N., Guldi, D. M., 2013. Toward multifunctional wet chemically oligomeric, molecular, and particulate building activity. *Acc. Chem. Res.* XXX, 53–64.

Mamidi, N., Zuníga, A. E., Villela-Castrejón, J., 2020. Engineering and evaluation of forcespun functionalized carbon nano-onions reinforced poly (ε-caprolactone) composite nanofibers for pH-responsive drug release. *Mater. Sci. Eng. C* 112, 110928. 10.1016/j.msec.2020.110928

Marcia, M., Hirsch, A., Hauke, F., 2017. Perylene-based non-covalent functionalization of 2D materials. *FlatChem* 1, 89–103. 10.1016/j.flatc.2017.01.001

Maślana, K., Kaleńczuk, R. J., Zielińska, B., Mijowska, E., 2020. Synthesis and characterization of nitrogen-doped carbon nanotubes derived from g-C3N4. *Materials (Basel)*. 13. 10.3390/ma13061349

Menaa, F., Fatemeh, Y., Vashist, S. K., Iqbal, H., Sharts, O. N., Menaa, B., 2021. Graphene, an interesting nanocarbon allotrope for biosensing applications: Advances, insights, and prospects. *Biomed. Eng. Comput. Biol.* 12, 117959722098382. 10.1177/1179597220983821

Montes, R., Sánchez, G., Zhao, J., Palet, C., Baeza, M., Bastos-Arrieta, J., 2020. Customized in situ functionalization of nanodiamonds with nanoparticles for composite carbon-paste electrodes. *Nanomaterials* 10, 1–15. 10.3390/nano10061179

Mountrichas, G., Tagmatarchis, N., Pispas, S., 2009. Functionalization of carbon nanohorns with polyethylene oxide: Synthesis and incorporation in a polymer matrix. *J. Nanosci. Nanotechnol.* 9, 3775–3779. 10.1166/jnn.2009.NS66

Naebe, M., Wang, J., Amini, A., Khayyam, H., Hameed, N., Li, L. H., Chen, Y., Fox, B., 2014. Mechanical property and structure of covalent functionalised graphene/epoxy nanocomposites. *Sci. Rep.* 4, 1–7. 10.1038/srep04375

Nimbalkar, S., Montgomery-Walsh, R., Bunnell, J., Galindo, S. L., Cariappa, B. K., Gautam, A., Arvizu, R., Yang, S., Kassegne, S., 2022. Carbon allotropes form a hybrid material: Synthesis, characterization, and molecular dynamics simulation of novel graphene-glassy carbon hybrid material. *Carbon N. Y.* 196, 1012–1023. 10.1016/j.carbon.2022.05.049

O'Bannon, E., Xia, G., Shi, F., Wirth, R., King, A., Dobrzhinetskaya, L., 2020. The transformation of diamond to graphite: Experiments reveal the presence of an intermediate linear carbon phase. *Diam. Relat. Mater.* 108, 107876. 10.1016/j.diamond.2020.107876

Park, S. J., Das, G. S., Schütt, F., Adelung, R., Mishra, Y. K., Tripathi, K. M., Kim, T. Y., 2019. Visible-light photocatalysis by carbon-nano-onion-functionalized ZnO tetrapods: Degradation of 2,4-dinitrophenol and a plant-model-based ecological assessment. *NPG Asia Mater.* 11. 10.1038/s41427-019-0107-0

Perumal, S., Atchudan, R., Immanuel Edison, T. N. J., Shim, J. J., Lee, Y. R., 2021. Exfoliation and noncovalent functionalization of graphene surface with Poly-N-Vinyl-2-pyrrolidone by in situ polymerization. *Molecules* 26(6), 1534. 10.3390/molecules26061534

Qu, S., Li, M., Xie, L., Huang, X., Yang, J., Wang, N., Yang, S., 2013. Noncovalent functionalization of graphene attaching [6,6]-phenyl-C61- butyric acid methyl ester (PCBM) and application as electron extraction layer of polymer solar cells. *ACS Nano.* 7, 4070–4081. 10.1021/nn4001963

Saifuddin, N., Raziah, A. Z., Junizah, A. R., 2013. Carbon nanotubes: A review on structure and their interaction with proteins. *J. Chem.* 2013. 10.1155/2013/676815

Sakellariou, G., 2015. Editorial (Thematic Issue: Graphene: From Synthesis and Functionalization to Applications). *Curr. Org. Chem.* 19, 1756–1756. 10.2174/138527281918150820165633

Sathish Kumar, P., Prakash, P., Srinivasan, A., Karuppiah, C., 2021. A new highly powered supercapacitor electrode of advantageously united ferrous tungstate and functionalized multiwalled carbon nanotubes. *J. Power Sources* 482, 228892. 10.1016/j.jpowsour.2020.228892

Serban, B. C., Cobianu, C., Dumbravescu, N., Buiu, O., Bumbac, M., Nicolescu, C. M., Cobianu, C., Brezeanu, M., Pachiu, C., Serbanescu, M., 2021. Electrical percolation threshold and size effects in polyvinylpyrrolidone-oxidized single-wall carbon nanohorn nanocomposite: The impact for relative humidity resistive sensors design. *Sensors* 21, 1–16. 10.3390/s21041435

Serhan, M., Sprowls, M., Jackemeyer, D., Long, M., Perez, I. D., Maret, W., Tao, N., Forzani, E., 2019. Total iron measurement in human serum with a smartphone. AIChE Annu. Meet. Conf. Proc. November 2019. 10.1039/x0xx00000x

Speranza, G., 2021. Carbon nanomaterials: Synthesis, functionalization and sensing applications. *Nanomaterials* 11(4), 967. 10.3390/nano11040967

Thakur, A., Kaya, S., Abousalem, A. S., Kumar, A., 2022a. Experimental, DFT and MC simulation analysis of Vicia Sativa weed aerial extract as sustainable and eco-benign corrosion inhibitor for mild steel in acidic environment. *Sustain. Chem. Pharm.* 29, 100785. 10.1016/j.scp.2022.100785

Thakur, A., Kaya, S., Abousalem, A. S., Sharma, S., Ganjoo, R., Assad, H., Kumar, A., 2022b. Computational and experimental studies on the corrosion inhibition performance of an aerial extract of Cnicus Benedictus weed on the acidic corrosion of mild steel. *Process Saf. Environ. Prot.* 161, 801–818. 10.1016/j.psep.2022.03.082

Thakur, A., Kumar, A., 2022. Recent advances on rapid detection and remediation of environmental pollutants utilizing nanomaterials-based (bio)sensors. *Sci. Total Environ.* 834, 155219. 10.1016/j.scitotenv.2022.155219

Thakur, A., Kumar, A., 2021. Sustainable inhibitors for corrosion mitigation in aggressive corrosive media: A comprehensive study. *J. Bio- Tribo-Corrosion* 7, 1–48. 10.1007/s40735-021-00501-y

Thakur, A., Kumar, A., Kaya, S., Marzouki, R., Zhang, F., Guo, L., 2022c. Recent advancements in surface modification, characterization and functionalization for enhancing the biocompatibility and corrosion resistance of biomedical implants. *Coatings* 12, 1459. 10.3390/coatings12101459

Thakur, A., Kumar, A., Kaya, S., Vo, D. V. N., Sharma, A., 2022d. Suppressing inhibitory compounds by nanomaterials for highly efficient biofuel production: A review. *Fuel* 312, 122934. 10.1016/j.fuel.2021.122934

Thakur, A., Kumar, A., Sharma, S., Ganjoo, R., Assad, H., 2022e. Materials today: Proceedings computational and experimental studies on the efficiency of Sonchus arvensis as green corrosion inhibitor for mild steel in 0. 5 M HCl solution. *Mater. Today Proc.* 66(2), 609–621. 10.1016/j.matpr.2022.06.479

Torres, T., 2011. Carbon nanotubes and related structures. Synthesis, Characterization, Functionalization, and Applications. Edited by Dirk M. Guldi and Nazario Martín. *Angew. Chemie Int. Ed.* 50, 1473–1474. 10.1002/anie.201006930

Tudose, I. V., Mouratis, K., Ionescu, O. N., Romanitan, C., Pachiu, C., Pricop, E., Khomenko, V. H., Butenko, O., Chernysh, O., Barsukov, V. Z., Suchea, M. P., Koudoumas, E., 2022. Carbon Allotropes-based paints and their composite coatings for electromagnetic shielding applications. *Nanomaterials* 12, 1–18. 10.3390/nano12111839

Vasilescu, A., Hayat, A., Gáspár, S., Marty, J. L., 2018. Advantages of carbon nanomaterials in electrochemical aptasensors for food analysis. *Electroanalysis* 30, 2–19. 10.1002/elan.201700578

Veeman, D., Shree, M. V., Sureshkumar, P., Jagadeesha, T., Natrayan, L., Ravichandran, M., Paramasivam, P., 2021. Sustainable development of carbon nanocomposites: Synthesis and classification for environmental remediation. *J. Nanomater.* 2021, 21. 10.1155/2021/5840645

Walters, F., Munem Ali, M., Burwell, G., Rozhko, S., Tehrani, Z., Daghigh Ahmadi, E., Evans, J. E., Abbasi, H. Y., Bigham, R., John Mitchell, J., Kazakova, O., Devadoss, A., Guy, O. J., 2020. A facile method for the non-covalent amine functionalization of carbon-based surfaces for use in biosensor development. *Nanomaterials* 10(9), 1808. 10.3390/nano10091808

Zhang, X., Hou, L., Samorì, P., 2016. Coupling carbon nanomaterials with photochromic molecules for the generation of optically responsive materials. *Nat. Commun.* 7, 11118. 10.1038/ncomms11118

Zhang, Y., Chang, W., Liu, Y., Maruyama, T., Zhao, X., 2020. High-yield growth of multi-walled carbon nanowires by magnetic field controlled arc technique. *Carbon N. Y.* 158, 672–680. 10.1016/j.carbon.2019.11.039

Part 2

Carbon Allotropes

Advanced Materials for Anticorrosive Coatings (Recent Advancements)

6 Graphene and Graphene Oxides in Anticorrosive Coatings

Avni Berisha
Department of Chemistry, Faculty of Natural and Mathematics Science, University of Prishtina, Prishtina, Kosovo

Fetah Podvorica
Department of Chemistry, Faculty of Natural and Mathematics Science, University of Prishtina, Prishtina, Kosovo

Materials Science – Nanochemistry Research Group, NanoAlb – Unit of Albanian Nanoscience and Nanotechnology, Tirana, Albania

Kosova Academy of Sciences and Arts, Prishtina, Kosovo

CONTENTS

6.1 INTRODUCTION

Corrosion of metals is a spontaneous reaction of metallic surface atoms with aggressive agents in their environment, and this process, depending on the nature of the metal, may lead to their complete degradation. As metals, especially structural metals (steel, aluminium, nickel, titane, etc.) are core components of all engineering infrastructures like buildings, railways, bridges, and gas and oil pipelines to name some of them, corrosion may cause severe damage. The total cost of the damage and that of conversion layers needed for the maintenance of these infrastructures is very high and is estimated at about 2.5 trillion US dollars per year, which is equivalent to 3,4% of the world's GDP (Chauhan et al. 2020).

To prevent metal corrosion, metal surfaces were usually covered by conversion coatings, and for a long time mainly hexavalent chromium was used. But due to their carcinogenic effects and environmental toxicity, these protection layers are now forbidden (Nine et al. 2015). Coatings that are made of different moieties and normally are stable toward the environment can be applied to a variety of substrates and have the main role to retain and protect a material's primary purpose. Along with protective and ornamental properties, researchers have also focused on producing functional coatings that exhibit anticorrosive, superhydrophobic, antifouling, self-healing, heat resistant, and antibacterial properties (Li et al. 2020; Zhang et al. 2021).

DOI: 10.1201/9781003323976-8

Typically, organic coatings were developed as an effective way to safeguard metallic substrates by establishing a physical barrier between their surfaces and the environmental aggressive agents that enable the permeation of water and oxidant agent species and therefore protect the deterioration of coinage metals. At the same time, the conversion coating ensures the maintenance, or even reinforces the stability of, the protected structure.

Polymer coatings were frequently used for the protection of metals, and their efficacy increases with the thickness of the layer. This approach might be beneficiary for the object itself, but it leads to a higher weight, which means higher energy consumption. To reduce the weight of the material and lower the cost of production but to maintain the efficacy of the inhibitors, nano-composite materials are considered key players. Recent studies demonstrate that incorporating nanomaterials into metallic coatings can help prevent their corrosion while also prolonging the life of the coating.

Graphene and graphene-based materials have found a tremendous increase in their use due to their exceptional properties such as: extraordinary electric and thermal conductivity, mechanical strength, chemical inertness, unique optical properties, high surface-to-volume ratio, and impermeability for all fluids and gases (Kulyk et al. 2021).

Due to their corrosion-resistant qualities, two-dimensional carbon nanostructures such as graphene and graphene oxide (GO) have lately garnered considerable interest. These nanomaterials, once introduced to the metallic surface, exhibit exceptional performance as a barrier to corrosive ions and can serve as containers for adsorbed corrosion inhibitors (Castaldo et al. 2018; Kulyk et al. 2021; Yan et al. 2014).

The application of graphene and GO, as well as their numerous forms and functionalised derivatives, in corrosion protection, namely in the areas of anticorrosion coatings and corrosion inhibition, is discussed in this chapter.

As seen from a simple SCOPUS search in Figure 6.1 by using the keywords: "graphene" OR "graphene oxide" AND "corrosion inhibition", the interest in this relatively new field of nanomaterials application is gaining more and more scientific interest.

Graphene and graphene-based materials have shown tremendous utilisation since the discovery of graphene in 2004 (Novoselov et al. 2004).

In this chapter, we will give more insights into different corrosion inhibition action modes that graphene and graphene-based materials reveal when covering a metal surface. Different forms of graphene and GO are used, namely: single-layer graphene (SLG), multi-layer graphene (MLG), GO, reduced graphene oxide (rGO), graphene platelets, and hybrid coatings.

There are three main modes of the mechanism of action of anticorrosion coatings that graphene and GO protection layers afford, namely; physical barrier creation between substrate materials and environments, slowdown of the corrosion processes, and coating layers acting as sacrificial materials.

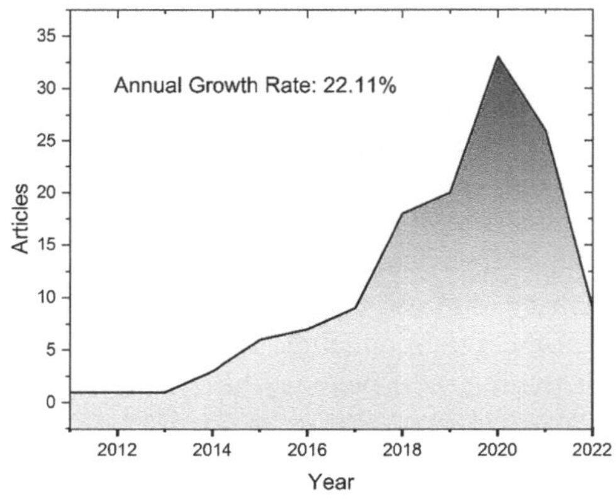

FIGURE 6.1 The number of published documents per year using SCOPUS search using the keywords: "graphene" OR "graphene oxide" AND "corrosion inhibition".

6.2 GRAPHENE AS A CORROSION INHIBITOR

The methods used for the production of graphene single and multi-layers and graphene-derived materials are described in the previous chapters of the book. Here, we will put in evidence only their use as corrosion inhibitors. The hydrophobic nature and impermeability of graphene have fueled the idea of the coverage of the metal surface with a graphene layer and its use as a protection barrier against corrosion. The growth of the graphene layer is made on transition metal surfaces and on magnesium and aluminium alloys by several methods such as chemical vapour deposition (CVD), electrophoretic deposition, laser irradiation, etc. (Kulyk et al. 2021). CVD is used more often and can be performed directly on the metal surface in the case of some transition metals or is transposed to another metal surface by a mechanical transfer technology (Li et al. 2009).

Graphene film that is almost free of any defect deposited on the metal surface does not only represent simple physical adsorption but is based on the nature of the metal and some unique properties of the graphene. It may have several interactions that stabilise the composite material. Therefore, the graphene layer can serve as an ideal physical barrier that prevents the corrosion of this entity in the aggressive environment (Dennis et al. 2016).

The mechanism of such action is depicted in Figure 6.2, where a pinhole-free graphene layer impedes the water and oxygen molecules and chloride ions to reach the metal surface (Ding et al. 2018).

Traditionally used polymer coatings to prevent metal corrosion are micrometric thick. They may alter optical and thermal properties of the protected system and do not resist at very high temperatures. Graphene single or few-layer films are chemically inert and stable at high temperatures and are very light due to the high ratio surface volume (Liu et al. 2008). It has now been more than one decade since Chen et al. published a paper that shows the capacity of the graphene layer to act as a corrosion protection barrier on copper and nickel/copper alloy surfaces against air oxidation under normal and high temperatures (Chen et al. 2011).

The graphene layer hinders corrosion degradation of the metal, for example, copper even at a high concentration of chloride ions (Singh Raman et al. 2012). Potentiodynamic and electrochemical impedance spectroscopy (EIS) measurements have shown a significant polarisation resistance of copper surface covered with a defect-free graphene film. Figure 6.3 shows the Bode plots for

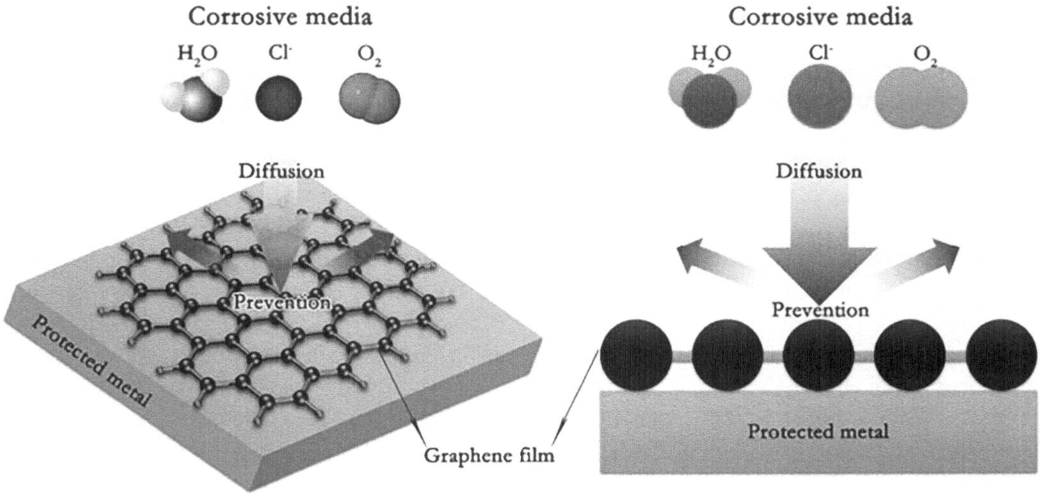

FIGURE 6.2 Graphene layer deposited on the metal surface hinders the diffusion of molecules and ions through the film (Ding et al. 2018).

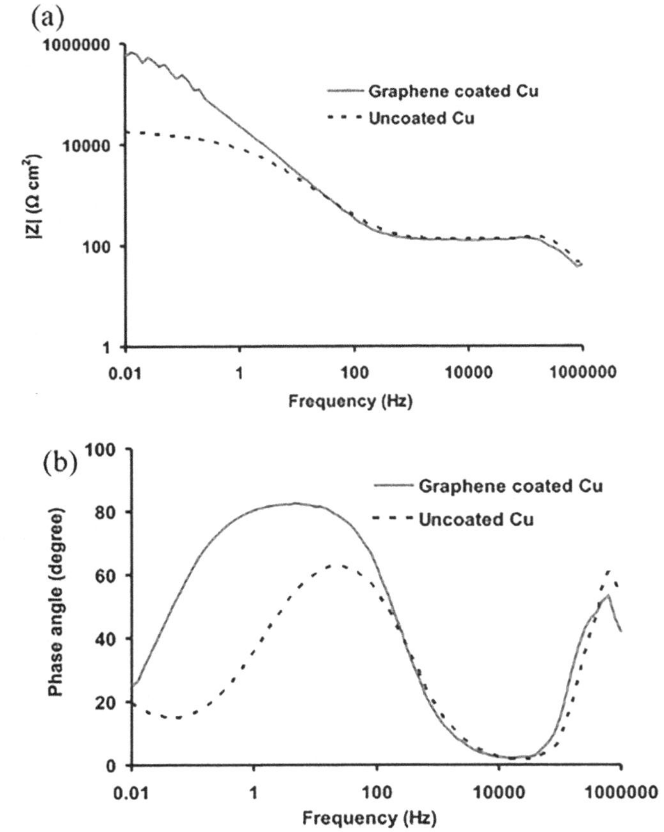

FIGURE 6.3 Bode plots of the copper surface covered with graphene and uncoated Cu: (a) Modulus and (b) phase angle plots (Singh Raman et al. 2012).

uncovered and copper surfaces with the graphene layer. One can see a much higher resistance for very low frequencies and a much broader phase angle for copper modified with graphene.

The barrier effect of the graphene layer depends also on the interfacial interactions between the surface atoms of the covered metal and the graphene layer. For example, graphene single layers to a great extent are most frequently prepared on the nickel and copper surface, and the interfacial distance is measured between the metal surface and graphene layer. It was 0.21 and 0.33 nm for Ni-G (Chen et al. 2011) and the Cu-G layer (Giovannetti et al. 2008). Cu and Ni surface interactions with the graphene single layer are theoretically calculated by DFT, and it is reported that there exist more covalent bonds between nickel and carbon atoms as a shorter distance enables them to share a higher number of electrons (Ma, Zeng, and Wang 2015). The impact of the graphene layer as a protective barrier on both metals was measured under an aggressive steam/air environment by potentiodynamic measurements (Chang et al. 2020). The corrosion inhibition efficiency was 66.3 times higher on the nickel directly grown on the graphene layer than on the transposed graphene layer and 361.6 higher than copper directly grown on the graphene film. These results and those of other teams put in evidence the importance of the interaction between the metal surface and graphene primary layer towards the ability to prevent their degradation in aggressive media (Weatherup et al. 2015). Given that nickel atoms have a specific interaction with the graphene layer and provide an efficient graphene layer growth, a thin alloy of nickel and iron is generated on a carbon steel surface using a high-power fiber laser (Ye et al. 2015). A multi-layer of graphene is grown on this alloy, and corrosion measurements were made in an aqueous solution of

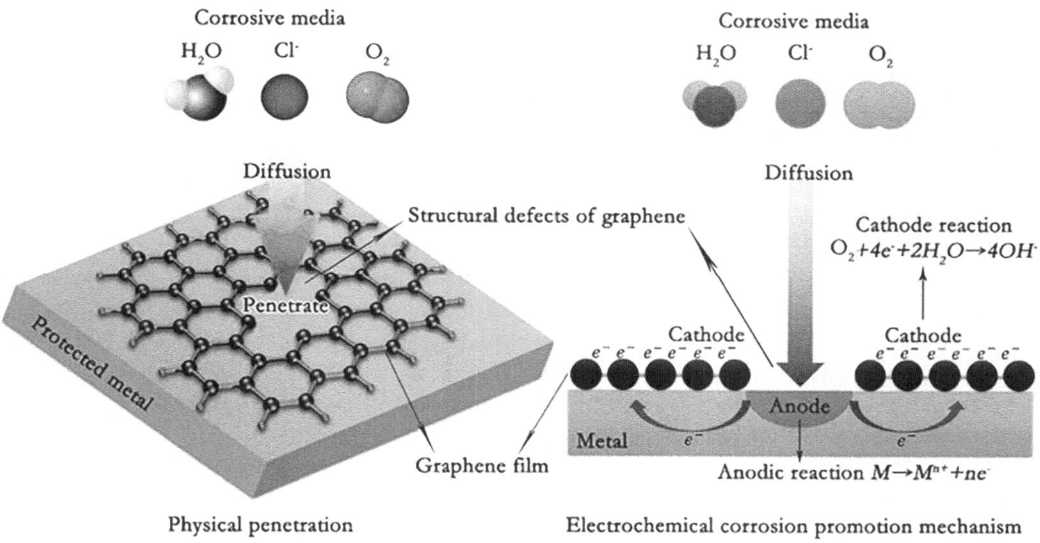

FIGURE 6.4 Graphene layer containing structural defects initiates the galvanic corrosion of the covered metal (Ding et al. 2018).

3.5% of NaCl by potentiodynamic and electrochemical impedance method. Experimental results gave a considerable diminution of corrosion rate for graphene-Ni-Fe alloy composite that is lower than that of stainless steel.

Apart from other unique properties, graphene is a very good electrical conductor. If its layer on the metal surface contains any defect – and knowing the fact that it is flexible and can easily create wrinkles (Kinloch et al. 2018), which is quite obvious, especially in case of high-surface-areas graphene films – it cannot prevent the penetration of solvent or gas molecules through the film, and the contact with the metallic surface induces galvanic corrosion when the graphene plays the role of the cathode, see Figure 6.4 (Ding et al. 2018). In this case, the graphene film plays the role of the cathode, and in the presence of aggressive agents, the oxidation of the covered metal is accelerated. Therefore, whenever graphene layers, deposited on the coinage metal surface, contain any defects, they not only fail to protect but increase the metal corrosion in the presence of oxidation agents.

The graphene layer is used also in the case of nanocomposites, as is shown in the case of graphene-nickel-phosphorous coating on the copper surface prepared by the electroless procedure (Kumari et al. 2019). The introduction of the graphene layer ameliorated the passive behavior of the coating by reducing the number of micropores and improved considerably the resistance against corrosion.

Graphene is used as a filler to form also nanocomposites with different polymers, but due to its hydrophobic nature, high electrical conductivity, poor solubility in different solvents, and the lack of heteroatoms that represent functional groups, graphene layers agglomerate, so it is quite difficult to have a strong interaction between the filler and the matrix and to protect the metals against corrosion (Kulyk et al. 2021).

Modification of the graphene surface with an insulator layer of a bio-based epoxy monomer and the formation of composite material in epoxy polymer has improved the corrosion resistance of carbon steel in the corrosive medium (Kulyk et al. 2021). An anticorrosive coating like a sandwich was prepared using the polymer, polyvinyl butyral (PVB), and CVD deposited graphene and is used to protect aluminium alloy used in aeroplanes (Kulyk et al. 2021). Two SLG were introduced between three PVB layers, and this hybrid coating has shown excellent efficacy to protect the corrosion of the metal exposed for 120 days in an aqueous solution of 3.5% NaCl. A multifunctional graphene composite made of nano-titan dioxide and graphene has enabled excellent protection of copper surfaces against corrosion (Gao et al. 2021).

6.3 GRAPHENE OXIDE AS A CORROSION INHIBITOR

GO is a 2D carbon nanomaterial that has attracted a great deal of interest in fields such as physics, chemistry, biology, and materials science in general. This nanomaterial is also an important building block and offers promising routes for the large-scale production of graphene, the "wonder material" (Ajala et al. 2022; Brisebois and Siaj 2020; Chen, Feng, and Li 2012). The protection of metals against corrosion is a relatively new use for this nanomaterial and its analogues (as protective films or as additives in coatings), which has lately gained traction (Ajala et al. 2022; Ding et al. 2020; Ding et al. 2018; Ghauri et al. 2017; Irfan, Bhat, and Ahmad 2018; Kulyk et al. 2021; Mehmeti and Podvorica 2018). Figure 6.5 shows some of the many state-of-the-art methods that are used to produce GO (Brisebois and Siaj 2020; Chen, Feng, and Li 2012).

There have been reports of functionalised GO being effective as a corrosion inhibitor in very acidic conditions (Ansari, Chauhan, Quraishi, Adesina, et al. 2020; Ansari, Chauhan, Quraishi, and Saleh 2020; Haruna, Alhems, and Saleh 2021; Haruna and Saleh 2021; Saleh, Haruna, and Mohammed 2021). The grafted groups on the GO structure included: Bis(2-aminoethyl)amine (Ansari, Chauhan, Quraishi, and Saleh 2020), polyethyleneimine (Ansari, Chauhan, Quraishi, Adesina, et al. 2020), amino- and diaminoazobenzene (Gupta et al. 2017), N,N′-bis-(2-aminoethyl) piperazine (Haruna and Saleh 2021), dopamine (Haruna, Alhems, and Saleh 2021), Octanoate groups (Saleh, Haruna, and Mohammed 2021), p-Aminophenol (Gupta et al. 2019), etc.

Ansari, Chauhan, Quraishi, and Saleh (2020) found that modified GO by bis(2-aminoethyl) amine moieties has an outstanding corrosion inhibition performance of 91% (at a concentration of 50 mg/L) towards carbon steel (in 15% HCl aqueous corrosion media) and that the inhibition can be further increased up to nearly 97% by using KI. In another study, the authors reached similar results by chemically grafting the GO via –COOH surface groups through the reaction with thionyl chloride in the first stage and polyethyleneimine in the second one (Ansari, Chauhan, Quraishi,

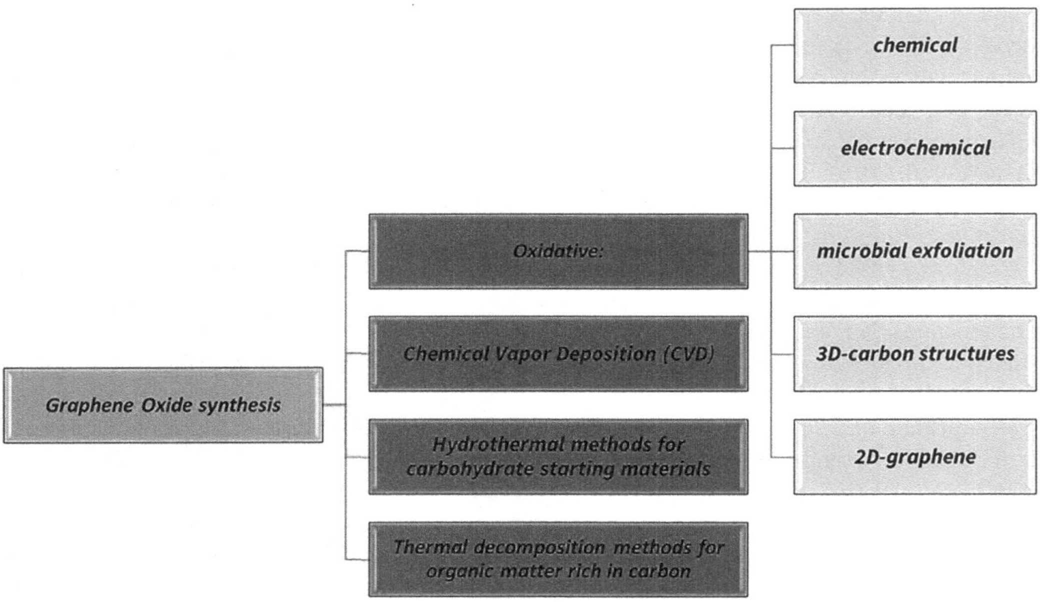

FIGURE 6.5 State-of-the-art methods that are used to produce graphene oxide.

Adesina, et al. 2020). The presence of the grafted groups in the modified GO structure was evaluated through FTIR-ATR spectroscopy.

As evidenced by the vast majority of studies conducted before their application in corrosion research, the attachment/surface modification of GO (covalent binding modification, noncovalent binding modification, and elemental doping) by various moieties is of the utmost importance (Berisha and Seydou 2022; Joshi et al. 2021; W. Yu et al. 2020).

This carbon nanomaterial is shown to strongly inhibit the corrosion of inhibition of niobium and tantalum surfaces (Mehmeti and Podvorica 2018), as was shown by Mehmeti et al. through the use of experimental measurements such as CV and EIS in combination with theoretical calculations methods such as Monte Carlo (MC) and molecular dynamics (MD).

Nb corrosion in 0.1 M H_2SO_4 aqueous solution may be inhibited by up to 70% in the presence of 100 ppm carboxylated GO, whereas the inhibition efficacy for Ta is 60%. In addition, by employing MD, as shown in Figure 6.6, they were able to demonstrate that this nanomaterial has a substantial adsorption propensity for these surfaces (Nb, Ta, and their corresponding oxidised surfaces).

In coating applications, GO, whether bare or grafted, has proven to be effective as filler in polymer matrices to generate high corrosion performance thin films. Such polymer matrices include: polyaniline (Yang et al. 2019; Yang, Zhu, and Hong 2020), epoxy resins (Liu et al. 2018; Pourhashem et al. 2017; Pourhashem, Vaezi, and Rashidi 2017), polystyrene (Yu et al. 2013), polyurethane (Liu et al. 2018), etc. When compared to other multilayer coating technologies, these polymer nanocomposites provide greater barrier properties at a lower cost and a simpler manufacturing approach (Kulyk et al. 2021). Metronidazole-modified GO nanocomposites were synthesised to prevent the presence of imperfections in epoxy coatings using a two-step method that involved grafting maleic anhydride on the GO surface followed by the covalent attachment of metronidazole via the esterification reaction (Yu et al. 2016).

It was demonstrated that the epoxy coating fillers manufactured from benzotriazole-loaded mesoporous silica nanocontainers mixed with GO displayed increased anticorrosion performance. These fillers were used in epoxy coatings. The presence of the peak signatures of the nanocomposite

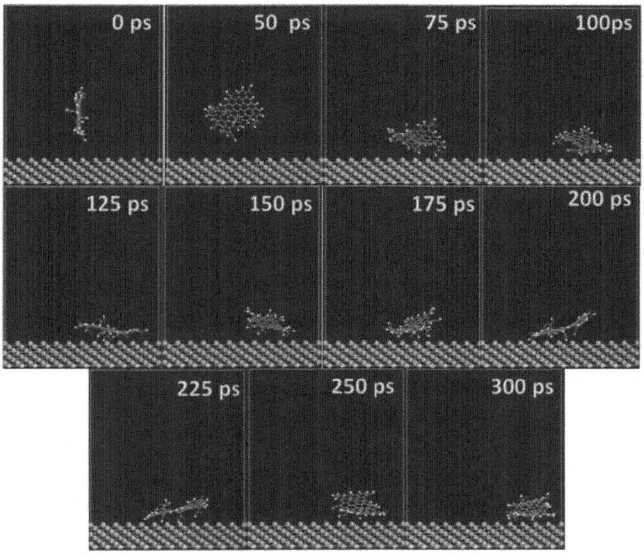

FIGURE 6.6 Time sequence interaction poses of the GO-COOH adsorbed on the Ta (111) surface (Mehmeti and Podvorica 2018).

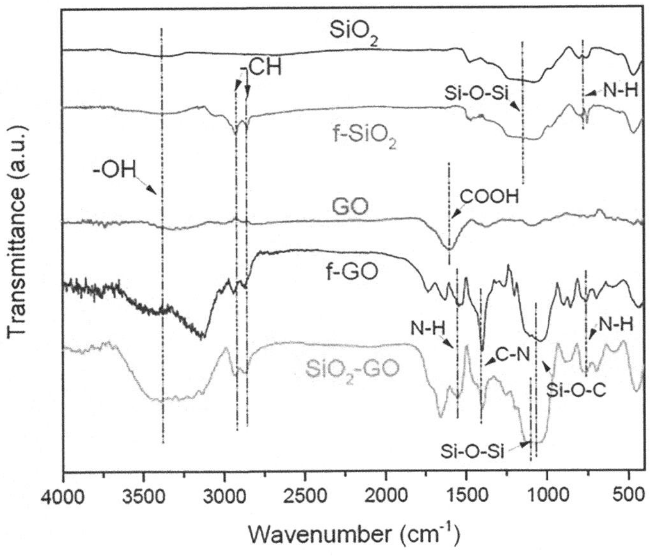

FIGURE 6.7 FTIR spectra of benzotriazole-loaded mesoporous silica nanocontainer and the GO-containing counterpart (Feng et al. 2022).

materials was demonstrated by infrared spectrum (FTIR) analyses, which can be seen in Figure 6.7. This was accomplished by meticulous investigation of the spectrum peaks.

An interesting approach for the construction of an anticorrosive GO material that shows self-healing properties was proposed by Zhang et al. (2022). In their approach, they employed the green liquid-phase synthetic technique to develop a covalent organic framework (COF) on GO sheets using 1,3,5-Triformylphloroglucinol and p-phenylenediamine as monomers. The newly produced nanocomposite served the purpose of a nanocontainer in which corrosion inhibitors could be loaded. Benzotriazole adsorption and release were controlled by local pH values and its ability to form complexes with iron(II) ions that are deposited in the metal-exposed area and prevent the diffusion of aggressive agents.

6.4 CONCLUSION AND PROSPECTS

Directly generated graphene coatings on metals can effectively prevent their corrosion by isolating them from corrosive surroundings. Due to its potential for mass production, CVD is largely used in laboratories to create graphene. During long-term exposures, however, graphene's flaws may aggravate the corrosion of metal substrates. The "barrier" effect of graphene and GO composite coatings and their unique properties, including very high chemical stability, make them more resistant to corrosion than traditional coatings. The protective mechanism of graphene and GO nanocomposite coatings requires more study at present. Future research should also investigate the subatomic scale distribution of metals under the protection of graphene and GO nanocomposite coatings. In addition, the failure derivation process of the graphene composite coating should be explained by artificial acceleration studies. As a consequence, the hunt for 2D barrier materials similar to graphene is a pressing concern. As a result of graphene's conductivity, galvanic corrosion is an unavoidable issue with graphene-based coatings. Emerging insulating 2D materials, such as GO (some extent) and other nanomaterials (TMD, BN, carbonitride MXenes), essentially prevent galvanic corrosion. In addition, it is worthwhile to create new materials and expand the category of graphene-like materials.

REFERENCES

Ajala, O. J., J. O. Tijani, M. T. Bankole, and A. S. Abdulkareem. 2022. "A Critical Review on Graphene Oxide Nanostructured Material: Properties, Synthesis, Characterization and Application in Water and Wastewater Treatment." *Environmental Nanotechnology, Monitoring & Management* 18 (December). Elsevier: 100673. doi:10.1016/J.ENMM.2022.100673.

Ansari, K. R., D. S. Chauhan, M. A. Quraishi, A. Y. Adesina, and T. A. Saleh. 2020. "The Synergistic Influence of Polyethyleneimine-Grafted Graphene Oxide and Iodide for the Protection of Steel in Acidizing Conditions." *RSC Advances* 10 (30). The Royal Society of Chemistry: 17739–17751. doi: 10.1039/D0RA00864H.

Ansari, K. R., D. S. Chauhan, M. A. Quraishi, and T. A. Saleh. 2020. "Bis(2-Aminoethyl)Amine-Modified Graphene Oxide Nanoemulsion for Carbon Steel Protection in 15% HCl: Effect of Temperature and Synergism with Iodide Ions." *Journal of Colloid and Interface Science* 564 (March). Academic Press: 124–133. doi:10.1016/J.JCIS.2019.12.125.

Berisha, A., and M. Seydou. 2022. "Grafting of Aryl Radicals onto Surfaces—A DFT Study." Springer, Cham, 121–135. doi:10.1007/978-3-031-04398-7_6.

Brisebois, P. P., and M. Siaj. 2020. "Harvesting Graphene Oxide – Years 1859 to 2019: A Review of Its Structure, Synthesis, Properties and Exfoliation." *Journal of Materials Chemistry C* 8 (5). The Royal Society of Chemistry: 1517–1547. doi:10.1039/C9TC03251G.

Castaldo, R., G. C. Lama, P. Aprea, G. Gentile, M. Lavorgna, V. Ambrogi, and P. Cerruti. 2018. "Effect of the Oxidation Degree on Self-Assembly, Adsorption and Barrier Properties of Nano-Graphene." *Microporous and Mesoporous Materials* 260 (April). Elsevier: 102–115. doi:10.1016/J.MICROMESO.2017.10.026.

Chang, W., P. Wang, Y. Zhao, C. Ren, B. N. Popov, and C. Li. 2020. "Characterizing Corrosion Properties of Graphene Barrier Layers Deposited on Polycrystalline Metals." *Surface and Coatings Technology* 398 (September). Elsevier: 126077. doi:10.1016/J.SURFCOAT.2020.126077.

Chauhan, D. S., M. A. Quraishi, K. R. Ansari, and T. A. Saleh. 2020. "Graphene and Graphene Oxide as New Class of Materials for Corrosion Control and Protection: Present Status and Future Scenario." *Progress in Organic Coatings* 147 (October). Elsevier: 105741. doi:10.1016/J.PORGCOAT.2020.105741.

Chen, D., H. Feng, and J. Li. 2012. "Graphene Oxide: Preparation, Functionalization, and Electrochemical Applications." *Chemical Reviews* 112 (11). American Chemical Society: 6027–6053. doi:10.1021/CR300115G/ASSET/IMAGES/CR300115G.SOCIAL.JPEG_V03.

Chen, S., L. Brown, M. Levendorf, W. Cai, S. Y. Ju, J. Edgeworth, X. Li, et al. 2011. "Oxidation Resistance of Graphene-Coated Cu and Cu/Ni Alloy." *ACS Nano* 5 (2). American Chemical Society: 1321–1327. doi:10.1021/NN103028D/ASSET/IMAGES/MEDIUM/NN-2010-03028D_0005.GIF.

Dennis, R. V., N. A. Fleer, R. D. Davidson, and S. Banerjee. 2016. "Graphene Coatings for the Corrosion Protection of Base Metals." *Graphene Technology: From Laboratory to Fabrication*, Edited by Soroush Nazarpour, and Stephen R. Waite August. John Wiley & Sons, 155–176. doi:10.1002/9783527687541.CH6.

Ding, J., P. Liu, M. Zhou, and H. Yu. 2020. "Nafion-Endowed Graphene Super-Anticorrosion Performance." *ACS Sustainable Chemistry and Engineering* 8 (40). American Chemical Society: 15344–15353. doi: 10.1021/ACSSUSCHEMENG.0C05679/SUPPL_FILE/SC0C05679_SI_001.PDF.

Ding, R., W. Li, X. Wang, T. Gui, B. Li, P. Han, H. Tian, et al. 2018. "A Brief Review of Corrosion Protective Films and Coatings Based on Graphene and Graphene Oxide." *Journal of Alloys and Compounds* 764 (October). Elsevier: 1039–1055. doi:10.1016/J.JALLCOM.2018.06.133.

Feng, C., L. Zhu, K. Cao, Z. Yu, and Y. Song. 2022. "Difunctional Silicon Dioxide Combined with Graphene Nanocomposite to Enhance the Anticorrosion Performance of Epoxy." *ACS Omega* 7 (28). American Chemical Society: 24134. doi:10.1021/ACSOMEGA.2C00494.

Gao, H., Y. Liu, G. Wang, S. Li, Z. Han, and L. Ren. 2021. "A Multifunctional Graphene Composite Coating with Switchable Wettability." *Chemical Engineering Journal* 415 (July). Elsevier: 128862. doi:10.1016/J.CEJ.2021.128862.

Ghauri, F. A., M. A. Raza, M. S. Baig, and S. Ibrahim. 2017. "Corrosion Study of the Graphene Oxide and Reduced Graphene Oxide-Based Epoxy Coatings." *Materials Research Express* 4 (12). IOP Publishing: 125601. doi:10.1088/2053-1591/AA9AAC.

Giovannetti, G., P. A. Khomyakov, G. Brocks, V. M. Karpan, J. Van Den Brink, and P. J. Kelly. 2008. "Doping Graphene with Metal Contacts." *Physical Review Letters* 101 (2). American Physical Society: 026803. doi:10.1103/PHYSREVLETT.101.026803/FIGURES/5/MEDIUM.

Gupta, R. K., M. Malviya, K. R. Ansari, H. Lgaz, D. S. Chauhan, and M. A. Quraishi. 2019. "Functionalized Graphene Oxide as a New Generation Corrosion Inhibitor for Industrial Pickling Process: DFT and

Experimental Approach." *Materials Chemistry and Physics* 236 (October). Elsevier: 121727. doi:10.1016/J.MATCHEMPHYS.2019.121727.

Gupta, R. K., M. Malviya, C. Verma, and M. A. Quraishi. 2017. "Aminoazobenzene and Diaminoazobenzene Functionalized Graphene Oxides as Novel Class of Corrosion Inhibitors for Mild Steel: Experimental and DFT Studies." *Materials Chemistry and Physics* 198 (September). Elsevier: 360–373. doi:10.1016/J.MATCHEMPHYS.2017.06.030.

Haruna, K., L. M. Alhems, and T. A. Saleh. 2021. "Graphene Oxide Grafted with Dopamine as an Efficient Corrosion Inhibitor for Oil Well Acidizing Environments." *Surfaces and Interfaces* 24 (June). Elsevier: 101046. doi:10.1016/J.SURFIN.2021.101046.

Haruna, K., and T. A. Saleh. 2021. "N,N′-Bis-(2-Aminoethyl)Piperazine Functionalized Graphene Oxide (NAEP-GO) as an Effective Green Corrosion Inhibitor for Simulated Acidizing Environment." *Journal of Environmental Chemical Engineering* 9 (1). Elsevier: 104967. doi:10.1016/J.JECE.2020.104967.

Irfan, M., S. I. Bhat, and S. Ahmad. 2018. "Reduced Graphene Oxide Reinforced Waterborne Soy Alkyd Nanocomposites: Formulation, Characterization, and Corrosion Inhibition Analysis." *ACS Sustainable Chemistry and Engineering* 6 (11). American Chemical Society: 14820–14830. doi:10.1021/ACSSUSCHEMENG.8B03349/SUPPL_FILE/SC8B03349_SI_001.PDF.

Joshi, D. J., J. R. Koduru, N. I. Malek, C. M. Hussain, and S. K. Kailasa. 2021. "Surface Modifications and Analytical Applications of Graphene Oxide: A Review." *TrAC Trends in Analytical Chemistry* 144 (November). Elsevier: 116448. doi:10.1016/J.TRAC.2021.116448.

Kinloch, I. A., J. Suhr, J. Lou, R. J. Young, and P. M. Ajayan. 2018. "Composites with Carbon Nanotubes and Graphene: An Outlook." *Science* 362 (6414). American Association for the Advancement of Science: 547–553. doi:10.1126/SCIENCE.AAT7439.

Kulyk, B., M. A. Freitas, N. F. Santos, F. Mohseni, A. F. Carvalho, K. Yasakau, A. J. S. Fernandes, et al. 2021. "A Critical Review on the Production and Application of Graphene and Graphene-Based Materials in Anti-Corrosion Coatings." *Critical Reviews in Solid State and Materials Sciences* 47 (3). Taylor & Francis: 309–355. doi:10.1080/10408436.2021.1886046. Https://Doi.Org/10.1080/10408436.2021.1886046.

Kumari, S., A. Panigrahi, S. K. Singh, M. Mohapatra, A. S. Khanna, S. K. Mishra, and S. K. Pradhan. 2019. "Electrochemical Behavior of Nanostructured Graphene Nickel Phosphorus Composite Coating on Copper." *Journal of Applied Electrochemistry 2019* 49 (12). Springer: 1157–1166. doi:10.1007/S10800-019-01333-Y.

Li, S., Y. Liu, Z. Tian, X. Liu, Z. Han, and L. Ren. 2020. "Biomimetic Superhydrophobic and Antibacterial Stainless-Steel Mesh via Double-Potentiostatic Electrodeposition and Modification." *Surface and Coatings Technology* 403 (December). Elsevier: 126355. doi:10.1016/J.SURFCOAT.2020.126355.

Li, X., Y. Zhu, W. Cai, M. Borysiak, B. Han, D. Chen, R. D. Piner, L. Colomba, and R. S. Ruoff. 2009. "Transfer of Large-Area Graphene Films for High-Performance Transparent Conductive Electrodes." *Nano Letters* 9 (12). American Chemical Society: 4359–4363. doi:10.1021/NL902623Y/ASSET/IMAGES/MEDIUM/NL-2009-02623Y_0006.GIF.

Liu, J., Q. Yu, M. Yu, S. Li, K. Zhao, B. Xue, and H. Zu. 2018. "Silane Modification of Titanium Dioxide-Decorated Graphene Oxide Nanocomposite for Enhancing Anticorrosion Performance of Epoxy Coatings on AA-2024." *Journal of Alloys and Compounds* 744 (May). Elsevier: 728–739. doi:10.1016/J.JALLCOM.2018.01.267.

Liu, L., S. Ryu, M. R. Tomasik, E. Stolyarova, N. Jung, M. S. Hybertsen, M. L. Steigerwald, L. E. Brus, and G. W. Flynn. 2008. "Graphene Oxidation: Thickness-Dependent Etching and Strong Chemical Doping." *Nano Letters* 8 (7). American Chemical Society: 1965–1970. doi:10.1021/NL0808684/SUPPL_FILE/NL0808684-FILE003.PDF.

Ma, L., X. C. Zeng, and J. Wang. 2015. "Oxygen Intercalation of Graphene on Transition Metal Substrate: An Edge-Limited Mechanism." *Journal of Physical Chemistry Letters* 6 (20). American Chemical Society: 4099–4105. doi:10.1021/ACS.JPCLETT.5B01841/SUPPL_FILE/JZ5B01841_SI_001.PDF.

Mehmeti, V., and F. I. Podvorica. 2018. "Experimental and Theoretical Studies on Corrosion Inhibition of Niobium and Tantalum Surfaces by Carboxylated Graphene Oxide." *Materials 2018, Vol. 11, Page 893* 11 (6). Multidisciplinary Digital Publishing Institute: 893. doi:10.3390/MA11060893.

Nine, M. J., M. A. Cole, D. N. H. Tran, and D. Losic. 2015. "Graphene: A Multipurpose Material for Protective Coatings." *Journal of Materials Chemistry A* 3 (24). The Royal Society of Chemistry: 12580–12602. doi:10.1039/C5TA01010A.

Novoselov, K. S., A. K. Geim, S. V. Morozov, D. Jiang, Y. Zhang, S. V. Dubonos, I. V. Grigorieva, and A. A. Firsov. 2004. "Electric Field in Atomically Thin Carbon Films." *Science* 306 (5696). American Association for the Advancement of Science: 666–669. doi:10.1126/SCIENCE.1102896/SUPPL_FILE/NOVOSELOV.SOM.PDF.

Pourhashem, S., A. Rashidi, M. R. Vaezi, and M. R. Bagherzadeh. 2017. "Excellent Corrosion Protection Performance of Epoxy Composite Coatings Filled with Amino-Silane Functionalized Graphene Oxide." *Surface and Coatings Technology* 317 (May). Elsevier: 1–9. doi:10.1016/J.SURFCOAT.2017.03.050.

Pourhashem, S., M. R. Vaezi, and A. Rashidi. 2017. "Investigating the Effect of SiO2-Graphene Oxide Hybrid as Inorganic Nanofiller on Corrosion Protection Properties of Epoxy Coatings." *Surface and Coatings Technology* 311 (February). Elsevier: 282–294. doi:10.1016/J.SURFCOAT.2017.01.013.

Saleh, T. A., K. Haruna, and A. R. I. Mohammed. 2021. "Octanoate Grafted Graphene Oxide as an Effective Inhibitor against Oil Well Acidizing Corrosion." *Journal of Molecular Liquids* 325 (March). Elsevier: 115060. doi:10.1016/J.MOLLIQ.2020.115060.

Singh Raman, R. K., P. C. Banerjee, D. E. Lobo, H. Gullapalli, M. Sumandasa, A. Kumar, L. Choudhary, R. Tkacz, P. M. Ajayan, and M. Majumder. 2012. "Protecting Copper from Electrochemical Degradation by Graphene Coating." *Carbon* 50 (11). Pergamon: 4040–4045. doi:10.1016/J.CARBON.2012.04.048.

Weatherup, R. S., L. D'Arsié, A. Cabrero-Vilatela, S. Caneva, R. Blume, J. Robertson, R. Schloegl, and S. Hofmann. 2015. "Long-Term Passivation of Strongly Interacting Metals with Single-Layer Graphene." *Journal of the American Chemical Society* 137 (45). American Chemical Society: 14358–14366. doi: 10.1021/JACS.5B08729/ASSET/IMAGES/LARGE/JA-2015-08729K_0005.JPEG.

Yan, N., G. Buonocore, M. Lavorgna, S. Kaciulis, S. K. Balijepalli, Y. Zhan, H. Xia, and L. Ambrosio. 2014. "The Role of Reduced Graphene Oxide on Chemical, Mechanical and Barrier Properties of Natural Rubber Composites." *Composites Science and Technology* 102 (1). Elsevier: 74–81. doi:10.1016/J.COMPSCITECH.2014.07.021.

Yang, N., T. Yang, W. Wang, H. Chen, and W. Li. 2019. "Polydopamine Modified Polyaniline-Graphene Oxide Composite for Enhancement of Corrosion Resistance." *Journal of Hazardous Materials* 377 (September). Elsevier: 142–151. doi:10.1016/J.JHAZMAT.2019.05.063.

Yang, S., S. Zhu, and R. Hong. 2020. "Graphene Oxide/Polyaniline Nanocomposites Used in Anticorrosive Coatings for Environmental Protection." *Coatings 2020, Vol. 10, Page 1215* 10 (12). Multidisciplinary Digital Publishing Institute: 1215. doi:10.3390/COATINGS10121215.

Ye, X., Z. Lin, H. Zhang, H. Zhu, Z. Liu, and M. Zhong. 2015. "Protecting Carbon Steel from Corrosion by Laser in Situ Grown Graphene Films." *Carbon* 94 (November). Pergamon: 326–334. doi:10.1016/J.CARBON.2015.06.080.

Yu, W., L. Sisi, Y. Haiyan, and L. Jie. 2020. "Progress in the Functional Modification of Graphene/Graphene Oxide: A Review." *RSC Advances* 10 (26). The Royal Society of Chemistry: 15328–15345. doi:10.1039/D0RA01068E.

Yu, Y. H., Y. Y. Lin, C. H. Lin, C. C. Chan, and Y. C. Huang. 2013. "High-Performance Polystyrene/Graphene-Based Nanocomposites with Excellent Anti-Corrosion Properties." *Polymer Chemistry* 5 (2). The Royal Society of Chemistry: 535–550. doi:10.1039/C3PY00825H.

Yu, Z., L. Lv, Y. Ma, H. Di, and Y. He. 2016. "Covalent Modification of Graphene Oxide by Metronidazole for Reinforced Anti-Corrosion Properties of Epoxy Coatings." *RSC Advances* 6 (22). The Royal Society of Chemistry: 18217–18226. doi:10.1039/C5RA23595B.

Zhang, C., W. Li, C. Liu, C. Zhang, L. Cao, D. Kong, W. Wang, and S. Chen. 2022. "Effect of Covalent Organic Framework Modified Graphene Oxide on Anticorrosion and Self-Healing Properties of Epoxy Resin Coatings." *Journal of Colloid and Interface Science* 608 (February). Academic Press: 1025–1039. doi:10.1016/J.JCIS.2021.10.024.

Zhang, Z., F. Xue, W. Bai, X. Shi, Y. Liu, and L. Feng. 2021. "Superhydrophobic Surface on Al Alloy with Robust Durability and Excellent Self-Healing Performance." *Surface and Coatings Technology* 410 (March). Elsevier: 126952. doi:10.1016/J.SURFCOAT.2021.126952.

7 Carbon Nanotubes (CNTs) in Anticorrosive Coatings

Omar Dagdag
Department of Mechanical Engineering, Gachon University, Seongnam, Republic of Korea

Rajesh Haldhar and Seong-Cheol Kim
School of Chemical Engineering, Yeungnam University, South Korea

Elyor Berdimurodov
Faculty of Chemistry, National University of Uzbekistan, Tashkent, Uzbekistan

Medical School, Central Asian University, Tashkent, Uzbekistan

Ekemini D. Akpan and Eno E. Ebenso
Centre for Materials Science, College of Science, Engineering and Technology, University of South Africa, Johannesburg, South Africa

CONTENTS

7.1 INTRODUCTION

Corrosion protection of metallic materials requires a big budget; around 3–4% of the world's gross domestic product (GDP) is spent on corrosion protection. Nearly 90% of metallic materials are defended by protective coatings [1–3]. The use of metal protection methods is one of the most common ways to prevent the interaction of the metal with the environment. Many studies have been done to find and develop more durable and stable materials, such as carbon nanotubes (CNTs) and their compounds, to prevent corrosion [4]. The data show that the addition of CNTs to the organic matrix of the composite improves skin absorption and improves absorption resistance. Carbonated materials of nanofillers are used in many fields such as technology, chemistry, biomedicine, and other interdisciplinary fields [1]. There are many types of carbon compounds in nature, and carbon compounds play an important role in the innovation of carbon-based advanced materials. Figure 7.1 shows the molecular structure of main C-based nanomaterials. From conventional 3D graphite [2] to the development of carbon-based nanomaterials such as fullerenes [3], 1D CNTs [4], and 2D graphene [5], the simple exploration of new carbon materials and their use is an important topic in the fields of material science, engineering, and physics, etc.

In recent years, there has been a search for CNTs that have good supporting additive excellence for the polymer matrix because of their unique and excellent thermal and chemical stability, electrical conductivity, and super mechanical strength (high aspect ratio, ultra-high tensile strength, etc.) [5–8]. Some studies have shown that the shielding ability of existing CNTs has improved, suggesting the ability to dissipate the electrical energy of the metal against localised attacks. CNT

DOI: 10.1201/9781003323976-9

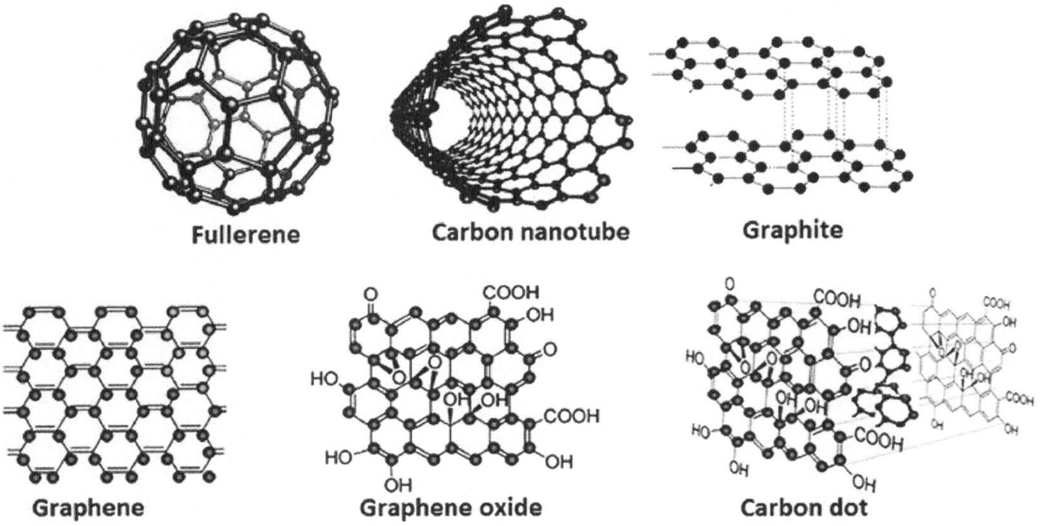

FIGURE 7.1 Molecular structures of main C-based nanomaterials. Reprinted with permission from [12] © 2021 Elsevier Publications.

overlay structures can also fill voids and micropores to seal the hull, reducing crack growth in the hull. For example, CNTs can provide an effective protective barrier in coatings. In addition, CNTs can increase the bond between rubber and metallic materials [9–11].

7.2 CARBON NANOTUBES (CNTs) IN ANTICORROSIVE COATINGS

CNTs that are shown in Figure 7.2 [13] are called good electronic materials due to their unique properties, high dynamic range, and good electrical properties [14,15]. In addition to high thermal stability, CNTs also have excellent mechanical properties, tear strength, stiffness, and Young's modulus (between 270 and 950 GPa). The addition of small amounts of CNTs to the biological system can improve the electrical, thermal, and mechanical properties [16,17].

Corrosion is well known to be one of the main causes of the degradation of metals in many industries. Therefore, ways to reduce wear and corrosion are being intensively researched. CNTs have many applications in several fields because of their specific multifunctional, electrical, and heat-stable performance [18,19]. CNT-doped composites are characterised by high weight, strength, and fatigue resistance [20]. In addition, CNTs are used as effective materials in hydrogen storage, as well as in chemical and electrical devices [21,22]. CNTs are also used as emulsifiers and are coated with a layer of polyaniline to reduce oxygen consumption and oxidation [23] and an epoxy coating to increase the adhesion and stickiness of rubber and steel [11].

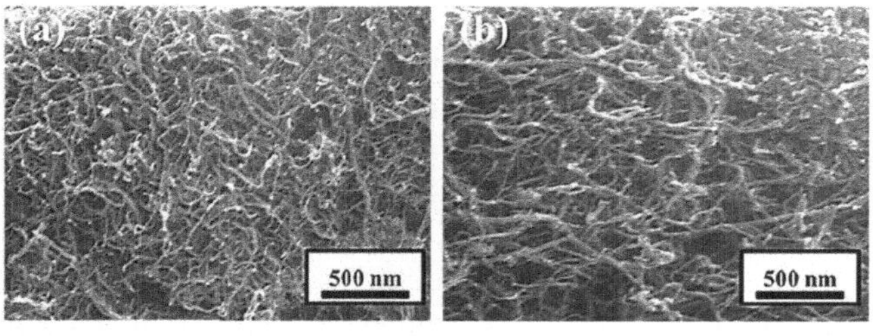

FIGURE 7.2 FE-SEM images of CNT nanoparticles. Reprinted with permission from [13] © 2015 Elsevier Publications.

Since their discovery in 1991, CNTs have been investigated and have been the subject of numerous studies in many fields [24,25]. In fact, in the last decade, CNTs were employed in many industries because of their good multifunctional, chemical, and physical characteristics. Some scientists have begun to use CNTs to protect steel from corrosion.

In the following research work [26], the graphene oxide (GO)/epoxy mixture was prepared as a suitable candidate for protection in the nuclear industry. In other research, the new CNT materials were modified with nanoparticles of silver, copper, and alumina to create a new generation of protective coating [27]. For example, adding CNTs to nanomaterials has been shown to increase the corrosion rate due to the poor chemical properties of CNTs by filling voids in the layer or modifying the process that protects against solubility changes such as polymeric materials. Several phosphonic acid surfactants are included in MWCNTs and BPA, HPA, and DPA. The DPA-modified alkyd MWCNT coating has better corrosion resistance than the additive-free coating [28]. A polyaniline-based nanotube is coated with ammonium peroxodisulfate as a disinfectant in aquatic systems. It is also indicated that the polyaniline nanotube was modified with the DPA and BPA in the water system; as a result, the antibacterial or disinfectant performance of the targeted nanotube increased. Therefore, these nanomaterials are mostly used in the bacterial or microbial corrosion of metallic materials. The adsorption performance and high density of polyaniline-based nanotubes were excellent [29]. In the next research work [30], a new multifunctional CNT was synthesised from β-cyclodextrin and MWCNTs. EIS and characterisation tests were used to test its corrosive and healing properties on the metal surface, confirming its high anticorrosion performance.

Nicola *et al.* [31] reported the development of a superhydrophobic coating based on MWCNTs on stainless steel (SS) substrates. The internal voids and metal content of SS were used to form a random pattern of MWCNTs by high-temperature chemical vapour deposition below 1000°C without further intervention or further processing. The superhydrophobic MWCNT film formed on the SS substrate had a peak of 154° with a large hysteresis angle. The superhydrophobic coating has been shown to stabilise the magnet and can be used to prevent corrosion.

Yang *et al.* [32] reported creating a new type of carbon nanomaterial that has effective corrosion resistance materials for metal. Cu was employed as a material to create a superhydrophobic material. An epoxy resin layer with random micro/nanostructures was made with low-energy materials such as X-ray-filtered nanoplates of various sizes and shapes in the liquid phase. The structure of the installation promotes the best protection, high mechanical protection, chemical stability, and self-cleaning of these materials in acidic and alkaline water. They provide new ideas and better protection of materials, including the protection of special and superhydrophobic compounds, which is very important for the development of special superhydrophobic materials in the chemical industry.

Zhu *et al.* [33] described a superhydrophobic carbon-based MWCNT-Co/a-C: H composite for self-cleaning. A stable and safe process called electrochemical deposition is used to create the superhydrophobic layer. The nano-performance of the surface was enhanced by MWCNT and cone nanoparticles with a nanocrystalline/amorphous microstructure. The highest water level is about 158° based on the lowest water level at 3°, that is, residual water. The compact design can be a good choice for corrosion prevention and self-cleaning. The superhydrophobicity of the enhanced part was also maintained after belt grinding and sand sampling, indicating that it remains due to the good relationship between MWCNTs and Co nanoparticles for the panel. The MWCNT-Co/a-C: H film is very effective in self-cleaning compared to pure carbon like a diamond. In other research work [34], the superhydrophobic carbon layer was developed with the MWCNTs and Ni nanoparticles, which were encapsulated in diamond-like carbon films, and MWCNT-Ni/a-C: H nanocomposite films were obtained by one-step electrochemical coating. The maximum water absorption of −159° is obtained from the lowest water absorption of −2°, which gives the film a hierarchical micro-nano-roughness. In addition, the developed superhydrophobic carbon layer was

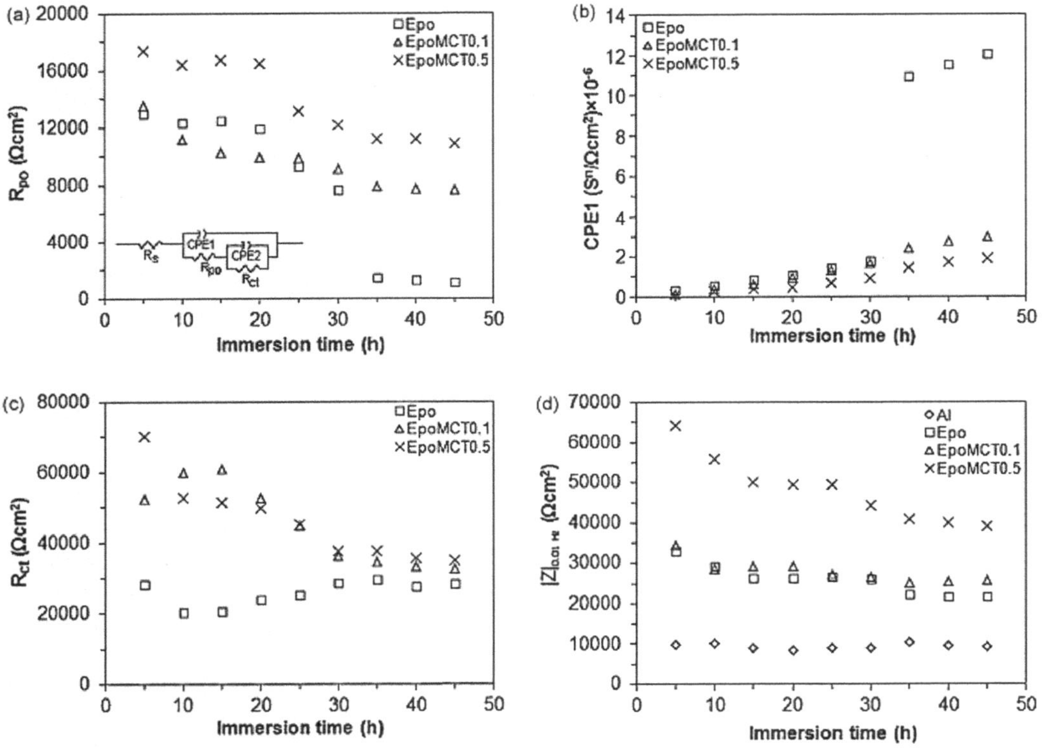

FIGURE 7.3 Metal protection performance of coating on AA2024-T (the MWCNT is addictive). Reprinted with permission from [35] © 2014 Elsevier Publications.

physical or chemical stable, had good adsorbent on the metal surface, was highly protective, and had excellent corrosion resistance.

Figure 7.3 shows the metal protection performance of the coating on AA2024-T (the MWCNT is addictive) [30]. It is clear from the obtained results that the inclusion of MWCNTs in the epoxy matrix led to a 29% increase in compressive strength, which determines the strength of the bond between the substrate and the epoxy resin. Uniform weight loss was divided by 0.5% MWCNT content. This confirmed that MWCNT cannot improve the multifunctional performances because of the negative interaction between CNT and epoxy structure, resulting in insufficient strength.

Wei *et al.* prepared polyurethane (PU) nanocomposite layers containing MWCNT (5%) by surface-initiated in situ polymerisation, which provides high elasticity of metal materials [36]. They found that the corrosion resistance of the nanocomposite layer was enhanced by the rise in micro and nano surface properties. These materials are more stable in aggressive saline solutions (Figure 7.4).

In the next study, Kumar *et al.* [37] examined the influence of the dispersion method on the protection performance of MWCNT epoxy nanocomposites deposited on the metal surface based on the parameters of ultrasound and shear force. They concluded that a homogeneous dispersion without agglomerates is 0.75 wt%. NanoMCNT in epoxy composites reduced the corrosion rate by 2.5×10^{-3} MPa per year with a low corrosion density (4.55 nA) and higher protection performance. Immunity is greater than 99% (Figure 7.5).

Based on the data obtained on the combined MWCNT samples, it can be seen that the higher the ratio, the larger the composite network and the higher the tensile strength. The diffuse polymer layer of MWCNTs effectively interconnected with the metal surface through the active functional groups [38].

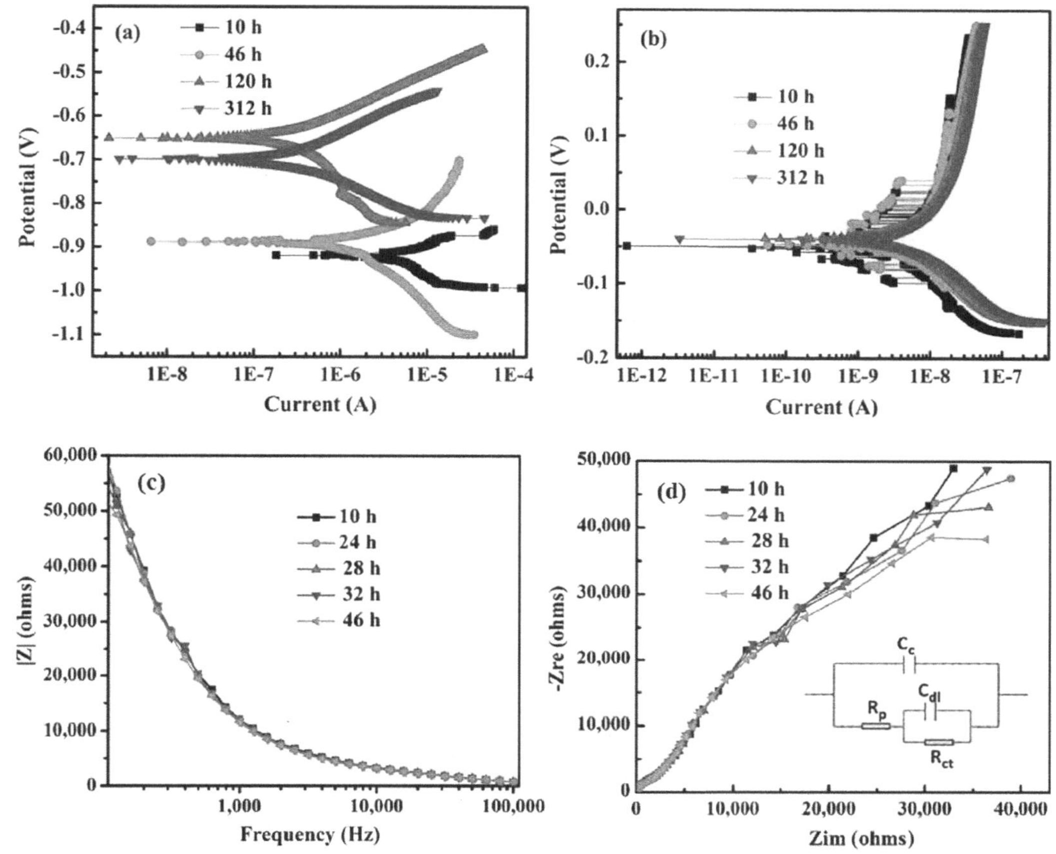

FIGURE 7.4 Electrochemical properties of PU/MWCNT coated stainless steel (3% NaCl). Reprinted with permission from [12] © 2021 Elsevier Publications.

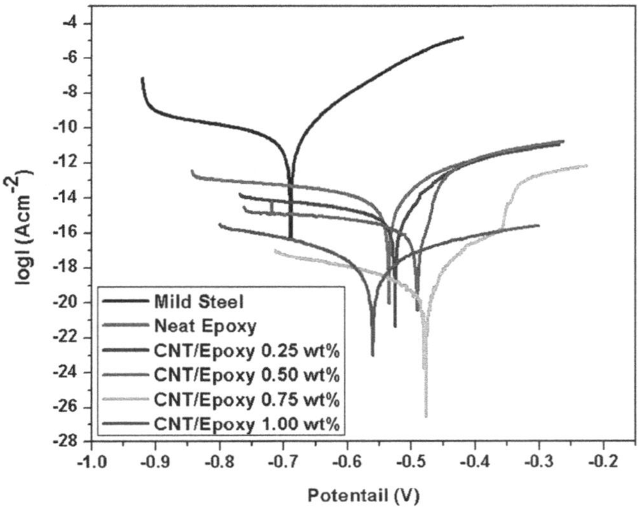

FIGURE 7.5 Polarisation curves of epoxy nanocomposites (MWCNT addition; steel surface). Reprinted with permission from [37] © 2017 Elsevier Publications.

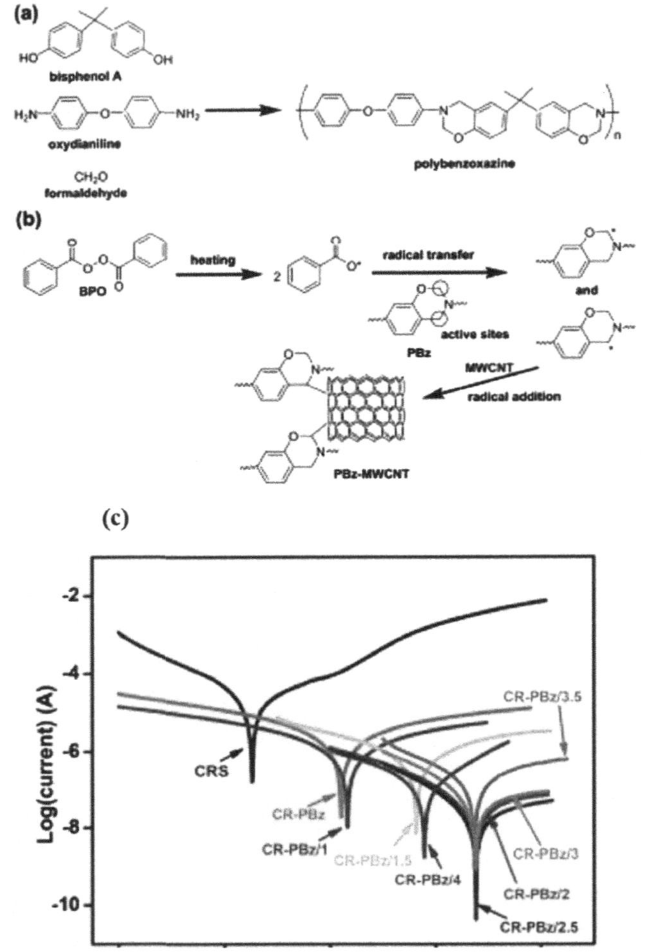

FIGURE 7.6 Tafel results of various nanohybrid protective coatings on Fe surface. Reprinted with permission from [39] © 2020 Elsevier Publications.

Researchers investigated the thermal properties of a polymer matrix doped with MWCNTs. Compared with the actual MWCNT aggregation in the PBz-MWCNT structure, the coating coated with PBz/PBz-MWCNT solution has the same morphology and good degradation. As a result, the contact area between PBz resin and MWCNTs increases. When using PBz as the polymer matrix, as shown in Figure 7.6(a,b), the addition of suboptimal MWCNT has a negative effect on the ability to form a floating plant due to aggregation. As can been seen from the obtained results in Figure 7.6c, the introduction of protective coating is mainly responsible for the dramatic decrease in the corrosion rate. It is also clear that the anti-microbial corrosion properties of this protective coating are high [39].

Hammer *et al.* synthesised MWCNTs functionalised with -CO_2H groups and bound to PMMA-based coatings. The nanocomposite coating showed higher strength compared to the non-permeable coating due to the high barrier of the coating because of the high interaction of the siloxane with the MWCNT functional group [40]. They also evaluated long-term corrosion protection and found that the nanocomposite layer provided long-term protection. The incorporation of non-functional materials, such as the CNT polymeric matrix used with -CO_2H groups, changes the ion transport process to a cation-specific one.

The prepared (PoPD)-CNT nanocomposite obtained had good strength when treated with pure CNT acid by cyclic voltammetry. They found that the addition of CNTs increased polymer deposition because CNTs have more surface area and more adsorption sites. Furthermore, the presence

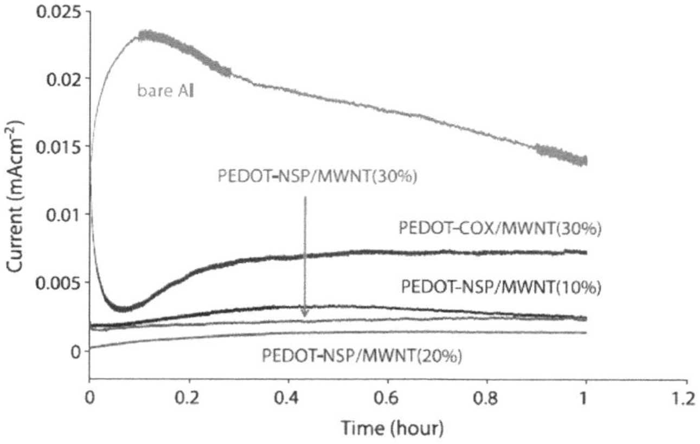

FIGURE 7.7 Chronoamperograms results for PEDOT-NSP/MWNT and PEDOT-COX/ MWNT coatings (Al metal samples). Reprinted with permission from [41] © 2012 Elsevier Publications.

of phenolic groups and -CO_2H in the CNT leads to an increase in the adsorption capacity of specific substances. As a result, the polymerisation of PoPD and its copolymerisation with CNTs increases, leading to the increase of the SS layer in the acid solution. They found that higher polymer deposition was observed in SWCNTs compared to MWCNTs [40].

In the next research work, PEDOT-based nanomaterials were introduced as a good corrosion inhibitor. The obtained results confirmed that this nanomaterial can protect Al metal samples from the corrosion in Li-ion batteries [84]. Figure 7.7 shows the frequency comparison of uncoated and coated aluminium samples and shows that PEDOT-NSP/MWCNTs reduce the corrosion of aluminium with increased electrostatic resistance. As shown in Figure 7.8, the PF6 anion can reduce the material to aluminium, which reduces corrosion. The protective behaviour of this coating depends on the formation of nanospheres on the metal surface. This is because the nanospheres effectively insulate the metal surface from the aquatic phase. Therefore, a small amount of MWCNTs cannot contribute much to the reduction of corrosion. By using effective MWCNTs, it is possible to stabilise the application and increase corrosion resistance [41].

In another study, the polyurethane layer was coated with PDA@CNT. Rapid tests such as EIS, salt spray, UV, and immersion showed corrosion and adhesion of the polyurethane polymer coating. In addition, a high CNT concentration (2.0 wt. %) increases the porosity and porosity of the layer, which reduces PU degradation due to the poor dispersion of CNTs. In contrast, CNT-wrapped PDA improved the distribution of CNTs in the polyurethane matrix because PDA was more compatible with the polyurethane polymer than CNTs. The melting temperature of the polyurethane coating also increased to 10^7 $\Omega.cm^2$ after 7 cycles. The presence of dopamine in CNTs prevents the formation and propagation of microdefects in polyurethane during long-term aging studies [42].

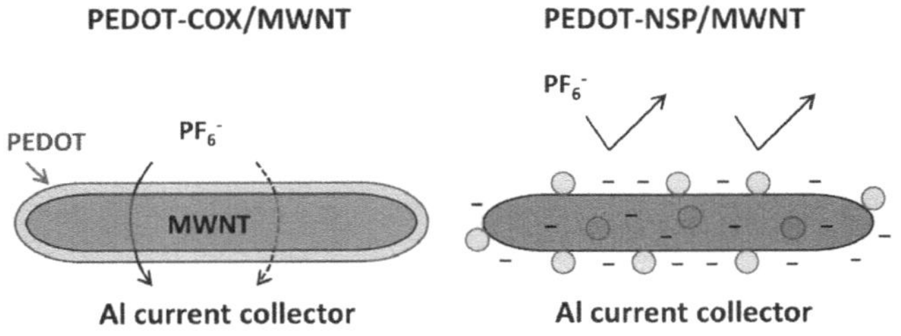

FIGURE 7.8 The pictorial representations of synergism for MWNT and PEDOT-NSP on the Al surface. Reprinted with permission from [41] © 2012 Elsevier Publications.

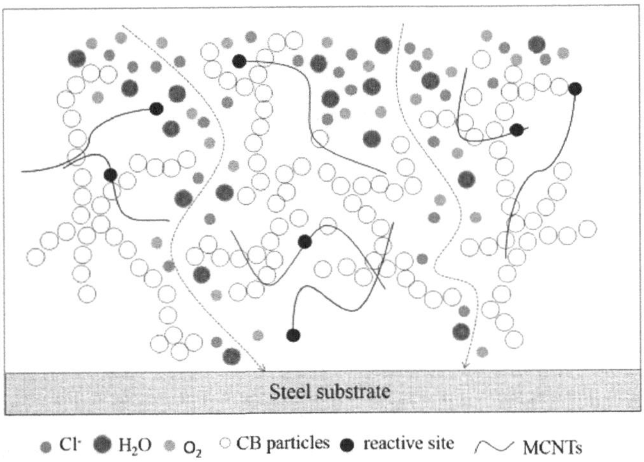

FIGURE 7.9 Interactions between the studied protective coating (MCNTs) and metal surface Reprinted with permission from [43] © 2018 Elsevier Publications.

In another investigation, Hou *et al.* [43] found that, with the participation of the reinforcement material, the interaction between the stiffness of the polyurethane chain and the filler is achieved. Complete dispersion of the composite increases the T_g and thermal stability of the nanocomposite. The degree of contact and the contact strength between the filler and the matrix was confirmed by SEM and TEM microphotographs. Adding filler increases the yield and improves the utilisation of the mixture. The absorption mechanism involves an increase in the circulation of ions. An increase in the interface also reduces the ion concentration and improves the barrier, as shown in Figure 7.9. It is indicated that an increase in corrosion is also responsible for the rise of corrosion current density.

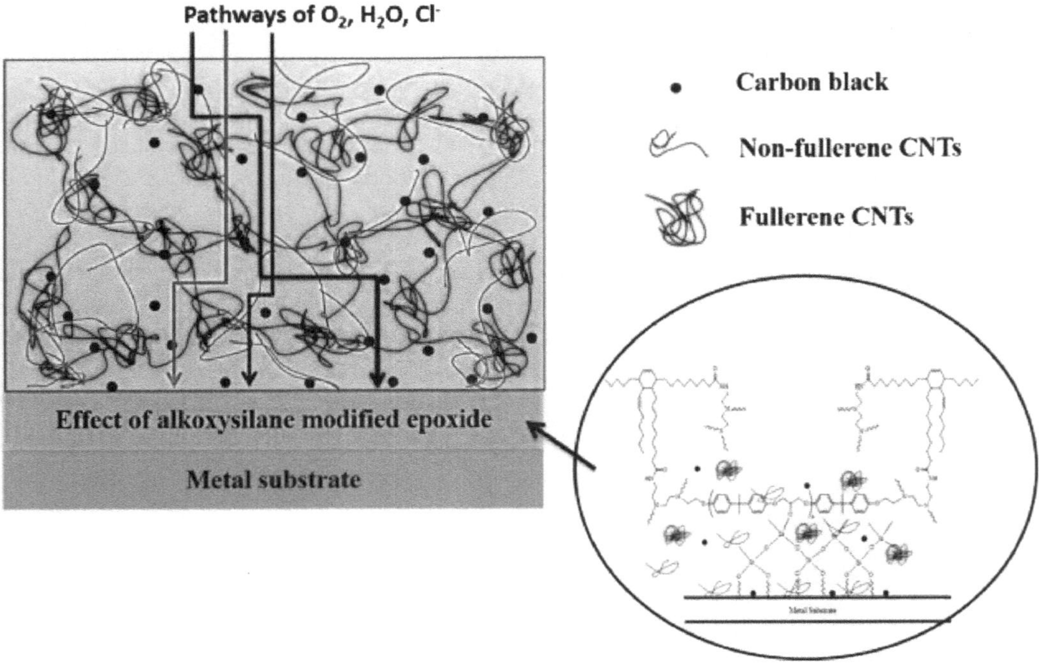

FIGURE 7.10 Pictorial representation of chemical and physical interactions between the protective coating and metal surface. Reprinted with permission from [44] © 2019 Elsevier Publications.

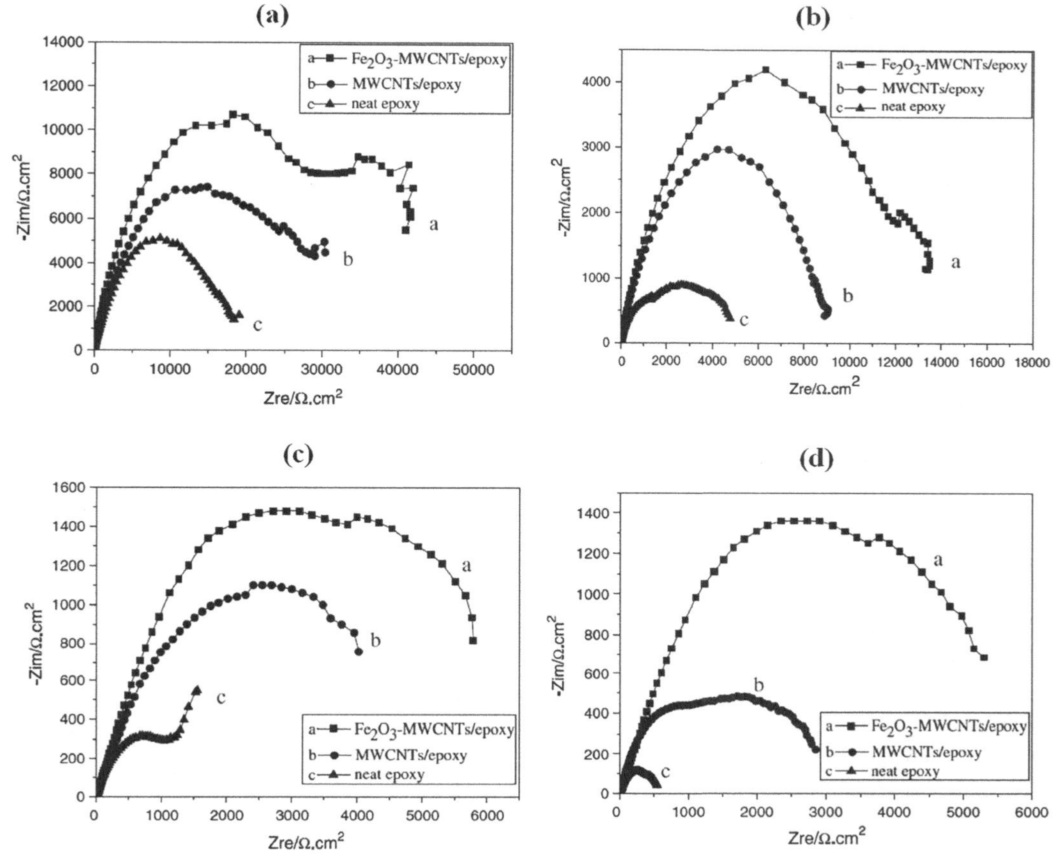

FIGURE 7.11 The obtained EIS results: (a) Fe_2O_3-MWCNTs/epoxy; (b) MWCNTs/epoxy; (c) Neat epoxy. Reprinted with permission from [12] © 2021 Elsevier Publications.

In the next research, the CBs, CNTs and blends of CBs, pure fullerenes, a mix of fullerenes and CNTs were introduced. Several parameters were used, including viscosity, mechanical, thermal, and electrical properties. The combination of fullerenes and CNTs was considered the best in terms of electrical properties and oxidation protection of the nanocomposite, and its effect on water use was also evident. In general, the anticorrosion process of the nanocomposite layer can be seen in Figure 7.10. In general, carbon composites are hydrophobic and have resistance to absorption, 1% by weight. This filler is highly resistant to water absorption and other damage [44].

The Fe_2O_3-MWCNTs was productively synthesised in [45]. The corrosion properties of this epoxy rise (ER) were explored by the various electrochemical and surface-morphological experiments [45]. It indicated a good difference of 2% by weight. Fe_2O_3-MWNT ES improves the coating's mechanical properties and corrosion resistance after different holding times (Figure 7.11(a-d)). In comparison, the results in Figure 7.11 suggested that the protective properties of the composite improved after Fe_2O_3-MWCNT coating. At the beginning of the immersion, the flow of electrolytes from the layer and pores of the EA layer was slower than the flow of aggressive ions between the EA layer and the metal of the EA layer.

7.3 CONCLUSION

CNTs are innovative materials for metal protection. Currently, CNTs are used in coatings and films, microelectronics, energy storage and environment, and biotechnology for heavy metals. Many CNTs have recently been tested for composition, material, and base. This chapter introduced some of the important aspects of the recent progress in using CNTs in anticorrosive coatings.

Author Contribution Statement: Equal contribution by all the authors.

REFERENCES

1. D. Song, Z. Yin, F. Liu, H. Wan, J. Gao, D. Zhang, *et al.*, "Effect of carbon nanotubes on the corrosion resistance of water-borne acrylic coatings," *Progress in Organic Coatings*, vol. 110, pp. 182–186, 2017.
2. H. Vakili, B. Ramezanzadeh, and R. Amini, "The corrosion performance and adhesion properties of the epoxy coating applied on the steel substrates treated by cerium-based conversion coatings," *Corrosion Science*, vol. 94, pp. 466–475, 2015.
3. M. Ghaffari, R. Naderi, and M. Sayehbani, "The effect of mixture of mercaptobenzimidazole and zinc phosphate on the corrosion protection of epoxy/polyamide coating," *Progress in Organic Coatings*, vol. 86, pp. 117–124, 2015.
4. D. Schmidt, D. Shah, and E. P. Giannelis, "New advances in polymer/layered silicate nanocomposites," *Current Opinion in Solid State and Materials Science*, vol. 6, pp. 205–212, 2002.
5. P. Calvert, "A recipe for strength," *Nature*, vol. 399, pp. 210–211, 1999.
6. M. J. Treacy, T. W. Ebbesen, and J. M. Gibson, "Exceptionally high Young's modulus observed for individual carbon nanotubes," *Nature*, vol. 381, pp. 678–680, 1996.
7. C. J. Lee, J. Park, S. Y. Kang, and J. H. Lee, "Growth and field electron emission of vertically aligned multiwalled carbon nanotubes," *Chemical Physics Letters*, vol. 326, pp. 175–180, 2000.
8. T. M. McCoy, P. Brown, J. Eastoe, and R. F. Tabor, "Noncovalent magnetic control and reversible recovery of graphene oxide using iron oxide and magnetic surfactants," *ACS Applied Materials & Interfaces*, vol. 7, pp. 2124–2133, 2015.
9. H. Jeon, J. Park, and M. Shon, "Corrosion protection by epoxy coating containing multi-walled carbon nanotubes," *Journal of Industrial and Engineering Chemistry*, vol. 19, pp. 849–853, 2013.
10. L. Yang, F. Liu, and E. Han, "Effects of P/B on the properties of anticorrosive coatings with different particle size," *Progress in Organic Coatings*, vol. 53, pp. 91–98, 2005.
11. S. Ganguli, H. Aglan, P. Dennig, and G. Irvin, "Effect of loading and surface modification of MWCNTs on the fracture behavior of epoxy nanocomposites," *Journal of Reinforced Plastics and Composites*, vol. 25, pp. 175–188, 2006.
12. A. Hosseinpour, M. R. Abadchi, M. Mirzaee, F. A. Tabar, and B. Ramezanzadeh, "Recent advances and future perspectives for carbon nanostructures reinforced organic coating for anti-corrosion application," *Surfaces and Interfaces*, vol. 23, p. 100994, 2021.
13. A. M. Kumar and Z. M. Gasem, "Effect of functionalization of carbon nanotubes on mechanical and electrochemical behavior of polyaniline nanocomposite coatings," *Surface and Coatings Technology*, vol. 276, pp. 416–423, 2015.
14. Z. Xiong, Y. S. Yun, and H.-J. Jin, "Applications of carbon nanotubes for lithium ion battery anodes," *Materials*, vol. 6, pp. 1138–1158, 2013.
15. Z. Han and A. Fina, "Thermal conductivity of carbon nanotubes and their polymer nanocomposites: A review," *Progress in Polymer Science*, vol. 36, pp. 914–944, 2011.
16. F. Gojny, M. Wichmann, U. Köpke, B. Fiedler, and K. Schulte, "Carbon nanotube-reinforced epoxy-composites: Enhanced stiffness and fracture toughness at low nanotube content," *Composites Science and Technology*, vol. 64, pp. 2363–2371, 2004.
17. D. Eder, "Carbon nanotube–inorganic hybrids," *Chemical Reviews*, vol. 110, pp. 1348–1385, 2010.
18. S. C. Canobre, D. A. Almeida, C. P. Fonseca, and S. Neves, "Synthesis and characterization of hybrid composites based on carbon nanotubes," *Electrochimica Acta*, vol. 54, pp. 6383–6388, 2009.
19. W. Jarernboon, S. Pimanpang, S. Maensiri, E. Swatsitang, and V. Amornkitbamrung, "Effects of multiwall carbon nanotubes in reducing microcrack formation on electrophoretically deposited TiO_2 film," *Journal of Alloys and Compounds*, vol. 476, pp. 840–846, 2009.

20. S. H. S. Zein, L.-C. Yeoh, S.-P. Chai, A. R. Mohamed, and M. E. M. Mahayuddin, "Synthesis of manganese oxide/carbon nanotube nanocomposites using wet chemical method," *Journal of Materials Processing Technology*, vol. 190, pp. 402–405, 2007.
21. A. C. Dillon, K. Jones, T. Bekkedahl, C. Kiang, D. Bethune, and M. Heben, "Storage of hydrogen in single-walled carbon nanotubes," *Nature*, vol. 386, pp. 377–379, 1997.
22. P. Serp, M. Corrias, and P. Kalck, "Carbon nanotubes and nanofibers in catalysis," *Applied Catalysis A: General*, vol. 253, pp. 337–358, 2003.
23. M. Deyab, "Corrosion protection of aluminum bipolar plates with polyaniline coating containing carbon nanotubes in acidic medium inside the polymer electrolyte membrane fuel cell," *Journal of Power Sources*, vol. 268, pp. 50–55, 2014.
24. V. Sharma, M. Goyat, A. Hooda, J. K. Pandey, A. Kumar, R. Gupta, *et al.*, "Recent progress in nano-oxides and CNTs based corrosion resistant superhydrophobic coatings: A critical review," *Progress in Organic Coatings*, vol. 140, p. 105512, 2020.
25. M. Goyat, V. Jaglan, V. Tomar, G. Louchaert, A. Kumar, K. Kumar, *et al.*, "Superior thermo-mechanical and wetting properties of ultrasonic dual mode mixing assisted epoxy-CNT nano-composites," *High Performance Polymers*, vol. 31, pp. 32–42, 2019.
26. W. Xia, H. Xue, J. Wang, T. Wang, L. Song, H. Guo, *et al.*, "Functionlized graphene serving as free radical scavenger and corrosion protection in gamma-irradiated epoxy composites," *Carbon*, vol. 101, pp. 315–323, 2016.
27. V. Srinivas, C. Moorthy, V. Dedeepya, P. Manikanta, and V. Satish, "Nanofluids with CNTs for automotive applications," *Heat and Mass Transfer*, vol. 52, pp. 701–712, 2016.
28. C. Oueiny, S. Berlioz, L. Patout, and F. Perrin, "Aqueous dispersion of multiwall carbon nanotubes with phosphonic acid derivatives," *Colloids and Surfaces A: Physicochemical and Engineering Aspects*, vol. 493, pp. 41–51, 2016.
29. C. Oueiny, S. Berlioz, and F. Perrin, "Assembly of polyaniline nanotubes by interfacial polymerization for corrosion protection," *Physical Chemistry Chemical Physics*, vol. 18, pp. 3504–3509, 2016.
30. Y. He, C. Zhang, F. Wu, and Z. Xu, "Fabrication study of a new anticorrosion coating based on supramolecular nanocontainer," *Synthetic Metals*, vol. 212, pp. 186–194, 2016.
31. F. De Nicola, P. Castrucci, M. Scarselli, F. Nanni, I. Cacciotti, and M. De Crescenzi, "Super-hydrophobic multi-walled carbon nanotube coatings for stainless steel," *Nanotechnology*, vol. 26, p. 145701, 2015.
32. Z. Yang, L. Wang, W. Sun, S. Li, T. Zhu, W. Liu, *et al.*, "Superhydrophobic epoxy coating modified by fluorographene used for anti-corrosion and self-cleaning," *Applied Surface Science*, vol. 401, pp. 146–155, 2017.
33. X. Zhu, S. Zhou, Q. Yan, and S. Wang, "Multi-walled carbon nanotubes enhanced superhydrophobic MWCNTs-Co/aC: H carbon-based film for excellent self-cleaning and corrosion resistance," *Diamond and Related Materials*, vol. 86, pp. 87–97, 2018.
34. S. Zhou, X. Zhu, L. Ma, Q. Yan, and S. Wang, "Outstanding superhydrophobicity and corrosion resistance on carbon-based film surfaces coupled with multi-walled carbon nanotubes and nickel nano-particles," *Surface Science*, vol. 677, pp. 193–202, 2018.
35. N. Khun, B. R. Troconis, and G. Frankel, "Effects of carbon nanotube content on adhesion strength and wear and corrosion resistance of epoxy composite coatings on AA2024-T3," *Progress in Organic Coatings*, vol. 77, pp. 72–80, 2014.
36. H. Wei, D. Ding, S. Wei, and Z. Guo, "Anticorrosive conductive polyurethane multiwalled carbon nanotube nanocomposites," *Journal of Materials Chemistry A*, vol. 1, pp. 10805–10813, 2013.
37. A. Kumar, P. Ghosh, K. Yadav, and K. Kumar, "Thermo-mechanical and anti-corrosive properties of MWCNT/epoxy nanocomposite fabricated by innovative dispersion technique," *Composites Part B: Engineering*, vol. 113, pp. 291–299, 2017.
38. J. Jeevanandam, A. Barhoum, Y. S. Chan, A. Dufresne, and M. K. Danquah, "Review on nanoparticles and nanostructured materials: History, sources, toxicity and regulations," *Beilstein Journal of Nanotechnology*, vol. 9, pp. 1050–1074, 2018.
39. S. Zachariah and Y.-L. Liu, "Nanocomposites of polybenzoxazine-functionalized multiwalled carbon nanotubes and polybenzoxazine for anticorrosion application," *Composites Science and Technology*, vol. 194, p. 108169, 2020.
40. P. Hammer, F. Dos Santos, B. Cerrutti, S. H. Pulcinelli, and C. V. Santilli, "Carbon nanotube-reinforced siloxane-PMMA hybrid coatings with high corrosion resistance," *Progress in Organic Coatings*, vol. 76, pp. 601–608, 2013.

41. S. R. Prabakar and M. Pyo, "Corrosion protection of aluminum in LiPF6 by poly (3, 4-ethylenedioxythiophene) nanosphere-coated multiwalled carbon nanotube," *Corrosion Science*, vol. 57, pp. 42–48, 2012.
42. G. Cai, J. Hou, D. Jiang, and Z. Dong, "Polydopamine-wrapped carbon nanotubes to improve the corrosion barrier of polyurethane coating," *RSC Advances*, vol. 8, pp. 23727–23741, 2018.
43. L. Hou, M. Zhou, and S. Wang, "Synthesis, thermal and anticorrosion performance of WPU nano-composites with low carbon-black content by adding amine-modified multiwall carbon nanotube," *Diamond and Related Materials*, vol. 90, pp. 166–171, 2018.
44. L. Meng and M. D. Soucek, "Comparison of the carbon additives on the conductivity, thermo-mechanical, and corrosion properties for TEOS oligomer modified epoxy-amine coating systems," *Progress in Organic Coatings*, vol. 130, pp. 168–181, 2019.
45. Y. He, C. Chen, F. Zhong, H. Chen, and D. Qing, "Synthesis and properties of iron oxide coated carbon nanotubes hybrid materials and their use in epoxy coatings," *Polymers for Advanced Technologies*, vol. 26, pp. 414–421, 2015.

ABBREVIATIONS

CNTs	Carbon nanotubes
MWCNTs	Multi-walled carbon nanotubes
BPA	Benzylphosphonic acid
HPA	Hexylphosphonic acid
DPA	Decylphosphonic acid
PANI	Polyaniline
EIS	Electrochemical impedance spectroscopy
ITO	Indium tin oxide
SS	Stainless steel
PBz	Polybenzoxazine
PMMA	Siloxane-poly(methyl methacrylate)
RTAR	Radical transfer and addition reaction.
PoPD	Poly (o-phenylenediamine)
PEDOT	Poly(ethylene dioxythiophene)
SS	Stainless steel
PDAs@CNTs	Polydopamine-wrapped carbon nanotubes
EIS	Electrochemical impedance spectroscopy
CB	Carbon black

8 Quantum Dots (QDs) in Anticorrosive Coatings

Taiwo W. Quadri and Eno E. Ebenso
Centre for Material Science, College of Science, Engineering and Technology, University of South Africa, Johannesburg, South Africa

CONTENTS

8.1 CORROSION AND ANTICORROSIVE COATINGS

Metallic corrosion, which has become one of the biggest challenges in the metal-based industry, is a phenomenon usually triggered by the contact of metals with corrosive species in the environment, such as oxygen, water and chloride ions. A report from the National Association of Corrosion Engineers (NACE) estimated that the yearly direct cost of metallic deterioration is about $2.5 trillion globally [1]. Annually, South Africa and the United States of America are reported to lose about $10 and $300 billion, respectively, due to material corrosion [2]. In addition, corrosion is known to have dire implications on human health, public infrastructures, the natural environment and national heritage. Therefore, reliable and robust combat strategies are urgently needed to address the problem of metallic degradation. Different methods have been proposed to combat metallic degradation, which include cathodic protection, anodic protection, protective coatings, inhibitors and equipment design [3]. Among these methods, inhibitors and protective coatings have been widely embraced because they are cost-effective, high-performing and practical. In particular, anticorrosive coatings, including conversion, metallic, organic and inorganic coatings, have been developed to offer effective corrosion resistance to metals by preventing contact of metallic substrates with aggressive ions [4,5]. The commonly adopted coating technologies are often based on heavy metals, including copper, zinc, chromium and phosphate, which are environmentally deleterious. Modern trends in the development of anticorrosive coatings are in favour of suitable eco-friendly materials with less harmful environmental impacts and superior corrosion protection properties. Therefore, the search for eco-friendly and effective materials has continued leading to the discovery, design and development of novel anticorrosive materials.

Over the years, polymers have gained a wide reputation as effective coating materials because they possess exceptional features such as good hydrophobic property, eco-friendliness, inherent stability and wide effectiveness. Owing to their high chemical resistance, strong adsorption property, high abrasion resistance and outstanding flexibility, several conventional polymers such as polyolefin, polyurethane (PU) and epoxy (EP) have been adopted as anticorrosive materials in

DOI: 10.1201/9781003323976-10

different corrosive media. Nevertheless, the protection properties provided by these conventional polymer coatings can wear out with time, leading to exposure to aggressive surroundings. Therefore, it becomes expedient to augment the corrosion protection abilities of these conventional polymers to meet stringent requirements for use as anticorrosive agents [6,7]. Conducting polymers such as polyaniline, polyindole and polypyrrole, which have the advantage of superior mechanical, electrical and optical properties, have shown better performance as anticorrosive agents. Over the last three decades, numerous reports have emerged on the anticorrosive potential of conductive polymer coatings [8–10]. Despite their wide application, they have certain deficiencies and limitations as well, leading to coating failure and corrosion acceleration. In recent years, the advancement of nanotechnology has provided effective alternatives to the challenges encountered with previous anticorrosive systems. Nowadays, nanofillers such as ZnO, TiO_2, SiO_2, graphene, carbon nanotubes, etc., are being incorporated into polymers to form polymer nanocomposite coatings, which have shown improved mechanical and anticorrosive properties [7,9,11–13]. Research has shown that protective coatings impede corrosion reaction by forming an exterior film to act as a shield to prevent the percolation of aggressive ions into the metal substrate [14,15]. Investigations on the corrosion protection of metals via anticorrosive coatings are often conducted using different electrochemical tests such as electrochemical impedance spectra (EIS), local electrochemical impedance spectra (LEIS), cyclic voltammetry (CV) and potentiodynamic polarisation (PDP). Furthermore, salt spray tests are also utilised to understand the mechanism of protective coatings in aggressive electrolytes [16–18].

8.2 SYNTHESIS, CHARACTERISATION, PROPERTIES AND APPLICATIONS OF CDs

Fondly called the rising star of carbon nanomaterials, carbon dots (CDs), which were first discovered by Xu et al. in a work published in 2004 [19] and Sun et al. in 2006 [20], have gained considerable interest in several fields of application because of their fascinating characteristics. CDs are novel zero-dimensional carbon nano-allotropes with a typical size less than 10 nm and comprising sp^2 and sp^3 carbon atoms. They are quasi-spherical nanomaterials possessing crystalline and amorphous parts and are popularly known for their strong fluorescence properties. Based on their structures, CDs are generally classified into four broad categories, namely: carbonised polymer dots (CPDs), carbon nanodots (CNDs), carbon quantum dots (CQDs) and graphene quantum dots (GQDs) [21]. A general structure and classification of CDs is presented in Figure 8.1.

The different structures of CDs are fabricated through various synthesis methods, which are largely classified into "bottom-up" and "top-down" techniques, as depicted in Figure 8.2 [24]. Bottom-up processes are carried out by pyrolysis or carbonisation of small sets of organic molecules through chemical reaction, whereas top-down techniques entail the cracking/decomposition of bigger carbon molecules into smaller forms via electrochemical, physical or chemical techniques. Generally, hydrothermal and solvothermal methods are preferred techniques in the development of CDs for application as anticorrosive coatings because they are simple, cost-effective and non-toxic.

Generally, the successful synthesis of CDs is often confirmed using several modern and sophisticated analytical methods. These techniques provide insights into the composition, size, shape and structure of the synthesised CDs, which have corresponding influence on their special properties. The widely reported microscopic methods available in the literature on CDs in anticorrosive coatings include transmission electron microscopy (TEM), scanning probe microscopy (SPM), scanning electron microscopy (SEM), atomic force microscopy (AFM), X-ray diffraction (XRD) and high-resolution TEM (HRTEM). Spectroscopic techniques such as nuclear magnetic resonance spectroscopy (NMR), X-ray fluorescence (XRF), Fourier transform infrared spectroscopy (FTIR), X-ray photoelectron spectroscopy (XPS), energy dispersive X-ray spectroscopy (EDS), ultraviolet-visible spectroscopy (UV–vis), photoluminescence spectroscopy (PL) and Raman spectroscopy (RS) have also been reported in the description of the synthetic features of prepared CDs [25].

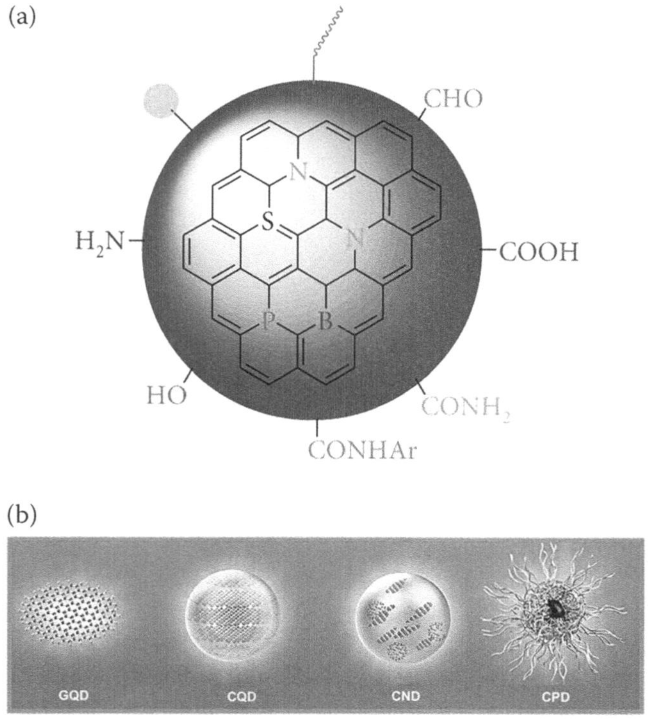

FIGURE 8.1 (a) A general structure [22] and (b) classification of carbon dots (CDs) [23].

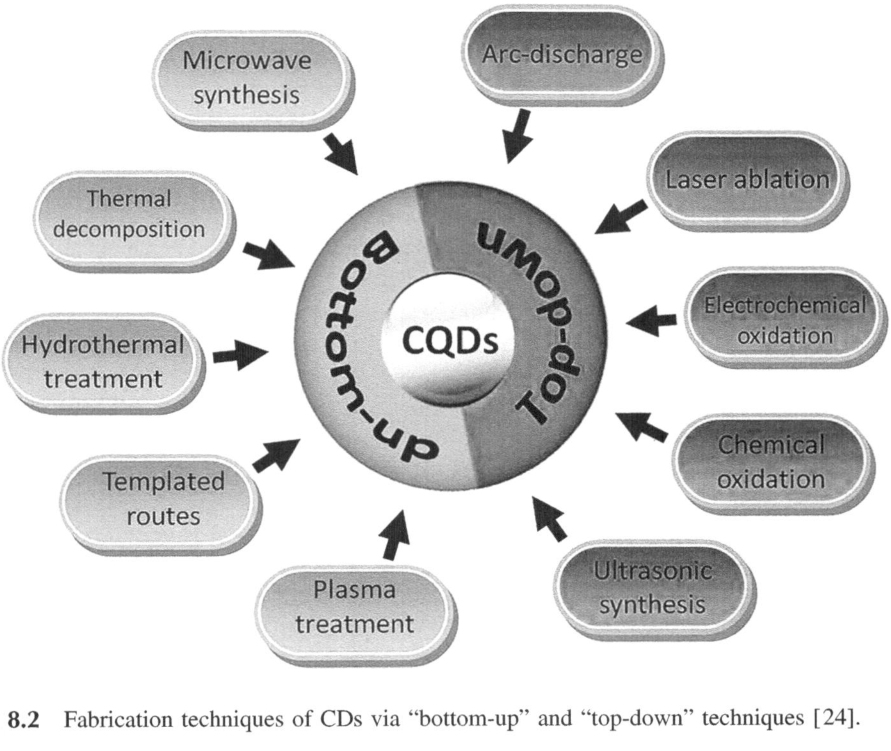

FIGURE 8.2 Fabrication techniques of CDs via "bottom-up" and "top-down" techniques [24].

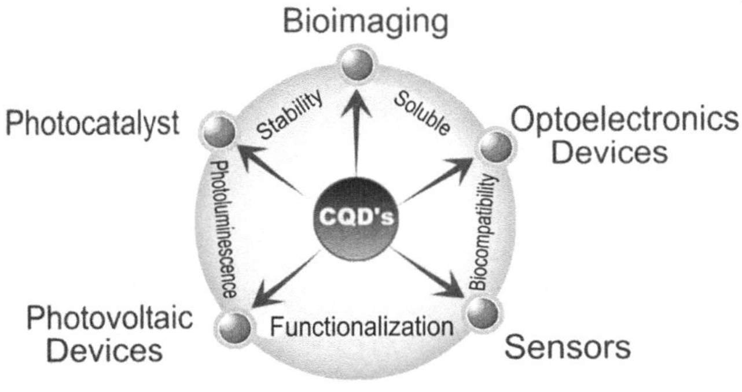

FIGURE 8.3 Properties and applications of carbon dots [26].

CDs are preferred materials of interest in science and engineering because of their superior features, such as unique optical properties, ease of surface functionalisation, good water solubility, outstanding biocompatibility, high chemical and photostability, facile synthesis, large-scale production, high adsorption ability, relatively low toxicity and tunable emission [22,23,26]. These special properties can be altered by the control of their morphology. Numerous studies have reported the excellent utilisation of CDs in several fields such as biomedicine, biosensors, bioimaging, drug/gene delivery, theranostics, explosive detection, optoelectronics, food safety, photocatalysis, and energy conversion and storage (Figure 8.3) [23,26].

Robust and comprehensive information on the syntheses, properties and applications of CDs, which are beyond the scope of this chapter, can be found elsewhere [27–33]. The focus of this chapter is to summarise the applications of CDs as reinforcements for corrosion resistance of different metal samples in common electrolytic media. The chapter focuses on available reports on CDs, otherwise known as CQDs, and GQDs utilised as nanofillers in anticorrosive coatings.

8.3 APPLICATIONS OF cds in anticorrosive coatings

8.3.1 CQDs and CNDs in Anticorrosive Coatings

In the last six years, CDs have gained wide interest in the corrosion protection of metals in different electrolytes. Corrosion researchers have explored these novel carbon-based nanomaterials as chemical inhibitors on account of their special properties, chief among which are their facile synthesis, excellent solubility, high adsorption ability and eco-friendliness. Recently emerging studies are showing that CDs synthesised from cheap and eco-friendly starting materials, such as citric acid, thiourea, urea and dopamine, can act as effective barriers to corrosion processes in different tested media [2,34]. Table 8.1 presents a summary of some recent studies on CDs prepared from citric acid as efficient aqueous inhibitors of corrosion in metals. The table clearly shows that both doped (nitrogen-doped, N-CDs or nitrogen, sulphur co-doped, N,S-CDs) and undoped CDs exhibit excellent inhibitory efficiencies (%IE) at minute concentrations for different metal types in common tested media.

The earliest available report on the application of CDs as anticorrosive nanofillers was published by Zhu et al. in 2017 [45]. The report focused on the self-healing and anticorrosive coating properties of four different polymer matrices intercalated with small-sized CDs ranging from 1 to 5 nm. The study found that CDs-5/polymethacrylate (PMMA) produced the strongest self-healing and anticorrosion property, which was attributed to the interfacial bonding occurring between PMMA and the CDs. The existence of different functional moieties such as amidogen, hydroxyl and carboxyl in the CDs fillers also contributed to improving the anticorrosive and healing properties of the polymer-CDs composite as compared to the pure polymer coatings.

TABLE 8.1
Summary of Some Recent Studies on Carbon Dots as Sustainable Chemical Inhibitors

No	Carbon Dots	Precursors	Metal/Medium	Dosage (mg/L)	Maximum %IE	References
1.	CDs	Citric acid and hexylamine	N80 steel/ 1 M HCl	150	92.90%	[35]
2.	CDs	Imidazole and citric acid	Q235 steel/ 1 M HCl	200	94.00%	[36]
3.	CDs	Citric acid monohydrate melamine	Q235 steel/ 3.5 wt% NaCl	200	93.18%	[37]
4.	N-CDs	Citric acid and ethylenediamine	Q235 steel/ 1 M HCl	100	93.70%	[38]
5.	N-CDs	Natural citric acid	Q235 steel/ 1 M HCl	200	97.43%	[39]
6.	N-CDs	Citric acid and aminoguanidine hydrochloride	Q235 steel/ 0.5 M H_2SO_4	200	92.20%	[40]
7.	N-CDs	Citric acid and thiourea	X80 steel/ 0.1 M HCl	10	92.80%	[41]
8.	N-CDs	Citric acid and urea	Carbon steel/ 0.5 M H_2SO_4	30	97.80%	[42]
9.	N-CDs	Citric acid and polyethyleneimine	Copper/ 0.5 M H_2SO_4	50	96.10%	[43]
10.	N,S-CDs	Citric acid, isoniazid and thiourea	Mild steel/ 15% HCl	200	96.30%	[44]

Besides the applications of CDs as nanofillers in EP matrix to boost the corrosion suppression of coated metals, they also play active roles as dispersants. Taking advantage of the special barrier and electrical properties of hexagonal boron-nitride (h-BN), coupled with the established dispersibility and self-healing effect of CDs, Yu et al. [46] and Wan et al. [47] in two different recent studies examined the anticorrosive prospects of waterborne EP (WEP) coating by the functionalisation of h-BN by CDs. These CDs were employed as eco-friendly dispersing agents to disperse h-BN nanosheets in the waterborne EP matrix based on their π-π interactions. The synthetic process and the obtained structural analyses from modern analytical methods for the functionalised carbon dots (FCDs) and the FCDs/boron nitride nanosheets composite (BNNS) are presented in Figure 8.4 and 8.5, respectively [47].

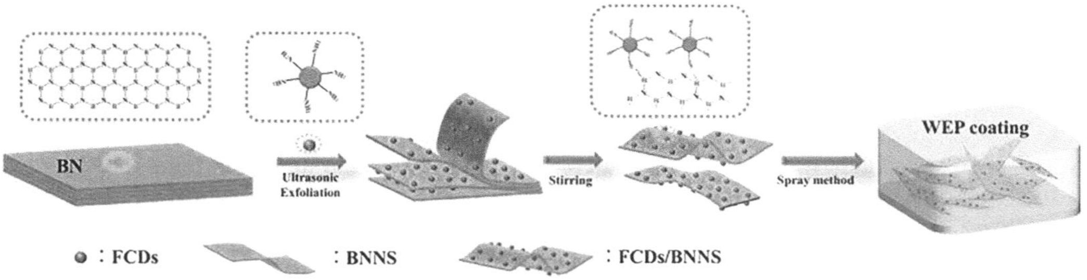

FIGURE 8.4 Schematic route for the fabrication of WEP FCDs/BNNS coating [47].

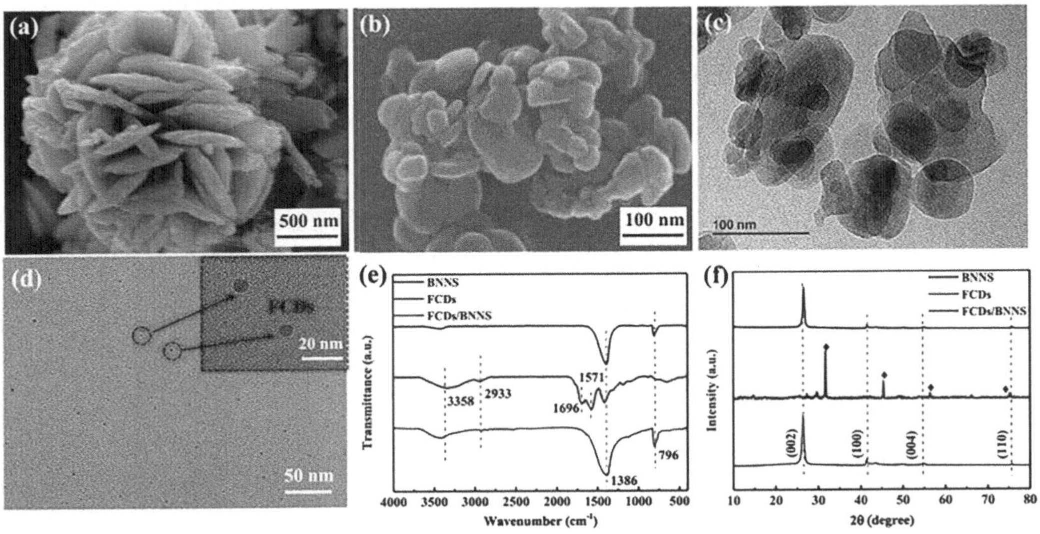

FIGURE 8.5 SEM images of boron nitride (a) and FCDs/BNNS (b); TEM images of FCDs/BNNS (c) and FCDs (d); FTIR (e) and XRD (f) spectra of BNNS, FCDs and FCDs/BNNS [47].

Saikia et al. [48] assessed the anticorrosive abilities of bio-based EP nanocomposites comprising different compositions of polyaniline nanofiber (PANI) and CDs in 3.5 wt% NaCl. Synthesised nanomaterials were characterised using FTIR, XRD, NMR, TEM and RS. The authors reported an impressive improvement in the elongation, scratch resistance, tensile strength, chemical resistance, thermal stability and impact resistance of the fabricated coating composites. Most importantly, the study revealed an excellent anticorrosive property in terms of corrosion rate than the pristine EP system. In another study, the group also reported the mechanical, anticorrosive, chemical resistance and thermal effectiveness of MWCNT/PANI/CDs embedded in EP resin [49]. The study established that the presence of CDs aided the dispersibility of the nanohybrid in the EP, which reflects the concentration-dependent enhancement of the studied properties.

The examination of the effect of solvothermally prepared NCDs on the corrosion suppression of waterborne EP in NaCl solution was carried out by Ren et al. [50]. Using SPM, the synthesised NCDs were found to range from 4 to 6 nm. Nanocomposite coatings were thereafter prepared using various compositions of NCDs (0, 0.5%, 2% and 4%) and analyzed for their anticorrosive properties using electrochemical methods and SEM. The degradation status as revealed by SEM showed that after 800 h of exposure in 3.5% NaCl, the bare EP displayed poor anticorrosion property evidenced by the percolation of aggressive species within the coating, unlike NCDs/EP. EIS studies revealed that 2.0% NCDs/EP yielded an impedance modulus that is 364 times greater than bare EP after 800 h exposure. The anticorrosive property of NCDs/EP was ascribed to the ability of the surface functional moieties on NCDs to repress pore enlargement and minimise the distribution of oxygen in the nanocomposite coating. In another report, the authors demonstrated that an anticorrosive microcapsule filled with NCDs as the main material can effectively synergise with ZnO as the wall material to provide long-term coating effect in the presence of various amounts of NaCl solution. Overall, impedance experiments revealed that waterborne polyacrylate (WPA)/NCD@ZnO composite coating offered a superior corrosion suppression than blank WPA and WPA/ZnO coatings by resisting the infiltration of corrosive ions [51].

N-CDs obtained from solvothermal techniques with size range of 3–5 nm were applied as nanofillers to intensify the corrosion suppression in waterborne EP resin coatings [52]. The authors remarked that anticorrosive activities were promoted by bond interactions such as hydrogen bonds, covalent bonds as well as van der Waals forces between the polymer matrix and the N-CDs. As nanofillers, the N-CDs were also found to suppress the defects and pores caused during curing and adhered to the sample via physisorption and chemical adsorption. A recent study explored the anticorrosive potential as well as the viscoelastic and UV-shielding properties of CDs incorporated into EP/PU coatings [53]. In alignment with previously conducted studies, the author established from salt spray tests that increment in the concentration of CDs in the coatings will improve the corrosion suppression abilities of the coatings. The study also revealed improved tensile strength, viscoelasticity properties and 85–98% protection against UV rays in the presence of 3 wt% CDs content in EP/PU. In another work, Bankole-Ojo et al. [54] demonstrated via salt spray test that steel coated with waterborne PU/silane modified CQDs (WPU/SCQDs) possesses enhanced corrosion protection properties than WPU/CQDs. Di et al. assessed the anticorrosive, fluorescence and mechanical properties of N-CQDs prepared via hydrothermal reaction in a recent report [55]. Successful synthesis of N-CQDs was confirmed with the aid of UV–vis, FTIR, XRD and TEM. An in-depth investigation on the long-term anticorrosive and mechanical properties of different concentrations of synthesised N-CQDs in WEP coating on Q235 steel in seawater was conducted via impedance, adhesion and salt spray tests. From the study, it can be seen that 0.3 N-CQDs/EP coating demonstrated stronger corrosion protection on Q235 steel than bare EP coating and other tested N-CQDs concentrations for the 30 days testing period. The exceptional protection offered by N-CQDs/EP coating was traceable to the ability of N-CQDS to fill the pores and thus boost compactness as well as their inhibitory potential to adhere to the metal substrate, thereby circumventing contact between NaCl and Q235 steel.

FCDs prepared from citric acid were incorporated into graphene (G) and dispersed into pure EP to fabricate CDs-G/EP coatings. The nanocomposite coating was found to exhibit superior anticorrosive potential on steel owing to the self-healing property of CDs and the shielding effect of dispersed graphene [56]. After 50 days immersion, it was observed that the optimised composite coating (CDs-G0.5%/EP) yielded a tangible decrease in the value of the steel coarseness, which was ascribed to the dative bond between the steel sample and the heteroatom. However, it was noticed that excess graphene content agglomerated and resulted in reducing the protective and barrier ability of the composite coating.

The ability of N-CDs and polydopamine (PDA) to provide effective corrosion protection to Mg alloy was demonstrated in a study conducted by Zhang et al. [57]. The fabrication of N-CDs/PDA composite was done via hydrothermal method, and the synthesised composite was confirmed using SEM, TEM, XPS and RS. The authors observed that an upsurge in the size of the N-CDs significantly increased the anticorrosive property of the coating (N-CDs-8 > N-CDs-6 > N-CDs-3). Electrochemical analysis via PDP revealed that the corrosion current density (i_{corr}) depreciated tremendously as the N-CDs of greater particle size were introduced. N-CDs-8/PDA displayed the lowest i_{corr} value of 1.6×10^{-7} Acm^{-2} and the highest charge transfer resistance (R_{ct}) value of 4568 Ωcm^2. Results obtained from the salt spray tests and self-healing performance test also revealed the superiority of the N-CDs-8/PDA on the AZ91D Mg alloy. In another study, hydrothermally-prepared N-CQDS were utilised for their compatibility and dispersibility to enhance the corrosion suppression of a novel waterborne resin in the presence of graphene [58]. Spin-coated N-CQDs@Gr/WEP coating displayed long-term (90 days) anticorrosion abilities and yielded an impedance modulus almost 200 times > WEP. The improved corrosion protection property was ascribed to the synergistic impact of graphene and N-CQDs in the WEP coating. Salt spray tests (Figure 8.6) revealed massive rust and a large presence of corrosion spots and blisters upon scratching Gr/WEP and WEP coatings after 13 days. The increased rust on Gr/WEP was correlated to the agglomeration of graphene, which resulted in corrosion acceleration. However, for N-CQDSs/WEP coating (Figure 8.6c2), a lesser number of corrosion spots and blisters was observed, which correlates to the production of barrier film on the metal substrate. This result corresponded to the obtained EIS results. The effectiveness of novel CD-Ti3C2Tx MXene hybrids prepared via covalent functionalisation has also been reported. The study investigated the microstructure, anticorrosion activity and passivation characteristics of the manufactured composite coatings after 35 days of exposure in NaCl solution

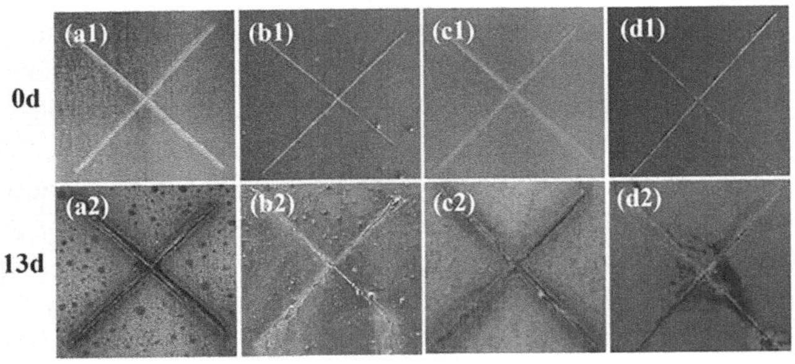

FIGURE 8.6 Images of (a1, a2) WEP, (b1, b2) Gr/WEP, (c1, c2) N-CQDs/WEP and (d1, d2) N-CQDs@Gr/WEP coatings prior and post 13 days immersion in salt spray test environment [58].

and obtained an impedance modulus having four orders of magnitude more than the bare EP [59]. A presentation of several obtained results (PDP, EIS and SEM) and the corrosion mechanism is displayed in Figure 8.7. Peng et al. [60] have also established that 0.5 mg/ml organic CDs coatings exhibit superior anticorrosive and self-cleaning properties for Al alloy than untreated or poorly treated surface in NaCl solution.

A recent report by Li et al. [61] furnished information on the performance of novel N-doped CNDs as effective nanofillers capable of improving the protection performance of EP coating in NaCl solution. The group fabricated CNDs via hydrothermal technique from citric acid owing to its facile synthesis, low cost, abundant functional groups and high yield, and afterwards characterised the product using FTIR, XPS, UV–vis, TEM and SEM. From impedance analysis, authors achieved a low-frequency impedance of 1.85×10^{10} Ω cm^2 with the optimal introduction of

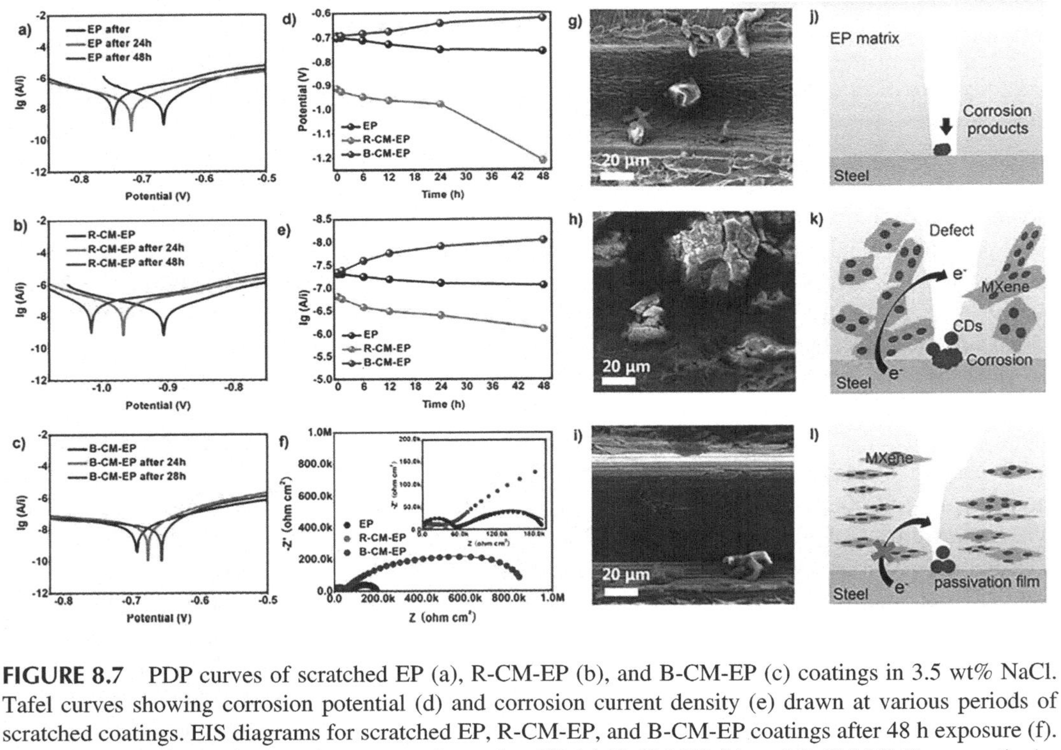

FIGURE 8.7 PDP curves of scratched EP (a), R-CM-EP (b), and B-CM-EP (c) coatings in 3.5 wt% NaCl. Tafel curves showing corrosion potential (d) and corrosion current density (e) drawn at various periods of scratched coatings. EIS diagrams for scratched EP, R-CM-EP, and B-CM-EP coatings after 48 h exposure (f). SEM diagrams of scratched coating protected samples: EP (g), R-CM-EP (h), and B-CM-EP (i), respectively. Mechanistic pathways of EP (j), R-CM-EP (k), and B-CM-EP (l) coatings [59].

0.15 wt% CNDs after 20 days immersion in NaCl solution. Furthermore, the introduction of 0.15 wt% significantly reduced the coating porosity from 2.09 × 10^{-3} cm^{-2} (pure EP) to 6.23 × 10^{-9} cm^{-2} after 20 days of immersion in the corrosive medium. The decreased porosity, which is indicative of an improved protection performance, was attributed to the formation of covalent and hydrogen bonds owing to the abundance of surface functional groups in CNDs.

8.3.2 GQDs and Graphene Oxide Quantum Dots in Anticorrosive Coatings

Despite the numerous attractive properties of graphene, they have not been extensively used in corrosion science and engineering because of their associated limitations, one of which is insolubility. To circumvent this limitation, GQDs and graphene oxide quantum dots (GOQDs) have been considered as potential substitutes. Besides, GQDs and GOQDs are reputed to reinforce the anticorrosion potential of EP coatings than pure EP, EP/graphene and EP/graphene oxide (GO) owing to their low cost, easy fabrication, relatively low toxicity and better corrosion suppression [62]. Jiang et al. [63] identified three properties of GQDs that makes them a preferred material for corrosion protection. Firstly, the high dispersibility of GQDs in water coupled with their ability to form good films on metal substrates. Secondly, the presence of several functional moieties such as carbonyl and hydroxyl groups, which can easily bond with other coating substances or the metal substrate. Thirdly, the properties of GQDs are tunable by adopting elemental doping. In their work, they utilised N-GQDs incorporated in polymethyltrimethoxysilane (PMTMS) composite coating to strengthen the corrosion resistance of AZ91D Mg.

Motivated by the need to overcome the deficiencies of low dispersibility of graphene and GO for use in protective coatings, Ramezanzadeh et al. [64] fabricated GOQDs via a bottom-up technique using citric acid as a starting material (Figure 8.8). A direct pyrolysis of citric acid was

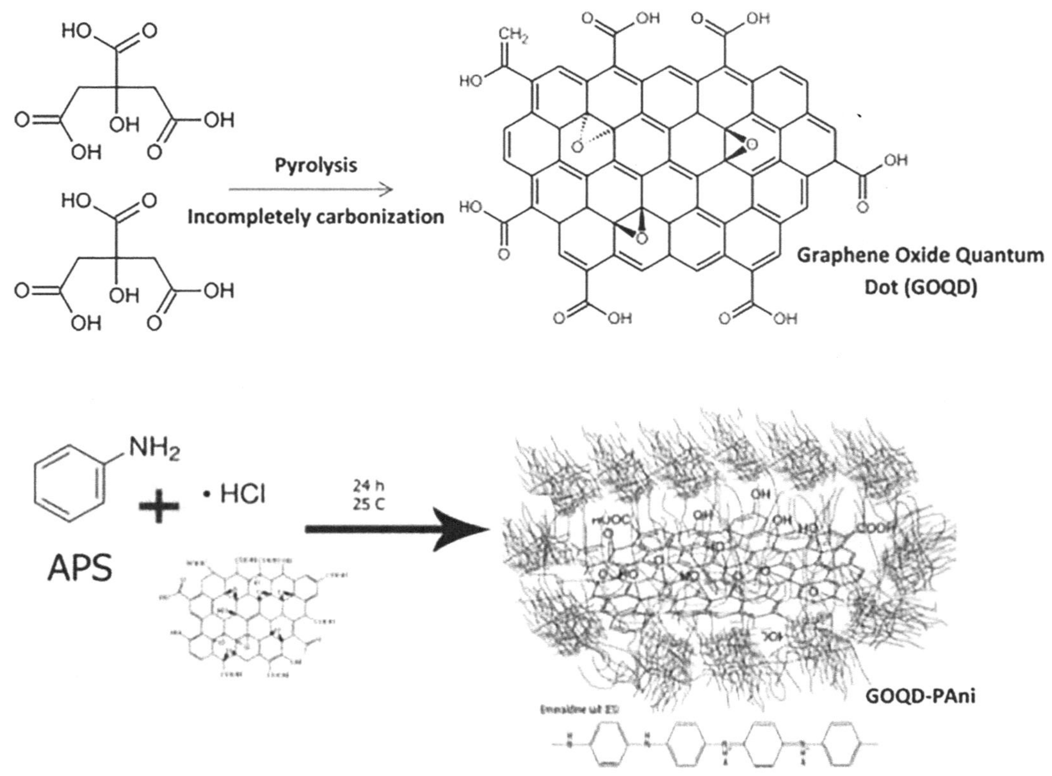

FIGURE 8.8 Schematic illustration of the preparation of GOQD-PANI [64].

performed under heat and after colour change to orange after 30 min, GOQD solution was added to the 100 mL NaOH solution. The produced GOQDs was functionalised by PANI through a rapid-mixture polymerisation technique and a purified dark green GOQD-PANI was obtained. Subsequently, GOQD/EP and GOQD-PANI/EP were prepared by addition of the dispersed products to 66g EP resin. The obtained composites were characterised using XPS, XRD, PL, FTIR, HR-TEM and thermogravimetric analysis (TGA). From the accelerated salt spray tests coupled with EIS techniques, the researchers established that the addition of GOQD-PANI provided both active and passive protection for the coatings.

The reinforcing effects of GOQDs fabricated using bottom-up technique and GO nanosheets synthesised via top-down method on the mechanical qualities of an EP coating were juxtaposed. The synthesised materials were probed using several analytical methods such as XRD, FTIR, XPS, AFM, HRTEM, FESEM and TGA. Employing tensile test and dynamic mechanical thermal analysis, GOQDs was found to enhance the mechanical features better than the pure EP and the GO nanosheets composite [65]. In the same vein, another research team explored the contrast between the anticorrosive effect of graphene nanosheets and GQDs in EP coatings [66]. In this case, GQDs were synthesised via CVD and graphene nanosheets prepared by a top-down approach. Characterisation of the prepared nanomaterials were carried out with the aid of FTIR, TEM and XRD. The authors utilised impedance techniques in studying the anticorrosion potential of 0.1 wt% graphene and 0.1 wt% GQDs in EP coating after 2 weeks of dipping in 3.5 wt% NaCl solution. The study concluded that while both nanofillers meaningfully enhanced the protection of EP coatings, the EP/GQDs displayed a better barrier to corrosion due to the availability of functional groups, which promotes dispersion quality in the EP coating. Similarly, Mohammadpour and Zare assessed the influence of GO nanosheets and GOQDs on the anticorrosion of Ni-Fe composite coatings [67].

Bopp and Santhanam studied the anticorrosive potential of GQDs in the corrosion protection of monel alloy (67.05% Ni and 29.4% Cu). Studies were carried out using XRF, FTIR and RS. Corrosion rates obtained from Tafel plots were reported to be lower in coated monel alloy samples in the studied electrolytes (Na_2SO_4 and NaCl) when compared to the uncoated specimens [68]. GO and GOQDs produced by modified Hummer's method were also successfully employed to provide long-term anticorrosive activity to Q235 steel. The intercalation of GOQDs into the Zn@EP was found to tremendously boost the anticorrosion activity of the metal as it provided protection even after 792 h salt spray test [69]. In an attempt to enhance the interfacial interaction of EP coating with GQDs, (3-aminopropyl) triethoxysilane was used to modify GQDs and characterisation was carried out using FTIR, XRD, TEM and fluorescence emission. Using electrochemical tests, the study obtained a lower i_{corr} value for EP/0.5 f-GQDs protection [70].

In an effort to develop long-term anticorrosive coatings, EP/GQDs with TiO_2 nanotubes were fabricated with different compositions of GQD:TNT. The study established that H(1:1) yielded the best anticorrosive property after 90 days of exposure to NaCl solution [71]. Density functional theory (DFT) was further utilised to understand the interaction mechanisms at the interfaces, which revealed strong interactions at each interface, resulting in the suppression of electrochemical reactions. The anticorrosion mechanistic pathway is presented in Figure 8.9. The proposed mechanism suggests that maximum anticorrosion effect is achieved when the ratio of TNT to GQD is 1:1 (Figure 8.9d).

8.4 CONCLUSIONS AND FUTURE PERSPECTIVES

CDs belong to a family of new carbon-based nanomaterials with exciting prospects on account of their fascinating features. These identified features include their excellent solubility, low toxicity, facile synthesis, high quantum yield, adjustable structures, abundant raw materials and excellent biocompatibility. It is only in recent years that corrosion scientists and engineers have begun to explore their anticorrosive properties despite their wide usage in other fields like biosensors,

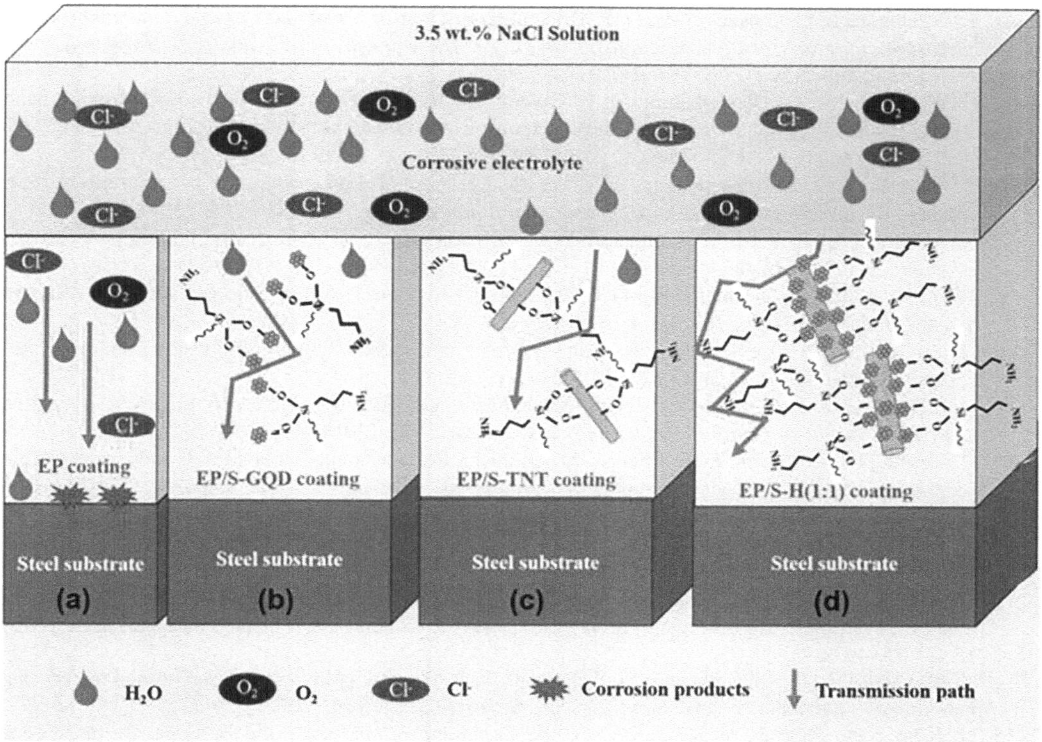

FIGURE 8.9 Anticorrosion mechanism of (a) EP, (b) EP/S-GQD, (c) EP/S-TNT, and (d) EP/S-H (1:1) coating specimens in NaCl solution [71].

nanomedicine, solar cells, supercapacitors and photocatalysis. The chapter presents an overview of available reports on the adoption of CDs in anticorrosive coatings for different media and metal types. From the chapter, it is obvious that there is a paucity of literature on the utilisation of CDs as nanofillers in the development of anticorrosive coatings. Furthermore, much work still needs to be done to understand the mechanism of protection of CD-based anticorrosive coatings. Looking forward, it is hoped that the investigations into the anticorrosive coating of CD-based materials will increase given its numerous special features and advantages.

ACKNOWLEDGEMENTS

TWQ acknowledges College of Science, Engineering and Technology, University of South Africa for the postdoctoral research fellowship.

REFERENCES

1. G. Koch, J. Varney, N. Thompson, O. Moghissi, M. Gould, J. Payer, International measures of prevention, application, and economics of corrosion technologies study, *NACE International*, (2016) 216.
2. E. Berdimurodov, D. K. Verma, A. Kholikov, K. Akbarov, L. Guo, The recent development of carbon dots as powerful green corrosion inhibitors: a prospective review, *Journal of Molecular Liquids*, (2021) 118124.
3. T. W. Quadri, E. D. Akpan, L. O. Olasunkanmi, O. E. Fayemi, E. E. Ebenso, *Fundamentals of corrosion chemistry, environmentally sustainable corrosion inhibitors*, Elsevier, 2022, pp. 25–45.

4. T. W. Quadri, L. O. Olasunkanmi, O. E. Fayemi, E. E. Ebenso, *Nanomaterials and nanocomposites as corrosion inhibitors, sustainable corrosion inhibitors II: synthesis, design, and practical applications*, ACS Publications, 2021, pp. 187–217.
5. T. W. Quadri, L. O. Olasunkanmi, O. E. Fayemi, M. M. Solomon, E. E. Ebenso, Zinc oxide nanocomposites of selected polymers: synthesis, characterization, and corrosion inhibition studies on mild steel in HCl solution, *ACS Omega*, 2 (2017) 8421–8437.
6. T. Peng, R. Xiao, Z. Rong, H. Liu, Q. Hu, S. Wang, X. Li, J. Zhang, Polymer nanocomposite-based coatings for corrosion protection, *Chemistry–An Asian Journal*, 15 (2020) 3915–3941.
7. A. A. Farag, Applications of nanomaterials in corrosion protection coatings and inhibitors, *Corrosion Reviews*, 38 (2020) 67–86.
8. M. Ates, A review on conducting polymer coatings for corrosion protection, *Journal of Adhesion Science and Technology*, 30 (2016) 1510–1536.
9. H. Xu, Y. Zhang, A review on conducting polymers and nanopolymer composite coatings for steel corrosion protection, *Coatings*, 9 (2019) 807.
10. P. P. Deshpande, N. G. Jadhav, V. J. Gelling, D. Sazou, Conducting polymers for corrosion protection: a review, *Journal of Coatings Technology and Research*, 11 (2014) 473–494.
11. C. Verma, T. W. Quadri, E. E. Ebenso, M. Quraishi, Polymer nanocomposites as industrially useful corrosion inhibitors: recent developments, *Handbook of Polymer Nanocomposites for Industrial Applications*, (2021) 419–435.
12. A. Kausar, Corrosion prevention prospects of polymeric nanocomposites: a review, *Journal of Plastic Film and Sheeting*, 35 (2019) 181–202.
13. H. Huang, X. Sheng, Y. Tian, L. Zhang, Y. Chen, X. Zhang, Two-dimensional nanomaterials for anticorrosive polymeric coatings: a review, *Industrial & Engineering Chemistry Research*, 59 (2020) 15424–15446.
14. T. W. Quadri, L. O. Olasunkanmi, O. E. Fayemi, E. E. Ebenso, Utilization of ZnO-based materials as anticorrosive agents: a review, *Inorganic Anticorrosive Materials*, (2022) 161–182.
15. N. Kumar, A. Sharma, *Surface coatings and functionalization strategies for corrosion mitigation, functionalized nanomaterials for corrosion mitigation: synthesis, characterization, and applications*, ACS Publications, 2022, pp. 291–316.
16. T. W. Quadri, L. O. Olasunkanmi, O. E. Fayemi, E. E. Ebenso, Industrial corrosion inhibitors: nanostructured carbon allotropes as ideal substitutes, *Carbon Allotropes: Nanostructured Anti-Corrosive Materials*, (2022) 327.
17. K. M. Amin, H. M. Amin, Carbon Nanoallotropes-based anticorrosive coatings: recent advances and future perspectives, *Corrosion Protection of Metals and Alloys Using Graphene and Biopolymer Based Nanocomposites*, (2021) 81–98.
18. C. Verma, M. Quraishi, E. E. Ebenso, C. M. Hussain, Recent advancements in corrosion inhibitor systems through carbon allotropes: past, present, and future, *Nano Select*, 2 (2021) 2237–2255.
19. X. Xu, R. Ray, Y. Gu, H. J. Ploehn, L. Gearheart, K. Raker, W. A. Scrivens, Electrophoretic analysis and purification of fluorescent single-walled carbon nanotube fragments, *Journal of the American Chemical Society*, 126 (2004) 12736–12737.
20. Y.-P. Sun, B. Zhou, Y. Lin, W. Wang, K. S. Fernando, P. Pathak, M. J. Meziani, B. A. Harruff, X. Wang, H. Wang, Quantum-sized carbon dots for bright and colorful photoluminescence, *Journal of the American Chemical Society*, 128 (2006) 7756–7757.
21. C. Xia, S. Zhu, T. Feng, M. Yang, B. Yang, Evolution and synthesis of carbon dots: from carbon dots to carbonized polymer dots, *Advanced Science*, 6 (2019) 1901316.
22. B. D. Mansuriya, Z. Altintas, Carbon dots: classification, properties, synthesis, characterization, and applications in health care—an updated review (2018–2021), *Nanomaterials*, 11 (2021) 2525.
23. B. Gayen, S. Palchoudhury, J. Chowdhury, Carbon dots: a mystic star in the world of nanoscience, *Journal of Nanomaterials*, 2019 (2019).
24. R. Wang, K.-Q. Lu, Z.-R. Tang, Y.-J. Xu, Recent progress in carbon quantum dots: synthesis, properties and applications in photocatalysis, *Journal of Materials Chemistry A*, 5 (2017) 3717–3734.
25. T. W. Quadri, L. O. Olasunkanmi, O. E. Fayemi, E. E. Ebenso, *Functionalized carbon allotropes as corrosion inhibitors, functionalized nanomaterials for corrosion mitigation: synthesis, characterization, and applications*, ACS Publications, 2022, pp. 87–114.
26. R. Souza da Costa, W. Ferreira da Cunha, N. Simenremis Pereira, A. Marti Ceschin, An alternative route to obtain carbon quantum dots from photoluminescent materials in peat, *Materials*, 11 (2018) 1492.

27. D. S. Chauhan, M. Quraishi, C. Verma, Carbon nanodots: recent advances in synthesis and applications, *Carbon Letters*, 32 (2022) 1–27.
28. L. Cui, X. Ren, M. Sun, H. Liu, L. Xia, Carbon dots: synthesis, properties and applications, *Nanomaterials*, 11 (2021) 3419.
29. A. M. El-Shafey, Carbon dots: discovery, structure, fluorescent properties, and applications, *Green Processing and Synthesis*, 10 (2021) 134–156.
30. J. Liu, R. Li, B. Yang, Carbon dots: a new type of carbon-based nanomaterial with wide applications, *ACS Central Science*, 6 (2020) 2179–2195.
31. W. Liu, C. Li, Y. Ren, X. Sun, W. Pan, Y. Li, J. Wang, W. Wang, Carbon dots: surface engineering and applications, *Journal of Materials Chemistry B*, 4 (2016) 5772–5788.
32. B. Wang, H. Song, X. Qu, J. Chang, B. Yang, S. Lu, Carbon dots as a new class of nanomedicines: opportunities and challenges, *Coordination Chemistry Reviews*, 442 (2021) 214010.
33. A. Himaja, P. Karthik, S. P. Singh, Carbon dots: the newest member of the carbon nanomaterials family, *The Chemical Record*, 15 (2015) 595–615.
34. C. Verma, A. Alfantazi, M. Quraishi, Quantum dots as ecofriendly and aqueous phase substitutes of carbon family for traditional corrosion inhibitors: a perspective, *Journal of Molecular Liquids*, 343 (2021) 117648.
35. Y. Yang, R. Lu, W. Chen, P. Mei, L. Lai, Amphiphilic carbon dots as high-efficiency corrosion inhibitor for N80 steel in HCl solution: performance and mechanism investigation, *Colloids and Surfaces A: Physicochemical and Engineering Aspects*, 649 (2022) 129457.
36. Y. Ye, D. Yang, H. Chen, A green and effective corrosion inhibitor of functionalized carbon dots, *Journal of Materials Science & Technology*, 35 (2019) 2243–2253.
37. Y. Zeng, L. Kang, Y. Wu, S. Wan, B. Liao, N. Li, X. Guo, Melamine modified carbon dots as high effective corrosion inhibitor for Q235 carbon steel in neutral 3.5 wt% NaCl solution, *Journal of Molecular Liquids*, 349 (2021) 118108.
38. M. Zhu, Z. He, L. Guo, R. Zhang, V. C. Anadebe, I. B. Obot, X. Zheng, Corrosion inhibition of eco-friendly nitrogen-doped carbon dots for carbon steel in acidic media: performance and mechanism investigation, *Journal of Molecular Liquids*, 342 (2021) 117583.
39. Z. Liu, Y. Ye, H. Chen, Corrosion inhibition behavior and mechanism of N-doped carbon dots for metal in acid environment, *Journal of Cleaner Production*, 270 (2020) 122458.
40. X. Xu, H. Wei, M. Liu, G. Shen, Q. Li, G. Hussain, F. Yang, R. Fathi, H. Chen, K. K. Ostrikov, Nitrogen-doped carbon quantum dots for effective corrosion inhibition of Q235 steel in concentrated sulphuric acid solution, *Materials Today Communications*, 29 (2021) 102872.
41. F. Niu, G. Zhou, J. Zhu, X. Zhang, Y. Shi, G. Lu, Z. Liu, Inhibition behavior of nitrogen-doped carbon dots on X80 carbon steel in acidic solution, *Journal of Molecular Liquids*, 339 (2021) 117171.
42. S. Cao, D. Liu, T. Wang, A. Ma, C. Liu, X. Zhuang, H. Ding, B. B. Mamba, J. Gui, Nitrogen-doped carbon dots as high-effective inhibitors for carbon steel in acidic medium, *Colloids and Surfaces A: Physicochemical and Engineering Aspects*, 616 (2021) 126280.
43. Q. Xu, K. Ge, S. Zhang, B. Tan, Understanding the adsorption and inhibitive properties of nitrogen-doped carbon dots for copper in 0.5 M H_2SO_4 solution, *Journal of the Taiwan Institute of Chemical Engineers*, 125 (2021) 23–34.
44. V. Saraswat, M. Yadav, Improved corrosion resistant performance of mild steel under acid environment by novel carbon dots as green corrosion inhibitor, *Colloids and Surfaces A: Physicochemical and Engineering Aspects*, 627 (2021) 127172.
45. C. Zhu, Y. Fu, C. Liu, Y. Liu, L. Hu, J. Liu, I. Bello, H. Li, N. Liu, S. Guo, Carbon dots as fillers inducing healing/self-healing and anticorrosion properties in polymers, *Advanced Materials*, 29 (2017) 1701399.
46. Y. Yu, M. Cui, W. Zheng, H. Zhao, Eco-friendly functionalization of hexagonal boron nitride nanosheets with carbon dots towards reinforcement of the protective performance of water-borne epoxy coatings, *New Journal of Chemistry*, 46 (2022) 6330–6342.
47. S. Wan, H. Chen, G. Cai, B. Liao, X. Guo, Functionalization of h-BN by the exfoliation and modification of carbon dots for enhancing corrosion resistance of waterborne epoxy coating, *Progress in Organic Coatings*, 165 (2022) 106757.
48. A. Saikia, D. Sarmah, A. Kumar, N. Karak, Bio-based epoxy/polyaniline nanofiber-carbon dot nanocomposites as advanced anticorrosive materials, *Journal of Applied Polymer Science*, 136 (2019) 47744.

49. A. Saikia, N. Karak, Fabrication of renewable resource based hyperbranched epoxy nanocomposites with MWCNT-polyaniline nanofiber-carbon dot nanohybrid as tough anticorrosive materials, *eXPRESS Polymer Letters*, 13 (2019).
50. S. Ren, M. Cui, H. Zhao, L. Wang, Effect of nitrogen-doped carbon dots on the anticorrosion properties of waterborne epoxy coatings, *Surface Topography: Metrology and Properties*, 6 (2018) 024003.
51. Y. Bao, Y. Yan, Y. Wei, J. Ma, W. Zhang, C. Liu, Salt-responsive ZnO microcapsules loaded with nitrogen-doped carbon dots for enhancement of corrosion durability, *Journal of Materials Science*, 56 (2021) 5143–5160.
52. J. Wang, P. Du, H. Zhao, J. Pu, C. Yu, Novel nitrogen doped carbon dots enhancing the anticorrosive performance of waterborne epoxy coatings, *Nanoscale Advances*, 1 (2019) 3443–3451.
53. A. Madhi, Smart epoxy/polyurethane/carbon quantum dots hybrid coatings: synthesis and study of UV-shielding, viscoelastic, and anti-corrosive properties, *Polymer-Plastics Technology and Materials*, 62 (2022) 403–418.
54. O. Bankole-Ojo, F. Oyedeji, K. Raju, N. Ramanuj, Salt spray resistance of superhydrophobic waterborne polyurethane/carbon dots nanocomposites, *Journal of Materials and Environmental Science*, 13 (2022) 814–824.
55. Z. Di, Q. Zhao, H. Xu, K. Wu, Y. Yan, X. Jin, H. S. Y. Lv, B. Ran, N-doped carbon quantum dots with mechanical enhancement and fluorescence properties toward anti-corrosion coating, *International Journal of Electrochemical Science*, 17 (2022) 2.
56. Y. Ye, H. Chen, Y. Zou, H. Zhao, Study on self-healing and corrosion resistance behaviors of functionalized carbon dot-intercalated graphene-based waterborne epoxy coating, *Journal of Materials Science & Technology*, 67 (2021) 226–236.
57. H. Zhang, A. Chen, B. Gan, H. Jiang, L. Gu, Corrosion protection investigations of carbon dots and polydopamine composite coating on magnesium alloy, *Journal of Magnesium and Alloys*, 10 (2021), 1358–1367.
58. X. Wang, C. Li, M. Zhang, D. Lin, S. Yuan, F. Xu, Y. Zhou, C. Wang, Y. Zhu, H. Wang, A novel waterborne epoxy coating with anti-corrosion performance under harsh oxygen environment, *Chemical Engineering Journal*, 430 (2022) 133156.
59. J. Ding, H. Zhao, H. Yu, Structure and performance insights in carbon dots-functionalized MXene-epoxy ultrathin anticorrosion coatings, *Chemical Engineering Journal*, 430 (2022) 132838.
60. H. Peng, L. Li, Q. Wang, Y. Zhang, T. Wang, B. Zheng, H. Zhou, Organic carbon dot coating for superhydrophobic aluminum alloy surfaces, *Journal of Coatings Technology and Research*, 18 (2021) 861–869.
61. S. Li, F. Du, Y. Lin, Y. Guan, W. Qu, J. Cheng, D. Wang, Excellent anti-corrosion performance of epoxy composite coatings filled with novel N-doped carbon nanodots, *European Polymer Journal*, 163 (2022) 110957.
62. A. Kausar, Epoxy and quantum dots-based nanocomposites: achievements and applications, *Materials Research Innovations*, 24 (2020) 235–243.
63. B. Jiang, A. Chen, J. Gu, J. Fan, Y. Liu, P. Wang, H. Li, H. Sun, J. Yang, X. Wang, Corrosion resistance enhancement of magnesium alloy by N-doped graphene quantum dots and polymethyltrimethoxysilane composite coating, *Carbon*, 157 (2020) 537–548.
64. B. Ramezanzadeh, B. Karimi, M. Ramezanzadeh, M. Rostami, Synthesis and characterization of polyaniline tailored graphene oxide quantum dot as an advance and highly crystalline carbon-based luminescent nanomaterial for fabrication of an effective anti-corrosion epoxy system on mild steel, *Journal of the Taiwan Institute of Chemical Engineers*, 95 (2019) 369–382.
65. B. Karimi, B. Ramezanzadeh, A comparative study on the effects of ultrathin luminescent graphene oxide quantum dot (GOQD) and graphene oxide (GO) nanosheets on the interfacial interactions and mechanical properties of an epoxy composite, *Journal of Colloid and Interface Science*, 493 (2017) 62–76.
66. S. Pourhashem, A. Rashidi, M. R. Vaezi, Comparing the corrosion protection performance of graphene nanosheets and graphene quantum dots as nanofiller in epoxy coatings, *Industrial Lubrication and Tribology*, 71 (2019) 653–656.
67. Z. Mohammadpour, H. R. Zare, The effect of graphene oxide nanosheets (GONSs) and graphene oxide quantum dots (GOQDs) on corrosion resistance enhancement of Ni–Fe nanocomposite coatings, *JOM*, 72 (2020) 4495–4504.
68. C. Bopp, K. Santhanam, Corrosion protection of monel alloy coated with graphene quantum dots starts with a surge, *ChemEngineering*, 3 (2019) 80.

69. H. Li, J. Pu, R. Zhang, Long-term corrosion protection of Q235 steel by graphene oxide composite coating, *Surface Topography: Metrology and Properties*, 7 (2019) 045022.
70. S. Pourhashem, E. Ghasemy, A. Rashidi, M. R. Vaezi, Corrosion protection properties of novel epoxy nanocomposite coatings containing silane functionalized graphene quantum dots, *Journal of Alloys and Compounds*, 731 (2018) 1112–1118.
71. X. Ji, A. Seif, J. Duan, A. Rashidi, Z. Zhou, S. Pourhashem, M. Mirzaee, X. Zhai, X. Zhao, B. Hou, Experimental and DFT studies on corrosion protection performance of epoxy/graphene quantum dots@ TiO_2 nanotubes coatings, *Construction and Building Materials*, 322 (2022) 126501.

9 Advanced Micro- and Nanocontainers in Smart Anticorrosive Coatings

Seyed Ali Rezaei
Research Department of Ceramic, Materials, and Energy Research Center, Alborz, Iran

Maryam Sirati Gohari
Research Department of Ceramic, Materials, and Energy Research Center, Alborz, Iran

Department of Materials Science and Engineering, School of Engineering, Shiraz University, Shiraz, Iran

Sirus Javadpour
Department of Materials Science and Engineering, School of Engineering, Shiraz University, Shiraz, Iran

Alimorad Rashidi
Nanotechnology Research Center, Research Institute of Petroleum Industry (RIPI), Tehran, Iran

CONTENTS

9.1 INTRODUCTION

Different kinds of strategies have been employed to avoid the corrosion reaction and decrease the costs of maintenance for metallic structures in an aggressive environment, including using corrosion resistance alloys, anticorrosive coatings, anodic/cathodic protection, and applying corrosion inhibitors (Haddadi et al. 2019; Kaghazchi, Naderi, and Ramezanzadeh 2021). Applying organic

DOI: 10.1201/9781003323976-11

coatings is considered the most common and economical technique among all strategies through which excellent barrier properties against corrosive ions can be provided as well as improving the surface features of metallic substrates such as strong adhesion (Alipanah et al. 2021; Habibiyan et al. 2020). However, the performance efficiency of these coating can be decreased due to their defects and the environmental condition, which can cause defects and serve as a proper site for corrosive ions (Alipanah et al. 2021).

Corrosion protection can be carried out through active strategies, such as corrosion inhibitors and passive strategies, such as anticorrosive coatings (Dehghani, Bahlakeh, and Ramezanzadeh 2020). As an active modifier for controlling corrosion reactions, inhibitors have been widely studied and developed (Gaidis 2004). Due to their chemical structure containing heterocyclic atoms, heteroatoms, π-electrons, and electron-rich elements (oxygen [O], nitrogen [N], sulfur [S], phosphorus [P]), inhibitors can absorb on the surface of metals (Haddadi et al. 2021). Due to their magnificent properties, inhibitors can play a leading role in corrosion protection of different sorts of metals and their alloys, including iron (Fe) (Bahlakeh, Ramezanzadeh, and Ramezanzadeh 2017; Mahdavian and Attar 2009; Amini et al. 2021; Bahremand, Shahrabi, and Ramezanzadeh 2021), zinc (Zn) (Custódio, Agostinho, and Simões 2010; Vimalanandan et al. 2013), copper (Cu) (Wang et al. 2018), magnesium (Mg) (Toorani et al. 2021; Ding et al. 2016), and aluminum (Al) (Changjean et al. 2014; Li et al. 2015). The presence of corrosion inhibitors in the protection process can alter the mechanism or rate of cathodic or/and anodic reactions, even at the first stage of the reaction. Moreover, they can prevent the corrosive ions from penetrating the surface of the substrate and, by the formation of a protective layer on the substrate's surface, improve the electrical resistance of the system (Akbarzadeh et al. 2019). Besides, the incorporation of inhibitors into the anticorrosive coating matrix is considered the most efficient and economical approach to enhance the coating properties, and it can also properly block the active areas on the surface of the metallic substrate (Fix et al. 2009). Nevertheless, some undesirable impacts can appear due to direct contact between the matrix and corrosion inhibitors, which can affect the protection performance anticorrosive layer (Yeganeh et al. 2020). More importantly, the corrosion inhibitors have some drawbacks, such as uncontrolled release rate, weathering, decomposition, and hydrolysis, which can lead to chemical waste, decrease in inhibitory efficiency and anticorrosive service time (Li et al. 2014).

Smart micro- and nanocontainers can be one of the best solutions for the above-mentioned issues, which have been widely developed and extensively used recently in different applications related to proper control of corrosion (Stimpfling et al. 2016; Nejman et al. 2014; Kulčar et al. 2010; Chuanjie, Juntao, and Xiaodong 2013). These containers are prepared to have the ability for a controlled release of the loaded contents depending on different environmental stimuli (Wei et al. 2014). The incorporation of stimuli-responsive capsules/containers into coating can be a promising method to provide self-healing and active corrosion protection features (Yeganeh et al. 2020). In this chapter, the research progress related to the anticorrosive application of corrosion inhibitor-based smart micro-and nanocapsules is reviewed. Stimuli-responsive release of such additives, including endogenous (pH fluctuation, electrochemical redox/potential, aggressive ions exchange) and exogenous ones (heat, magnetic field, light, and mechanical impact), followed by multistimuli-responsive containers, are summarised. In the end, some prospects associated with future studies and the development of corrosion inhibitor smart containers are proposed.

9.2 STIMULI-INFLUENCED CONTROLLED MECHANISM

Mainly, stimuli-influenced controlled releases are divided into two categories: endogenous and exogenous. The endogenous mechanisms originate from the materials and/or the parameters related to the electrochemical system and any changes that can lead to corrosion reaction occurrence, including pH fluctuation, electrochemical redox/potential, and aggressive ions exchange. These physiochemical parameters can play a significant role in the controlled release mechanism of inhibitors/healing agents from the containers to damaged areas, so they are referred to as the endogenous stimuli-influenced controlled release (Kazemi and Naghib 2020; Zhou et al. 2019). On the other hand, the mechanisms originating from the outside of the materials and electrochemical system, including heat, the magnetic field, light, and mechanical impact, which can also trigger the release of inhibitors, are referred to as the exogenous stimuli-influenced controlled release (Mercato et al. 2014).

9.2.1 pH Fluctuation

In the corrosion reaction, local pH fluctuation can be considered as the most well-known trigger due to pH variations in cathodic and anodic sites. Change of pH value can be employed for the controlled release of healing agents/inhibitors, which can improve the self-healing ability of smart anticorrosive coatings. When carbon steel is exposed to aggressive media, because of the reduction of oxygen, the pH value can increase in the cathodic zone, whereas Fe^{3+} or Fe^{2+} hydrolysis can occur due to the Fe dissolution, which can result in a decrease in pH in the anodic zone. The pH value fluctuation close to the corroded area can stimulate corrosion inhibitors' release, which is considered a parameter to control this action (Fu et al. 2013). Applying pH-sensitive polyelectrolyte (PE) polymers for container fabrication can be an effective strategy through which the inhibitors are released by pH fluctuation near the defects in the coating. Local pH variations can accelerate polymer molecules' hydrolytic degradation (Chen et al. 2014) or lead to protonating/deprotonating them for enhanced swelling (Cai, Wang, and Zhang 2019; Pirhady Tavandashti, Ghorbani, Shojaei, Mol et al. 2016; Pirhady Tavandashti, Ghorbani, Shojaei, Gonzalez-Garcia et al. 2016), resulting in the inhibitors release improvement. Fabricating a core-shell structure made from nanoparticles' core and multilayer PE film using the layer-by-layer (LBL) approach is considered a useful technique to produce pH-sensitive nanocontainers in which the inhibitors can be components of multilayers or can be loaded in the core (nanoparticles) (Skorb et al. 2009; Singh, Yeole, and Mhaske 2016; Feng, Chen, and Frank Cheng 2017). Since the cathodic and anodic reactions at the damaged sites in the coating lead to pH variations resulting in shell opening by changing PE conformation, the encapsulated inhibitors can be released. For instance, poly (methacrylic acid) (PMAA) has a shrunken conformation in an acidic environment, and it can swell when the pH has recovered (neutral condition). After the corrosion inhibition, the PMAA shell is closed to store the healing agents due to the large negative charge caused by the ionisation of carboxyl groups (COOH), and it loses a proton to form a negatively charged carboxylate ion (COO^-) (Wei et al. 2017).

Using inorganic nanocarriers based on silicon dioxide (SiO_2) coated with PE through the LBL assembly method can provide a controlled release of corrosion inhibitors (Zheludkevich et al. 2007; Shchukin, Zheludkevich, and Möhwald 2006). Incorporation of benzotriazole (BTA) between PE layers with the negative charge can provide pH-sensitive release of inhibitors by using the dependence of permeability of PE shells so that the inhibition process can be started when the corrosion process initiates. The compatibility of SiO_2 nanoparticles as the inorganic cores with sol-gel hybrid matrix makes it a suitable choice for the production of smart containers. To prepare such containers, negatively charged SiO_2 nanoparticles were coated with a poly(ethylene imine) (PEI) film with the positive charge and then adsorption of poly(styrene sulfonate) (PSS) with the negative charge, followed by BTA positively charged absorption. To load more cargo, the adsorption of PSS/BTA was repeated (Shchukin et al. 2006).

SiO_2/PEI/PSS/BTA suspension prepared by the mentioned procedure was added to organosiloxane and zirconium dioxide (ZrO_2) sols, followed by coating the film on aluminum alloy 2024 (AA2024) substrates (Zheludkevich et al. 2005; 2006). The effect of sol-gel layers doped with SiO_2/PEI/PSS/BTA, the films filled with BTA, and the ones without any additives on corrosion protection of AA2024 was investigated. Electrochemical impedance spectroscopy (EIS) measurement (frequency: 105–10-2 Hz, 0.005 M NaCl) showed that the highest impedance values

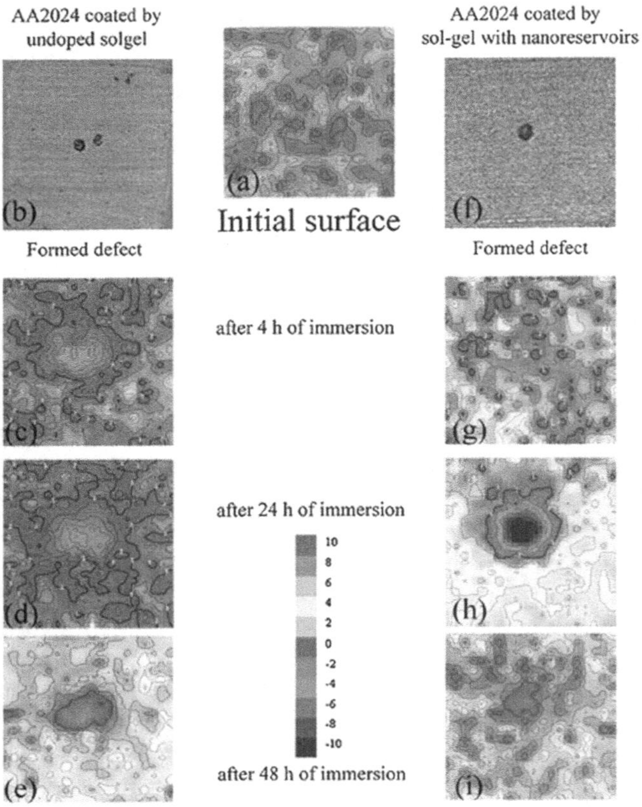

FIGURE 9.1 SVET maps of the ionic currents measured (scanned area: 2 × 2 mm, scale units: A cm^{-2}) above the aluminum surface (AA2024) coated with sol-gel without doping (a, c, d, e) and nanocontainers doped sol-gel (g, h, i). Adapted with permission from Zheludkevich et al. (2007).

related to the SiO_2/PEI/PSS/BTA doped layer, the sol-gel layer without additive, and the BTA doped layer, respectively. Based on the results, although BTA is considered an efficient inhibitor for such an alloy, the presence of BTA in the coating could be destructive for the hybrid matrix due to the interactions, which can lead to coating degradation. The scanning vibrating electrode techniques (SVET) were used to investigate the protection performance of anticorrosive layers. The SiO_2/PEI/PSS/BTA doped layer showed better self-healing and retained its barrier properties. SVET maps (Figure 9.1) show that the artificial defects can be observed on the surfaces of both unfilled ZrO_x/SiO_x hybrid coating and nanocapsule loaded coating in aggressive media (exposure in 0.05 M NaCl for 24 h). Clear-cut cathodic activity can be observed in the damaged area in the sample coated with the unloaded hybrid layer, and with the passage of exposure time, the activity becomes higher. On the other hand, there is no detected corrosion activity in the sample coated with nanocontainer-filled film 4 h after defect formation. After 24 h of exposure, the cathodic activity can be observed in the defected area, which healed by passivation and remained intact (after 48 h) because of the inhibitor (BTA) release and avoiding the corrosion reaction (Zheludkevich et al. 2007).

PMAA-coated silica nanotube (SiO_2NTs) was also used as the pH-responsive container filled with BTA as the corrosion inhibitor (Li et al. 2013). L-Carnosine (L-CAR) was the corrosion inhibitor, and hollow mesoporous zirconia nanospheres (HMZSs) was used as a container to prepare a pH-responsive release container (Figure 9.2). It was reported that the release mechanism was related to electrostatic interactions between the healing agent and the container. At neutral condition (pH=7),

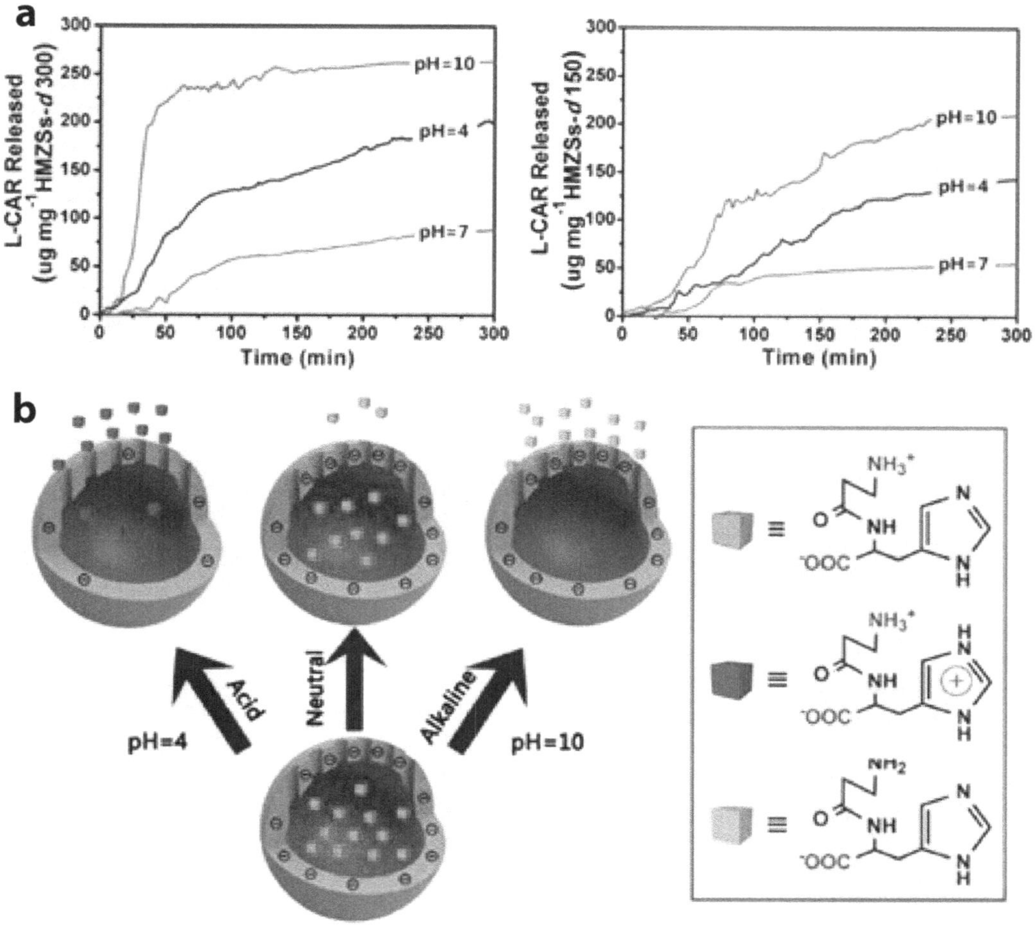

FIGURE 9.2 The released concentration of corrosion inhibitor (L-CAR) related to the variation of pH values (a), and schematic illustration of the pH-induced release mechanism of inhibitor (L-CAR) from the container (HMZSs) (b). Adapted with permission from Wang, Liu, and Fu (2015).

inhibitor molecules absorb into HMZSs cavities due to the weak positive charge of L-CAR. In alkaline conditions, amino group deprotonation forces negatively charged inhibitor molecules to release because of electrostatic attractions. Decreasing pH (acidic condition) causes a reduction of electrostatic attraction and positive charge accumulation leading to the release of L-CAR since their molecules can't pack inside the porosities of containers because of the intermolecular electrostatic repulsion (Wang, Liu, and Fu 2015). Encapsulation of 2-mercaptobenzimidazole (MBI) as the corrosion inhibitor in containers prepared from carbon hollow spheres (CHSs) and dispersed in anticorrosive coating showed remarkable protection of substrate (mild steel) in aggressive media due to smart release (MBI molecules deprotonation) of inhibitor in alkaline condition (Haddadi et al. 2018).

Taking advantage of supramolecular chemistry development, the production of micro- and nanocontainers such as mesoporous silica nanoparticles (MSNs) by installing supramolecular nanoswitches such as the pseudorotaxane or rotaxane nano-valves has been provided, which can be responsible for pH fluctuation (Ding et al. 2017). In a study, MSNs modified by axle molecules that contain a ferrocenecarboxylic acid ($C_{11}H_{10}FeO_2$) bonded via amide linkages and hexylammonium unit, the cucurbit[7]uril (CB[7]) macrocyclic molecular switches moving through the axle, through which inhibitor release can be controlled based on local pH variations (Fu et al. 2013). In another

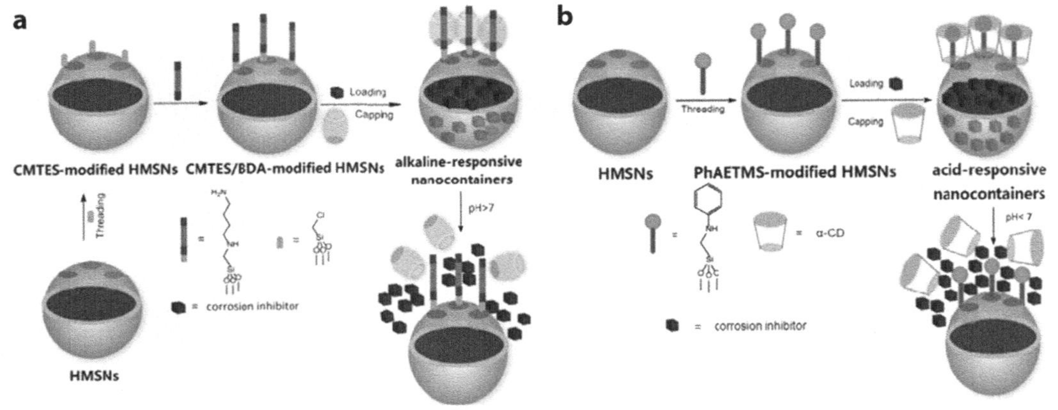

FIGURE 9.3 Schematic illustration of synthesis and inhibitor (BTA) release mechanism of alkaline-responsive containers (a) and acid-responsive containers (b). Adapted with permission from Chen and Fu (2012).

study, acid-sensitive nanoswitches based on α-cyclodextrin (α-CD)/aniline pseudorotaxane and alkali-responsive nanoswitches based on cucurbit[6]uril (CB[6])/bisammonium pseudorotaxane were covalently bonded to cavities of hollow mesoporous SiO_2 nanoparticles (HMSNs) (Figure 9.3). In alkaline conditions, deprotonation (completely or partially) of the initial protonated 1, 4-butanediamine units leads to the CB[6] rings bound dissociation via hydrophobic and ionic-dipole interactions. In acidic condition, the aniline stalks are surrounded by the α-CD rings via supramolecular interactions, whereas the aniline stalks' protonation reduces their binding affinity, followed by the dethreading of α-CD rings from the stalks (Chen and Fu 2012).

Recently, the use of natural materials to produce pH-sensitive gatekeepers has obtained increasing attention due to their properties, such as being eco-friendly and their low price. For instance, Tannic acid-Fe(III) complexes (TAC) was employed as the gatekeeper, which can be responsible for pH fluctuation in the development of containers made from MSNs and filled with BTA as the inhibitor (Qian, Michailidis et al. 2019). In another study, polydopamine (PDA) was coated on MSNs capsules filled with BTA to be used as the gatekeeper and incorporated into a water-based alkyd anticorrosive film on mild steel (Figure 9.4) (Qian, Zheng et al. 2019). MSNs coated with PDA to prepare pH-responsive drug release is another example (Zeng et al. 2017; Chang et al. 2016).

9.2.2 Aggressive Ions Exchange

Instead of preventing the corrosion process on the metallic substrate via passivation by corrosion inhibitors, reduction of aggressive ions concentration such as SO_4^{2-}, Cl^-, and O_2, which play an essential role in corrosion reaction, can be an efficient strategy to enhance protection efficiency of anticorrosive films. Taking advantage of the ion-exchange process, aggressive ions can be deactivated via entrapping into a layered structure, which is considered a technique to trigger the self-healing mechanism in a protective layer near the scratched substrate surfaces (Mohammadi et al. 2021a, 2021b). Employing cation-exchange clays as corrosion inhibitors containers (McMurray et al. 2003; Bohm et al. 2001) and layered materials, such as montmorillonite (MMT), which are constructed from negatively charged layers with exchangeable compensating cations, have been reported (Bohm et al. 2001). The gallery of Na-MMT clay particles was used for the exchange of Zn^{2+} and benzimidazole (BIA) cations by the Na^+ cations for an epoxy protective film on the steel substrate. By formation of insoluble complexes inside the defects due to the synergetic effect of Zn^{2+} and BIA^+ cations of the modified clay particles (Zn-MMT and BIA-MMT), the metallic substrate was perfectly protected from corrosion (Ghazi et al. 2015).

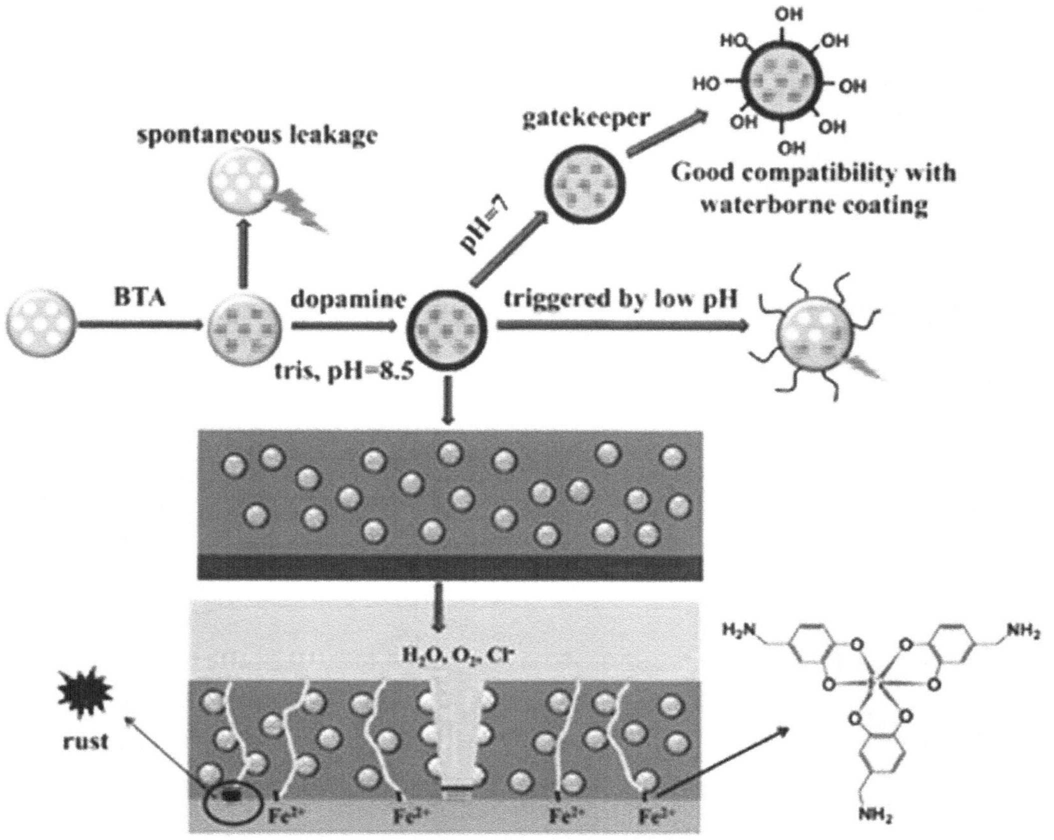

FIGURE 9.4 Schematic illustration of the self-healing process of mussel-inspired films. Adapted with permission from Qian, Zheng et al. (2019).

Layered double hydroxides (LDHs) are the composition of positively charged mixed metallic hydroxides multilayers, which can absorb the destructive chlorides and release the inhibitors via interaction with solvent molecules and anions (Zheludkevich et al. 2010). Hydroxides based on Mg-Al and Zn-Al loaded with different kinds of anionic inhibitors are known as the most common LDHs (Vega et al. 2011; Tedim et al. 2012). The protection performance of Zn-Al-based LDHs filled with three different inhibitors, including 2-mercaptobenzothiazole (MBT), phosphate, and vanadate incorporated in anticorrosive film coated on Al alloy, was investigated (Figure 9.5). The anion exchange with aggressive Cl^- triggered anionic inhibitors releases and capturing aggressive species (Tedim et al. 2010).

In a study, a soluble molybdate (MoO_4^{2-}) inhibitor intercalated Zn_2Al-LDH was employed in the anticorrosive film on a steel substrate. The release of MoO_4^{2-} anions was induced by the CO_3^{2-} anions in solution since the latter anions had a higher chemical affinity to the LDH than former anions (Shkirskiy et al. 2015). The investigation of the self-healing capacity of Mg-Al-based LDHs cerium molybdate (CeMo) corrosion inhibitor-based smart containers loaded with MBT showed that there was a synergistic impact on self-healing capacity and protection performance due to simulates using these two containers (Montemor et al. 2012). In another study, MgAl-LDHs/MBT composite was prepared for Mg alloy protection from corrosion. It was proposed that the release of CO_3^{2-} and OH^- anions by LDH phase entrapped Cl^- anions, which could reduce the concentration of Cl^- at the interface of coating solution. Simultaneously, the released anions could interact with free Mg^{2+} to create magnesium carbonate ($MgCO_3$) and magnesium hydroxide ($Mg(OH)_2$)

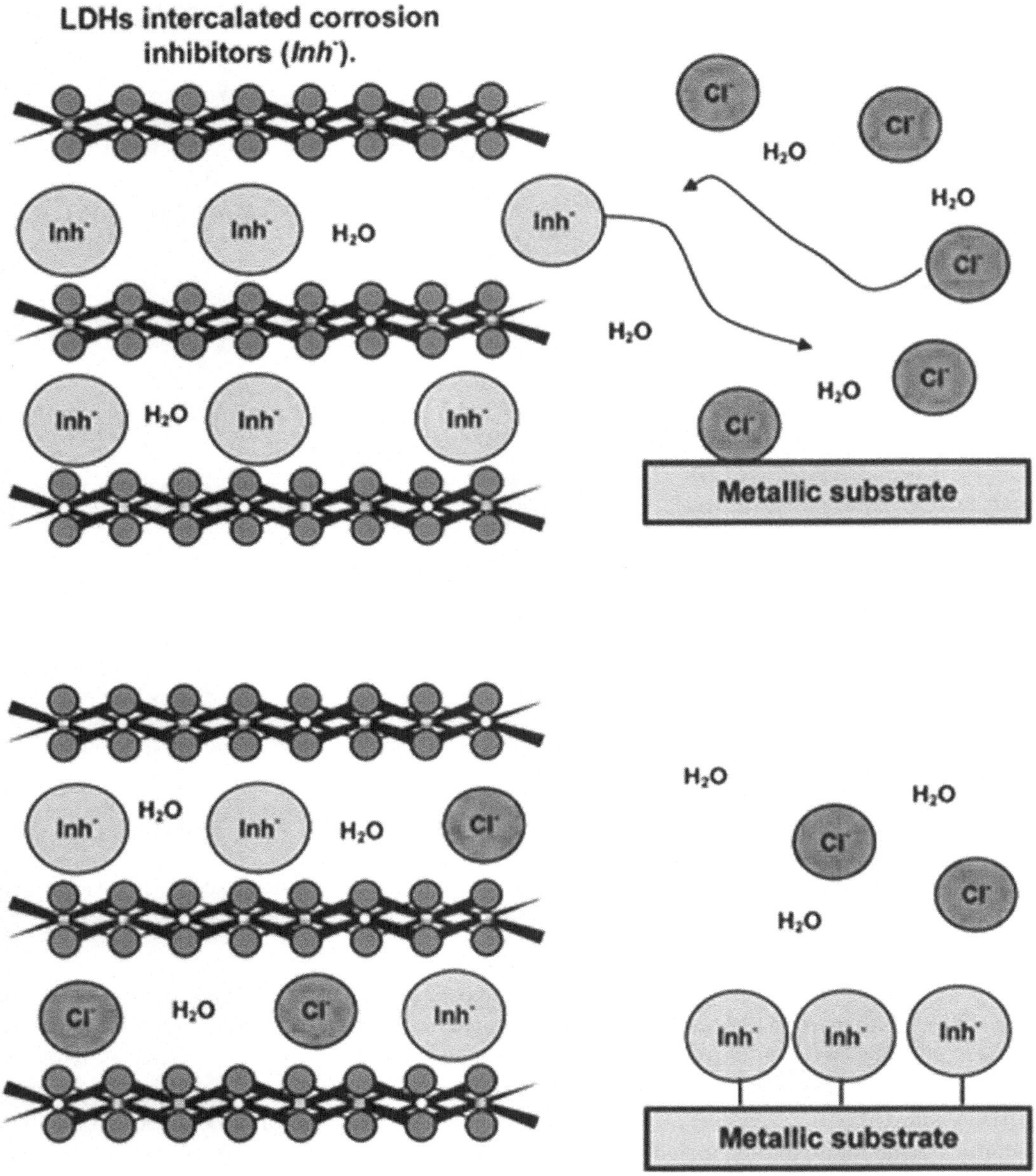

FIGURE 9.5 Schematic representation of inhibitor (LDHs) release mechanism for entrapping aggressive ions (Cl^-) and releasing anionic inhibitors (Inh^-) as anodic corrosion inhibitors on the substrate's surface for the self-healing process. Adapted with permission from Tedim et al. (2010).

precipitations, which can cover the damaged area and repair it (Hu et al. 2021). Recently, LDH-layer films have been widely investigated (Wang 2018; Li et al. 2020).

9.2.3 Heat

This stimulus is considered the most common one to be used as the trigger for the self-healing process. Shape memory polymers (SMPs) are known as intelligent materials, which can recover their initial form after they are deformed temporarily due to external heat (Xie 2011; Behl, Razzaq, and Lendlein 2010). SMPs can be used to produce smart anticorrosive coatings possessing thermo-responsive self-healing features. There are a large number of reports related to applying SMPs as anticorrosive coating with the ability of self-healing, such as polyurethane (PU) possessing poly(ε-caprolactone) (PCL) soft segments (Lutz et al. 2015; González-García et al. 2011b; 2011a; Jorcin et al. 2010). Based on the results from relevant studies, the shape memory effect (above 60°C) can narrow the cracks to a large extent, but the barrier features cannot be recovered entirely due to the absence of anticorrosion species with healing ability, which can seal the cracks. SMPs-based composite coatings were developed by providing the self-healing ability through which firstly, cracks can be closed through shape memory feature followed by filling them to recover the function of protective film (Nji and Li 2012; Li and Nettles 2010). Based on the mentioned strategy, PCL microfibers were applied on a metallic substrate via electrospinning technique as the healing agent, which can be melted by temperature increase, followed by coverage of the scaffolds with epoxy. By increasing temperature (from 25°C to 80°C), epoxy closed the cracks owing to the shape memory effect and melted PCL microfibers filled the cracks through which the barrier properties could be recovered entirely after cooling (Luo and Mather 2010).

Incorporating BTA as the corrosion inhibitor into a superhydrophobic epoxy possessing shape memory effect to produce dual action smart anticorrosive coatings for carbon steel corrosion protection was investigated. The results showed that the shape memory effect, which was thermo-responsive, could close the cracks and scratches and recover the surface's superhydrophobic morphology. The presence of BTA in the coating could provide corrosion inhibition in damaged sites and the SMP could improve the efficiency of the inhibition process via the size of the defects (cracks and scratches) in the protective coating. Based on the results, the BTA-doped protective coating showed better adhesion, strength, and self-healing properties (scratch repairing and crack sealing capacity) compared to the BTA-free film, even in the presence of pre-existing defects. It was demonstrated that the sunlight illumination could also enhance the performance of such smart protective coating by increasing the temperature from ambient temperature to 50°C, leading to motivating the shape memory effect (Qian et al. 2017).

It should be considered that the self-healing capacity of such a protective coating can be limited at the outside sites without the help of an external energy source to increase the temperature (Cai et al. 2018; Wang et al. 2016; Jorcin et al. 2010). To solve the mentioned problem, incorporation of photothermal (additives) fillers such as aniline black, carbon nanotubes (CNTs), carbon black (CB), and graphene (G) was suggested. The presence of such fillers in coatings can provide the required temperature through the transformation of light to heat energy to motivate the shape memory effects on polymers (Fang et al. 2017; Chen et al. 2016). A high-performance self-healing protective coating was prepared from shape-memory PU (SMPU) filled with two different types of containers, including a heat-generating agent (HGA) and alodine microcapsules. The reaction of released HGA (composition of vermiculite, carbon powder, malleable iron) with oxygen (O_2) led to motivating the shape memory effect of PU, and the inhibitors released from the alodine microcontainer caused the generation of a passive layer at the defective sites (Figure 9.6) (Fan et al. 2019).

SiO_2/polymer double-walled hybrid nanotubes were fabricated from a hollow cavity and a porous inner wall made of SiO_2 filled with poly(Nisopropylacrylamide-co-methylenebisacrylamide) (PNIPAM) as temperature-sensitive molecules. At 25°C, the outer wall (PNIPAM) can be swelled, whereas it can be shrunken at high temperature (50°C), resulting in facilitating the inhibitor (BTA) molecules diffusion in the cavities out of the wall (Li et al. 2013). PNIPAM, which is considered the common temperature-responsive material, has also been used for drug delivery by grafting onto the halloysite external surface (Cavallaro et al. 2015).

9.2.4 Light

Light energy (UV light or visible light) can also be used as an effective stimulus for controlling the release of healing agents or corrosion inhibitors from containers which are predictably,

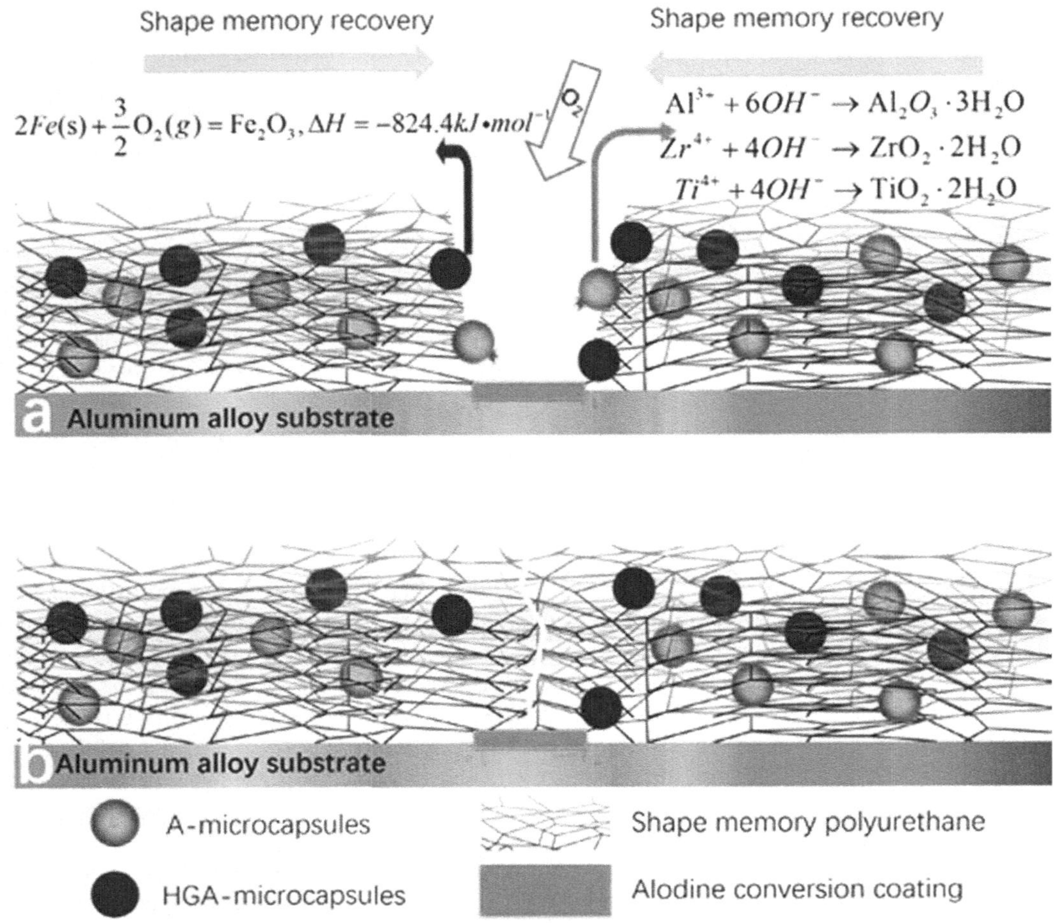

FIGURE 9.6 Self-healing mechanism via formation of a passivation layer on the damaged area and the generation of heat to motivate the shape memory effect of PU. Adapted with permission from Fan et al. (2019).

accurately, and can be controlled remotely (Habault, Zhang, and Zhao 2013). It is interesting to know that unexpected defects such as cracks can be recovered through periodic sunlight radiation without monitoring. Mainly the light-responsive self-healing polymeric coatings are categorised into two groups, including polymeric composites incorporated with micro- and nanocapsules filled with healing species (Park and Braun 2010) and dynamic polymers (Ying, Zhang, and Cheng 2014; Kuroki et al. 2013). The dynamic polymers mostly consist of a variety of self-healing processes. UV-responsive polymers with self-healing features have been expanded based on reversible reactions of photo-crosslinking, such as coumarin (Zhao et al. 2016), the cycloaddition of cinnamoyl (Song et al. 2013), or anthracene derivatives (Fang et al. 2017) and dynamic covalent attachment, such as allyl sulfide (AS) (Meng et al. 2016), trithiocarbonates (TTCs) (Amamoto et al. 2011), alkoxyamines (Telitel et al. 2014), and disulfides (Michal et al. 2013). Using polymeric composites has obtained increasing attention since there is no need to change the chemical structure of the polymers, and more importantly, defects with large sizes can be recovered through this strategy due to the presence of active containers and photo/catalyst initiator in the composites.

Photo-absorbing hybrid microcontainers (poly(urea-formaldehyde)/titania (TiO_2)) loaded with epoxy-based healing species to fabricate light-induced self-healing composites coating.

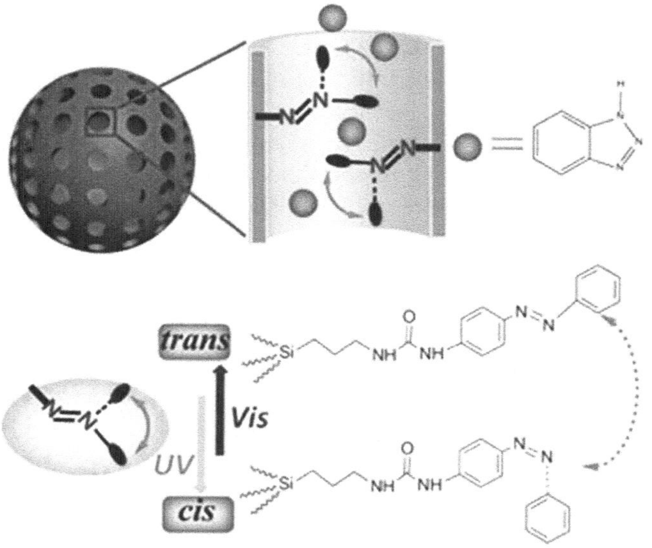

FIGURE 9.7 Schematic illustration related to mechanism reversible release mechanism using trans-cis photo-isomerisation of azobenzene molecules grafted in the mesopores structure of HMSs. Adapted with permission from Chen et al. (2015).

It was reported that the UV light could be absorbed with a wavelength of 330 nm owing to the presence of the TiO_2 nanoparticles, through which the photocatalytic performance could be induced and the epoxy-based healing species solidification could be accelerated (Gao et al. 2015). In another study, benzoin isobutyl ether (BIE) as photoinitiator and methacryloxypropyl-terminated polydimethylsiloxane (MAT-PDMS) as healing species were encapsulated into urea-formaldehyde (UF) polymeric microcontainers. Due to the presence of such microcapsules, the anticorrosive coating featured the light-induced self-healing mechanism through which controlled release of healing species from the broken microcontainers can be provided, resulting in filling the defects and then solidification by the photo-activatable crosslinking mechanism (Song et al. 2013). The effect of silver (Ag) nanoparticles decoration as photo-absorption sites on PE multilayers of mesoporous SiO_2 and TiO_2 on a self-healing system was investigated. The results depicted that the remote release of corrosion inhibitors from capsules can be perfectly controlled through a light-induced mechanism, leading to controlling the corrosion reaction at the right time and space (Skorb et al. 2009; Shchukin et al. 2009). A new release system was fabricated based on photo-responsiveness to improve the protection efficiency through the azobenzene species implanting into hollow mesoporous SiO_2 (HMSs) nanochannels (Figure 9.7). By UV radiation, azobenzene isomerisation impeller from thermodynamically stable trans isomer to metastable cis isomer opened the nanopores led to the simultaneous release of molecules out of the nanoparticles. As a result, the corrosion inhibitors are transferred to defect sites from the containers to prevent the diffusion of aggressive species and protect the surface of the substrate from major damage (Chen et al. 2015).

9.2.5 Electrochemical Redox/Potential

Electrochemical potential fluctuation is considered the most dependable trigger due to uniform distribution, in situ generation, and strong reducibility. More importantly, it always decreases when the corrosion reaction starts, which can be used for controlled self-healing (Zhao et al. 2015; Tran et al. 2015). In a study, to enhance the protection performance of magnesium alloy, potential sensitive nanocontainers (filled with 8-hydroxyquinoline (8-HQ) as the inhibitor) were prepared from $Fe_3O_4@mSiO_2$ as magnetic vehicles and arenes (BIPY ⊂ WP[5]- bipyridinium water-soluble pillar[5] as supramolecular switches and disulfide as chemical linkers). The corrosion process onset led to disulfide linkages breaking, nanoswitch removal, and release of an inhibitor. The incorporation of potential sensitive smart containers remarkably improved the anticorrosive performance of the coating as well as the self-healing feature (Ding et al. 2017).

Using conductive polymer (CP) to prepare nanocontainers is considered a great idea since the release of the filled corrosion inhibitor can be appropriately controlled by CP reduction caused by storage of negative ions as counter-charge till the polymer backbone oxidation (Kowalski, Ueda, and Ohtsuka 2010). Redox-sensitive nanocontainers were prepared from redox-responsive polyaniline (PANI) as capsules filled with 3-nitrosalicylic acid (3-NisA) as the inhibitor. To overcome the possible misalignment of fermi level, the nanocontainer modification was done by gold (Au) nanoparticles to provide a steady electronic contact by preventing direct contact between PANI and metallic substrate (Figure 9.8). The results demonstrated that the release of corrosion inhibitor (3-NisA) could be perfectly controlled by the gatekeeper (PANI). The onset of corrosion reaction led to shell reduction and increasing PANI permeability resulting in 3-NisA release. After the reoxidation process (metal passivation), the release of inhibitor was stopped due to decreasing shell permeability (Vimalanandan et al. 2013).

Multifunctional nanocapsules were prepared by the approach mentioned above and possessed two different inhibitors, including MBT, which was covalently attached to the polymeric shell of the container, and polydimethylsiloxane diglycidyl ether of which was packed in capsules. It was reported that the release of healing agents was done by polymeric shell reduction, in which release of MBT provided the packed inhibitor (polydimethylsiloxane diglycidyl ether), and the efficiency of release was controlled based on the properties of the shell structures and the inhibitors (Zhao et al. 2015).

To design supermolecule valves for containers to control healing agent release, macrocycle molecules can be utilised. Redox-sensitive MSNs with switch was added into sol-gel film (Fu et al. 2013), which was deposited on another coating made from sol-gel film doped with cerium(IV) (Ce(IV)). A layered nanocomposite coating for Al alloy protection was fabricated from a layer made of redox-triggered containers loaded with (p-coumaric acid [CA]) as a corrosion

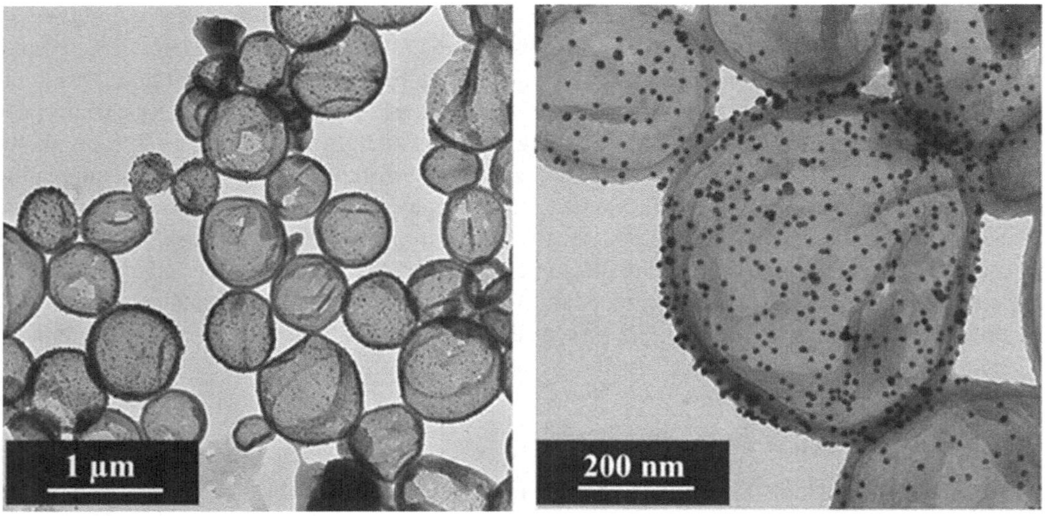

FIGURE 9.8 Transmission electron microscopy (TEM) images of AuNPs-decorated PANI containers. Adapted with permission from Vimalanandan et al. (2013).

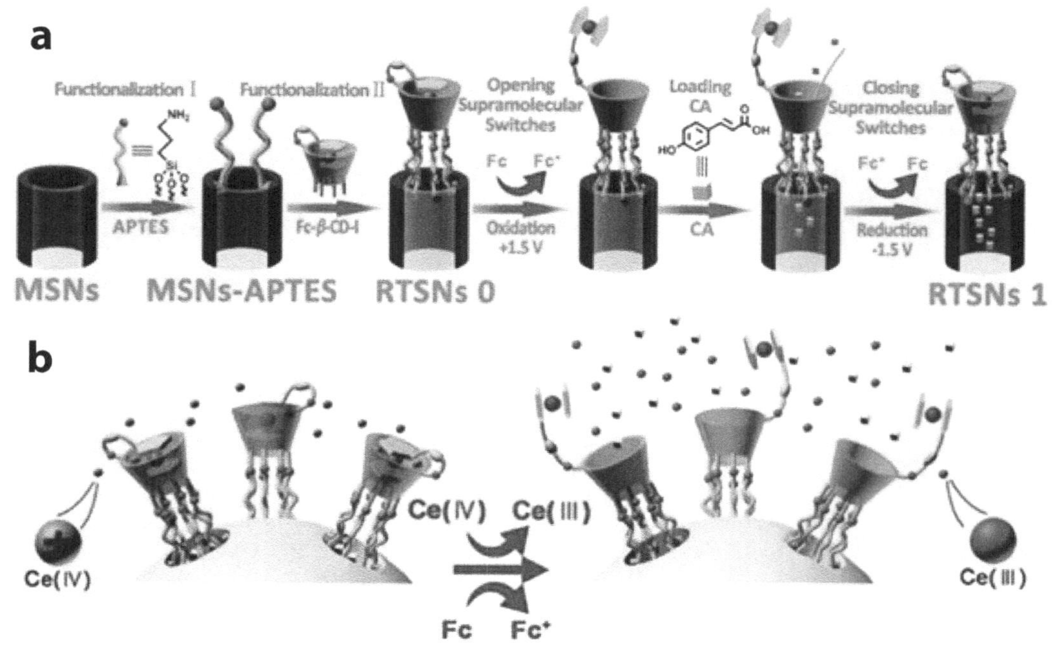

FIGURE 9.9 Schematic representation related to fabrication (a) and release mechanism (b) of redox-triggered smart nanocontainers. Adapted with permission from Wang et al. (2017).

inhibitor incorporated into ZrO_2–SiO_2 film, and a layer made of CeIV)-doped ZrO_2-SiO_2 film. β-cyclodextrin (β-CD) monofunctionalised with a ferrocene (Fc) moiety and pseudorotaxane was used as a supramolecular nanovalve to control the inhibitor release via Fc to Fc^+ transformation when the Ce^{4+} salt contacted with redox-triggered smart nanocontainers at the damaged sites (Figure 9.9) (T. Wang et al. 2017).

9.2.6 Magnetic Field

Owing to magnetic nanoparticles properties, incorporating such materials as an additive into nano/microcapsule, microgels, hydrogels, and polymer has obtained increasing attention (Kulbaba et al. 2002; Bhattacharya et al. 2007). The use of such additives to improve protection performance can also be effective. Conventional smart self-healing anticorrosive coatings are mostly related to inhibitor passivation incorporated with active containers as self-healing functional fillers, which are distributed spatially in the protective coatings. Such additives can react with the damaged site of the metallic substrate and form a passivation layer to stop the metal oxidation and control the corrosion process. Since the inhibitors distribute randomly in the protective coatings by the occurrence of corrosion reaction, there might not be required inhibitors nearby the defects for the passivation process. Based on the performance of targeted drug delivery (Kaur et al. 2014; Liu et al. 2011), self-healing containers are prepared and filled with magnetic materials to ease the transmission of released inhibitors and improve the inhibition efficiency via an external magnetic field. A novel corrosion inhibitor-based smart microcontainer made of poly(urea-formaldehyde) filled with BTA as a corrosion inhibitor and magnetic multiwall carbon nanotubes (MWCNTs) was prepared and incorporated into the polymeric coating for metallic substrate (copper [Cu]) protection in aggressive media. The presence of magnetic MWCNTs played a leading role in the efficiency improvement of self-healing magnetic gradient (SMG) anticorrosive coating. Due to the magnetic property of containers, the migration of such nanocapsules was accelerated, and they

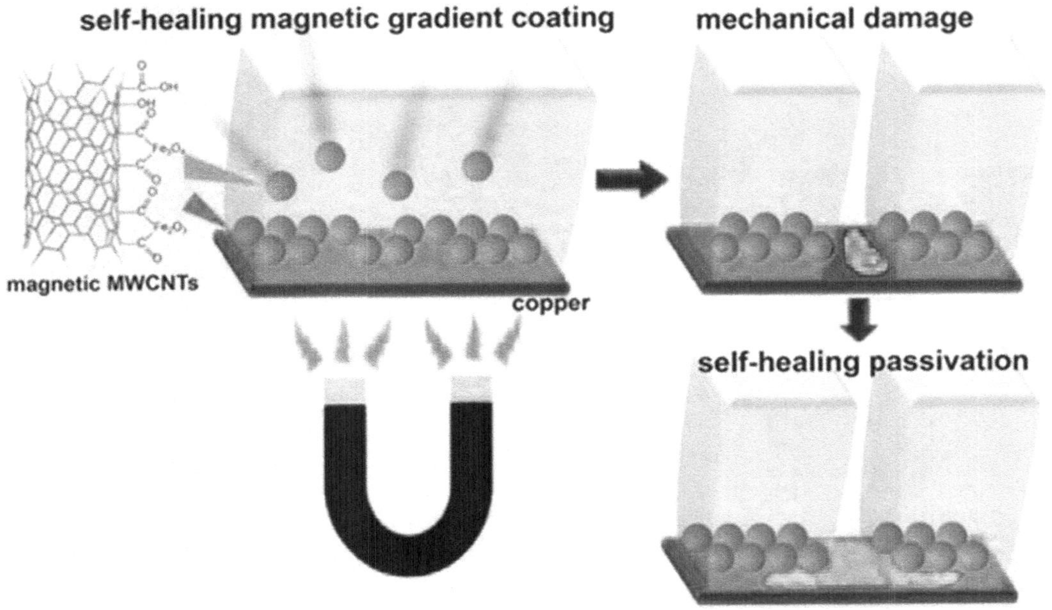

FIGURE 9.10 Schematic representation related to the self-healing acceleration process of the damaged protective coating possessing MWCNTs as functional fillers. Adapted with permission from Wang et al. (2018).

could be quickly and easily moved in the coating (epoxy resin) by using an external magnetic field to release the inhibitor and form a passive layer (Cu(I)-BTA) on the damaged sites of the substrate (Cu) (Figure 9.10) (Wang et al. 2018).

9.2.7 Mechanical Impact

Opposite to the mentioned stimulus, this one is not corrosion related, and the release of corrosion inhibitors/healing agents in self-healing coatings is under mechanical impact caused by the protective coating degradation, such as scratches. Therefore, it is known as the first external stimulus. Recently, the study of anticorrosive coatings, which can provide multiple-action self-healing, has obtained considerable attention, such as using healing agents with the ability of polymerisation and corrosion inhibitors, which can be encapsulated separately or together (Mahmoudian et al. 2018; Szabó, Telegdi, and Nyikos 2015). In a study, dicyclopentadiene (DCPD) was used as a corrosion inhibitor encapsulated in microcapsules made of polymer and embedded into an epoxy coating possessing Grubbs' catalyst. The results showed that due to the healing agent polymerisation, the cracks could be filled, and the barrier features of polymeric coating could be recovered. It was proposed that the containers can also be penetrated by microcracks initiation and development in the coating, which leads to DCPD release. Accordingly, the cracks are moistened and infiltrated, and due to the presence of Grubbs' catalyst, polymerisation is initiated and heals the defects (White et al. 2002). The studies showed the simultaneous incorporation of two containers filled with different healing agents (Keller, White, and Sottos 2007), encapsulation of oil (He and Shi 2009), and reactive epoxy resin (Caruso et al. 2008) can be effective in improving the corrosion inhibition process. LBL assembly technique was employed to produce microcapsule containers filled with linseed and covered with PE layers in which BTA was entrapped. As a result of linseed oil release under mechanical impact and BTA by pH fluctuation, the polymeric self-healing anticorrosive coated on carbon steel showed effective protection (Figure 9.11) (Leal et al. 2018).

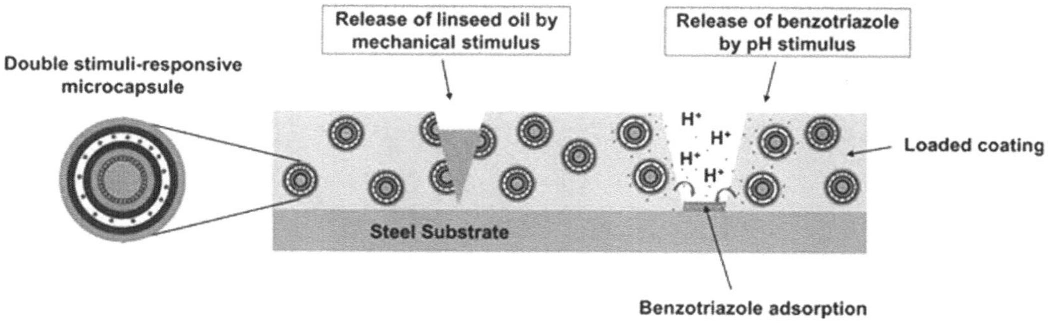

FIGURE 9.11 Schematic representation of the inner structure related to microcontainers and the self-healing mechanism of the protective film. Adapted with permission from Leal et al. (2018).

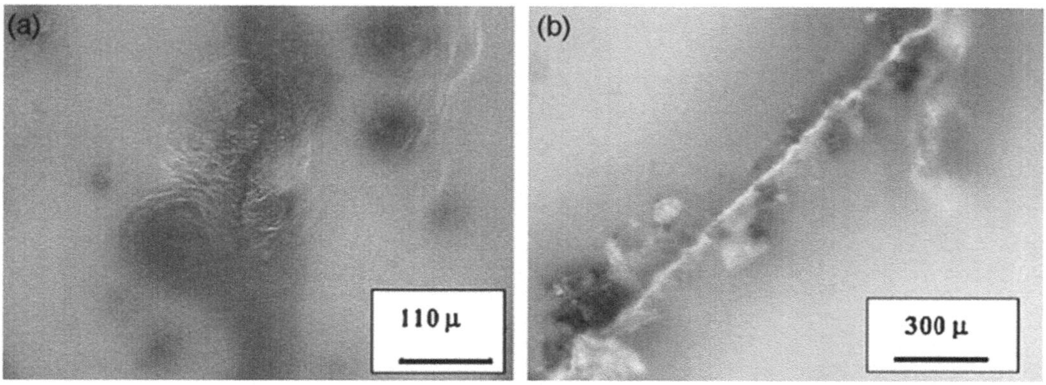

FIGURE 9.12 Dye release from UF microcontainer as a result of an induced crack in protective film (a), and caused by scribing the coated film. Adapted with permission from Kumar, Stephenson, and Murray (2006).

In another study, alkylammonium salt of (2-benzothiazolylthio)succinic acid ($C_{11}H_9NO_4S_2$) in xylene and camphor ($C_{10}H_{16}O$) were loaded in formaldehyde microcapsules as corrosion inhibitors to be embedded into polymeric coating in the presence of encapsulated spar varnish and tung oil as healing agents for corrosion protection of outdoor steel structures. Based on the results, the encapsulated inhibitors were released when the coating was damaged (Figure 9.12) and the corrosion propagation was controlled because of the healing agents' presence (Kumar, Stephenson, and Murray 2006).

9.2.8 Multistimuli-Responsive Containers

Developing the smart micro- and nanocontainers with multistimuli-responsive release mechanisms has recently obtained increasing attention by combining interior and/or exterior stimuli according to drug delivery smart containers (Wang et al. 2016; Chen, Yang, and Fu 2013). Utilising such containers provides a larger range of response mechanisms and triggers through which the damaged area can be healed and the performance efficiency of anticorrosive coating can be significantly improved (Fu et al. 2013). Mechanised silica nanoparticles (MSNPs) were incorporated as smart containers (guest) into assembled nanophase particles (SNAP) barrier film (host) to fabricate a novel host–guest feedback active anticorrosive film coated on the Mg alloy substrate (Figure 9.13). Pillar[5]arenes (WP[5])-based supramolecular nanoswitch controlled the release of inhibitor (2-hydroxy-4-methoxy-acetophenone [HMAP]). The inhibitor release mechanism is based on alkali or Mg^{2+} stimuli occurring on the corroded area of the substrate.

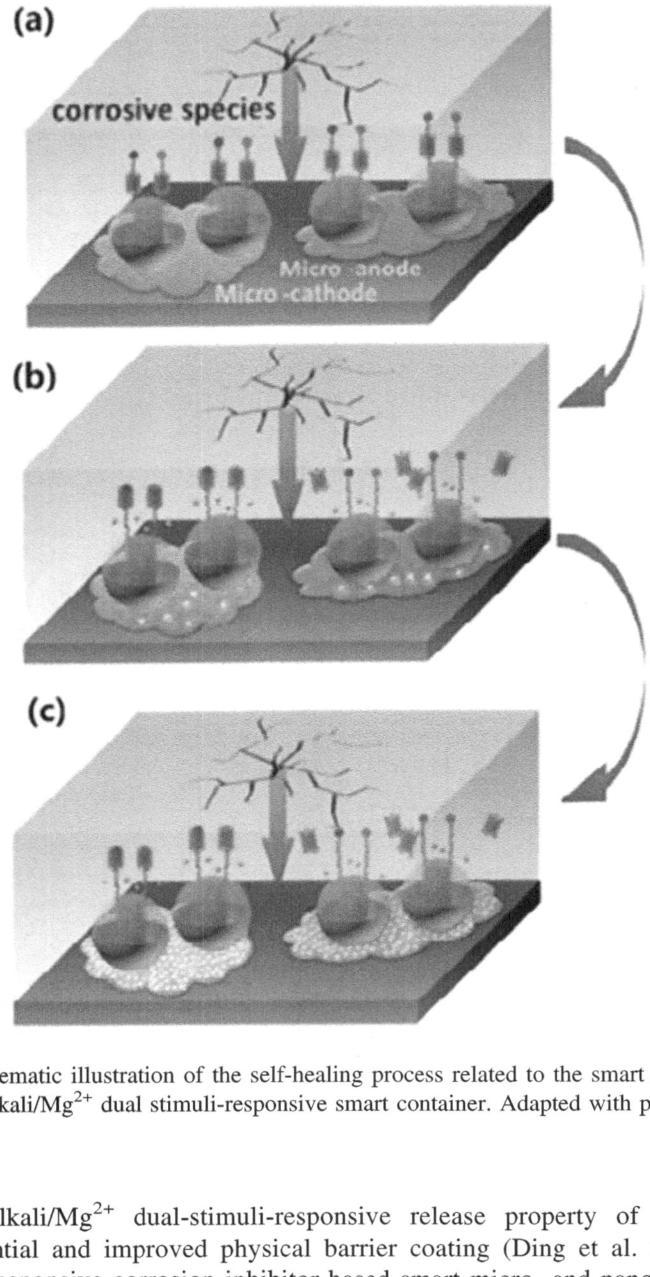

FIGURE 9.13 Schematic illustration of the self-healing process related to the smart anticorrosive coating immobilised with alkali/Mg^{2+} dual stimuli-responsive smart container. Adapted with permission from Ding et al. (2016).

The particular alkali/Mg^{2+} dual-stimuli-responsive release property of MSNPs provided self-healing potential and improved physical barrier coating (Ding et al. 2016). Fabrication of triple-stimuli-responsive corrosion inhibitor-based smart micro- and nanocontainers was reportedly due to the ability to be responsive to acid and alkali conditions and corrosion potential fluctuation (Wang et al. 2019; Ding, Tong, and Fu 2017).

Dual stimuli-responsive nanocontainers, which are responsive to reduction and pH fluctuation, were fabricated. The shell was prepared by PANI loaded with 3-NisA (3-nitrosalicylic acid) as the inhibitor, which is sensitive to pH value fluctuation and MBT (2-mercaptobenzothiazole), which can bond to Au and then reduce electrochemically (Figure 9.14). Incorporation of MBT-Au@PANI/3-NisA corrosion inhibitor smart containers into the anticorrosive could positively affect the protection performance in which coating corrosion inhibitors can be released independently and selectively based on the increase of solution pH value and/or by electrochemical reduction (Lv, Landfester, and Crespy 2014).

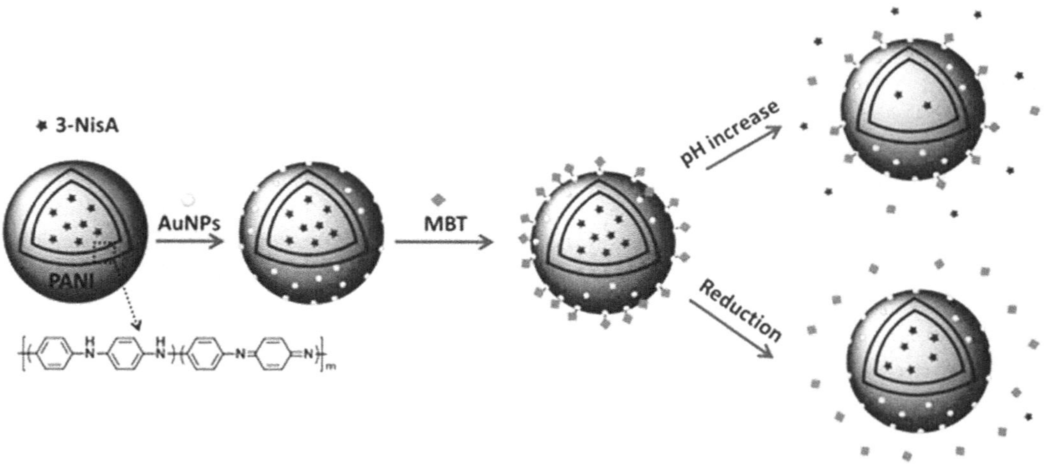

FIGURE 9.14 Schematic illustration of the dual-delivery nanocapsules fabrication and the release mechanisms of the loaded inhibitors depending on pH fluctuation or reduction. Adapted with permission from Lv, Landfester, and Crespy (2014).

9.3 OUTLOOK AND FUTURE STUDIES

Recently, a considerable number of new classes of micro- and nanocontainers with smart stimuli-responsive release mechanisms have been introduced in different fields, such as drug delivery and dye molecules, including electromagnetic irradiation (Lee et al. 2018; Hu et al. 2018), ultrasonic treatment (Sciurti et al. 2020; Zhou et al. 2019), enzymes (Patel et al. 2008; Sun et al. 2013), and reactive oxygen species (Saravanakumar, Kim, and Kim 2017; Lee et al. 2013), but these materials don't have the potential to be used for the corrosion inhibition process. Since controlling corrosion of metals in aggressive environments is a crucial task and it has been extended in different industries, it must be developed to introduce suitable stimuli-responsive based on the needs and harsh media (Hu et al. 2018). Based on the studies reviewed above, there are varying degrees of complexity in terms of fabricating and applying smart coatings to obtain an effective product so that all the proposed systems don't have the potential to be industrialised. Realising the full potential of corrosion inhibitor smart containers in the corrosion protection field requires a deeper understanding of various aspects, which can be done via extensive studies of this system through the cooperation of researchers from different fields, such as material science, corrosion science, chemical engineering, and electrochemistry. In summary, future research in this field will be conducted towards:

- Development of multifunctional (multistimuli) micro- and nanocontainers with the ability to be loaded with several inhibitors/healing agents and to be responsible for different triggers
- Development of shells with the smallest possible size and high loading capacity
- Development of facile, environmentally friendly, cost-efficient, reproducible fabrication methods
- Development of a multilevel self-healing/inhibition process

9.4 CONCLUSION

Corrosion is known as a complex process containing chemical reactions between the environment and the surface of materials through which chemical and physical features can be changed at the interface, leading to putting the material at risk of failure. Acidifying damaged sites is one of the most common changes that can self-accelerate corroding the material. Therefore, the local pH fluctuation is inevitable and due to the chemical valence changes in corrosion reaction, ion exchange and redox elevation may also be indicated. The mentioned parameters can be a suitable endogenous stimulus to control the release of inhibitor/healing agents from the smart containers to prevent the propagation of corrosion reactions. In fact, under these circumstances, corrosion inhibitor smart containers can perform as a sensor, which can track the reaction of corrosion and control this process by releasing of inhibitor/healing agent in the damaged area. There are some exogenous stimuli that are related to the environment, such as magnetic field, light, temperature, and mechanical impact, that can increase the susceptibility of the corroding process. Exogenous stimuli-responsive corrosion inhibitor smart containers can be a proper solution to control the possible corrosion attack caused by environmental changes. To enhance the protection efficiency of such additives, the idea of producing multistimuli-responsive containers and a multilevel self-healing process seems to be promising. By applying corrosion inhibitor smart containers into or between polymeric coating layers, according to their function, the active compounds can be targeted in the face of an aggressive environment. Although significant progress has been obtained in this field, many technological and scientific challenges remain, which can be a proper direction for future studies.

REFERENCES

Akbarzadeh, Sajjad, Bahram Ramezanzadeh, Ghasem Bahlakeh, and Mohammad Ramezanzadeh. 2019. "Molecular/Electronic/Atomic-Level Simulation and Experimental Exploration of the Corrosion Inhibiting Molecules Attraction at the Steel/Chloride-Containing Solution Interface." *Journal of Molecular Liquids* 296: 111809. 10.1016/j.molliq.2019.111809.

Alipanah, Nariman, Hossein Yari, Mohammad Mahdavian, Bahram Ramezanzadeh, and Ghasem Bahlakeh. 2021. "MIL-88A (Fe) Filler with Duplicate Corrosion Inhibitive/Barrier Effect for Epoxy Coatings: Electrochemical, Molecular Simulation, and Cathodic Delamination Studies." *Journal of Industrial and Engineering Chemistry* 97 (May): 200–215. 10.1016/j.jiec.2021.01.035.

Amamoto, Yoshifumi, Jun Kamada, Hideyuki Otsuka, Atsushi Takahara, and Krzysztof Matyjaszewski. 2011. "Repeatable Photoinduced Self-Healing of Covalently Cross-Linked Polymers through Reshuffling of Trithiocarbonate Units." *Angewandte Chemie - International Edition* 50 (7): 1660–1663. 10.1002/anie.201003888.

Amini, Misagh, Reza Naderi, Mohammad Mahdavian, and Alireza Badiei. 2021. "Release of Lanthanum Cations Loaded into Piperazine-Modified SBA-15 to Inhibit the Mild Steel Corrosion." *Microporous and Mesoporous Materials* 315 (December 2020): 110908. 10.1016/j.micromeso.2021.110908.

Bahlakeh, Ghasem, Mohammad Ramezanzadeh, and Bahram Ramezanzadeh. 2017. "Experimental and Theoretical Studies of the Synergistic Inhibition Effects between the Plant Leaves Extract (PLE) and Zinc Salt (ZS) in Corrosion Control of Carbon Steel in Chloride Solution." *Journal of Molecular Liquids* 248 (December): 854–870. 10.1016/j.molliq.2017.10.120.

Bahremand, Farshad, Taghi Shahrabi, and Bahram Ramezanzadeh. 2021. "Epoxy Coating Anti-Corrosion Properties Enhancement via the Steel Surface Treatment by Nanostructured Samarium Oxide-Poly-Dopamine Film." *Journal of Hazardous Materials* 403 (August 2020): 123722. 10.1016/j.jhazmat.2020.123722.

Behl, Marc, Muhammad Yasar Razzaq, and Andreas Lendlein. 2010. "Multifunctional Shape-Memory Polymers." *Advanced Materials* 22 (31): 3388–3410. 10.1002/adma.200904447.

Bhattacharya, Sanchita, Franziska Eckert, Volodymyr Boyko, and Andrij Pich. 2007. "Temperature-, PH-, and Magnetic-Field-Sensitive Hybrid Microgels." *Small* 3 (4): 650–657. 10.1002/smll.200600590.

Bohm, S., H. N. McMurray, S. M. Powell, and D. A. Worsley. 2001. "Novel Environment Friendly Corrosion Inhibitor Pigments Based on Naturally Occurring Clay Minerals." *Werkstoffe Und Korrosion* 52 (12): 896–903. 10.1002/1521-4176(200112)52:12<896::aid-maco896>3.0.co;2-8.

Cai, Haoyuan, Peng Wang, and Dun Zhang. 2019. "PH-Responsive Linkages-Enabled Layer-by-Layer Assembled Antibacterial and Antiadhesive Multilayer Films with Polyelectrolyte Nanocapsules as Biocide Delivery Vehicles." *Journal of Drug Delivery Science and Technology* 54 (June): 101251. 10.1016/j.jddst.2019.101251.

Cai, Tingting, Shuangjun Huo, Tao Wang, Weixiang Sun, and Zhen Tong. 2018. "Self-Healable Tough Supramolecular Hydrogels Crosslinked by Poly-Cyclodextrin through Host-Guest Interaction." *Carbohydrate Polymers* 193: 54–61. 10.1016/j.carbpol.2018.03.039.

Caruso, Mary M., Benjamin J. Blaiszik, Scott R. White, Nancy R. Sottos, and Jeffrey S. Moore. 2008. "Full Recovery of Fracture Toughness Using a Nontoxic Solvent-Based Self-Healing System." *Advanced Functional Materials* 18 (13): 1898–1904. 10.1002/adfm.200800300.

Cavallaro, Giuseppe, Giuseppe Lazzara, Marina Massaro, Stefana Milioto, Renato Noto, Filippo Parisi, and Serena Riela. 2015. "Biocompatible Poly(N -Isopropylacrylamide)-Halloysite Nanotubes for Thermoresponsive Curcumin Release." *Journal of Physical Chemistry C* 119 (16): 8944–8951. 10.1021/acs.jpcc.5b00991.

Chang, Danfeng, Yongfeng Gao, Lijun Wang, Gan Liu, Yuhan Chen, Teng Wang, Wei Tao, Lin Mei, Laiqiang Huang, and Xiaowei Zeng. 2016. "Polydopamine-Based Surface Modification of Mesoporous Silica Nanoparticles as PH-Sensitive Drug Delivery Vehicles for Cancer Therapy." *Journal of Colloid and Interface Science* 463: 279–287. 10.1016/j.jcis.2015.11.001.

Changjean, Wen Chyuan, Lin Yi Huang, Po Yi Liu, and Tseng Chang Tsai. 2014. "Repairable Mesoporous Silica Film with Replenishing Corrosion Inhibitor as Corrosion Protection Layer of Aluminum Alloy." *Microporous and Mesoporous Materials* 192: 82–88. 10.1016/j.micromeso.2014.02.030.

Chen, Jiamei, Liang Fang, Zhongzi Xu, and Chunhua Lu. 2016. "Self-Healing Epoxy Coatings Curing with Varied Ratios of Diamine and Monoamine Triggered via near-Infrared Light." *Progress in Organic Coatings* 101: 543–552. 10.1016/j.porgcoat.2016.09.020.

Chen, Jing Yu, Xiao Bo Chen, Jing Liang Li, Bin Tang, Nick Birbilis, and Xungai Wang. 2014. "Electrosprayed PLGA Smart Containers for Active Anti-Corrosion Coating on Magnesium Alloy AMlite." *Journal of Materials Chemistry A* 2 (16): 5738–5743. 10.1039/c3ta14999d.

Chen, Tao, Renpeng Chen, Zhong Jin, and Jie Liu. 2015. "Engineering Hollow Mesoporous Silica Nanocontainers with Molecular Switches for Continuous Self-Healing Anticorrosion Coating." *Journal of Materials Chemistry A* 3 (18): 9510–9516. 10.1039/C5TA01188D.

Chen, Tao, and Jiajun Fu. 2012. "An Intelligent Anticorrosion Coating Based on PH-Responsive Supramolecular Nanocontainers." *Nanotechnology* 23 (50). 10.1088/0957-4484/23/50/505705.

Chen, Tao, Nianwang Yang, and Jiajun Fu. 2013. "Controlled Release of Cargo Molecules from Hollow Mesoporous Silica Nanoparticles Based on Acid and Base Dual-Responsive Cucurbit[7]Uril Pseudorotaxanes." *Chemical Communications* 49 (58): 6555–6557. 10.1039/c3cc43221a.

Chuanjie, Fan, Tang Juntao, and Zhou Xiaodong. 2013. "Effects of Process Parameters on the Physical Properties of Poly (Urea-Formaldehyde) Microcapsules Prepared by a One-Step Method." *Iranian Polymer Journal (English Edition)* 22 (9): 665–675. 10.1007/s13726-013-0165-z.

Custódio, Juliana V., Sílvia M.L. Agostinho, and Alda M. P. Simões. 2010. "Electrochemistry and Surface Analysis of the Effect of Benzotriazole on the Cut Edge Corrosion of Galvanized Steel." *Electrochimica Acta* 55 (20): 5523–5531. 10.1016/j.electacta.2010.03.072.

Dehghani, Ali, Ghasem Bahlakeh, and Bahram Ramezanzadeh. 2020. "Synthesis of a Non-Hazardous/Smart Anti-Corrosion Nano-Carrier Based on Beta-Cyclodextrin-Zinc Acetylacetonate Inclusion Complex Decorated Graphene Oxide (β-CD-ZnA-MGO)." *Journal of Hazardous Materials* 398 (March): 122962. 10.1016/j.jhazmat.2020.122962.

Ding, Chen Di, Ling Tong, and Jia Jun Fu. 2017. "Quadruple Stimuli-Responsive Mechanized Silica Nanoparticles: A Promising Multifunctional Nanomaterial for Diverse Applications." *Chemistry – A European Journal* 23 (60): 15041–15045. 10.1002/chem.201704245.

Ding, Chendi, Ying Liu, Mingdong Wang, Ting Wang, and Jiajun Fu. 2016. "Self-Healing, Superhydrophobic Coating Based on Mechanized Silica Nanoparticles for Reliable Protection of Magnesium Alloys." *Journal of Materials Chemistry A* 4 (21): 8041–8052. 10.1039/c6ta02575g.

Ding, ChenDi, JianHua Xu, Ling Tong, GuangCai Gong, Wei Jiang, and Jiajun Fu. 2017. "Design and Fabrication of a Novel Stimulus-Feedback Anticorrosion Coating Featured by Rapid Self-Healing Functionality for the Protection of Magnesium Alloy." *ACS Applied Materials & Interfaces* 9 (24): 21034–21047. 10.1021/acsami.7b06347.

Fan, Weijie, Yong Zhang, Weihua Li, Wei Wang, Xiaodong Zhao, and Liying Song. 2019. "Multi-Level Self-Healing Ability of Shape Memory Polyurethane Coating with Microcapsules by Induction Heating." *Chemical Engineering Journal* 368 (January): 1033–1044. 10.1016/j.cej.2019.03.027.

Fang, Liang, Jiamei Chen, Yuting Zou, Shunping Chen, Tianyu Fang, Chunhua Lu, and Zhongzi Xu. 2017. "Self-Healing Epoxy Coatings via Focused Sunlight Based on Photothermal Effect." *Macromolecular Materials and Engineering* 302 (9): 1–13. 10.1002/mame.201700059.

Feng, Yuanchao, Shougang Chen, and Y. Frank Cheng. 2017. "Fabrication of SiO_2 Nanoparticle–Polyelectrolyte Nanocontainers with Preloaded Benzotriazole Inhibitors and Their Self-Releasing Mechanism and Kinetics." *Journal of Materials Science* 52 (14): 8576–8590. 10.1007/s10853-017-1074-x.

Fix, Dmitri, Daria V. Andreeva, Yuri M. Lvov, Dmitry G. Shchukin, and Helmuth Möhwald. 2009. "Application of Inhibitor-Loaded Halloysite Nanotubes in Active Anti-Corrosive Coatings." *Advanced Functional Materials* 19 (11): 1720–1727. 10.1002/adfm.200800946.

Fu, Jiajun, Tao Chen, Mingdong Wang, Nianwang Yang, Suning Li, Ying Wang, and Xiaodong Liu. 2013. "Acid and Alkaline Dual Stimuli-Responsive Mechanized Hollow Mesoporous Silica Nanoparticles as Smart Nanocontainers for Intelligent Anticorrosion Coatings." *ACS Nano* 7 (12): 11397–11408. 10.1021/nn4053233.

Gaidis, James M. 2004. "Chemistry of Corrosion Inhibitors." *Cement and Concrete Composites* 26 (3): 181–189. 10.1016/S0958-9465(03)00037-4.

Gao, Lei, Jinliang He, Jun Hu, and Chao Wang. 2015. "Photoresponsive Self-Healing Polymer Composite with Photoabsorbing Hybrid Microcapsules." *ACS Applied Materials and Interfaces* 7 (45): 25546–25552. 10.1021/acsami.5b09121.

Ghazi, A., E. Ghasemi, M. Mahdavian, B. Ramezanzadeh, and M. Rostami. 2015. "The Application of Benzimidazole and Zinc Cations Intercalated Sodium Montmorillonite as Smart Ion Exchange Inhibiting Pigments in the Epoxy Ester Coating." *Corrosion Science* 94 (3): 207–217. 10.1016/j.corsci.2015.02.007.

González-García, Y., J. M. C. Mol, T. Muselle, I. De Graeve, G. Van Assche, G. Scheltjens, B. Van Mele, and H. Terryn. 2011a. "A Combined Mechanical, Microscopic and Local Electrochemical Evaluation of Self-Healing Properties of Shape-Memory Polyurethane Coatings." *Electrochimica Acta* 56 (26): 9619–9626. 10.1016/j.electacta.2011.03.081.

González-García, Y., J. M. C. Mol, T. Muselle, I. De Graeve, G. Van Assche, G. Scheltjens, B. Van Mele, and H. Terryn. 2011b. "SECM Study of Defect Repair in Self-Healing Polymer Coatings on Metals." *Electrochemistry Communications* 13 (2): 169–173. 10.1016/j.elecom.2010.12.005.

Habault, Damien, Hongji Zhang, and Yue Zhao. 2013. "Light-Triggered Self-Healing and Shape-Memory Polymers." *Chemical Society Reviews* 42 (17): 7244–7256. 10.1039/c3cs35489j.

Habibiyan, Aylin, Bahram Ramezanzadeh, Mohammad Mahdavian, Ghasem Bahlakeh, and Mojtaba Kasaeian. 2020. "Rational Assembly of Mussel-Inspired Polydopamine (PDA)-Zn (II) Complex Nanospheres on Graphene Oxide Framework Tailored for Robust Self-Healing Anti-Corrosion Coatings Application." *Chemical Engineering Journal* 391 (Ii): 123630. 10.1016/j.cej.2019.123630.

Haddadi, S. A., S. A. A. Ramazani, M. Mahdavian, P. Taheri, and J. M. C. Mol. 2018. "Fabrication and Characterization of Graphene-Based Carbon Hollow Spheres for Encapsulation of Organic Corrosion Inhibitors." *Chemical Engineering Journal* 352: 909–922. 10.1016/j.cej.2018.06.063.

Haddadi, Seyyed Arash, Eiman Alibakhshi, Ghasem Bahlakeh, Bahram Ramezanzadeh, and Mohammad Mahdavian. 2019. "A Detailed Atomic Level Computational and Electrochemical Exploration of the Juglans Regia Green Fruit Shell Extract as a Sustainable and Highly Efficient Green Corrosion Inhibitor for Mild Steel in 3.5 Wt% NaCl Solution." *Journal of Molecular Liquids* 284: 682–699. 10.1016/j.molliq.2019.04.045.

Haddadi, Seyyed Arash, Saeed Ghaderi, Mohsen Sadeghi, Behzad Gorji, Farhad Ahmadijokani, Ahmad Ramazani S. A., Mohammad Mahdavian, and Mohammad Arjmand. 2021. "Enhanced Active/Barrier Corrosion Protective Properties of Epoxy Coatings Containing Eco-Friendly Green Inorganic/Organic Hybrid Pigments Based on Zinc Cations/Ferula Asafoetida Leaves." *Journal of Molecular Liquids* 323 (February): 114584. 10.1016/j.molliq.2020.114584.

He, Xiaodong, and Xianming Shi. 2009. "Self-Repairing Coating for Corrosion Protection of Aluminum Alloys." *Progress in Organic Coatings* 65 (1): 37–43. 10.1016/j.porgcoat.2008.09.003.

Hu, Haihua, Haopeng Liu, Dianjun Zhang, Jianjun Wang, Gaowu Qin, and Xuefeng Zhang. 2018. "PH and Electromagnetic Dual-Remoted Drug Delivery Based on Bimodal Superparamagnetic Fe_3O_4@Porous Silica Nanoparticles." *Engineered Science* 2: 43–48. 10.30919/es8d136.

Hu, Ting, Yuejun Ouyang, Zhi Hui Xie, and Liang Wu. 2021. "One-Pot Scalable in Situ Growth of Highly Corrosion-Resistant MgAl-LDH/MBT Composite Coating on Magnesium Alloy under Mild Conditions." *Journal of Materials Science and Technology* 92: 225–235. 10.1016/j.jmst.2021.03.021.

Jorcin, Jean Baptiste, Gill Scheltjens, Yves Van Ingelgem, Els Tourwé, Guy Van Assche, Iris De Graeve, Bruno Van Mele, Herman Terryn, and Annick Hubin. 2010. "Investigation of the Self-Healing Properties of Shape Memory Polyurethane Coatings with the 'odd Random Phase Multisine'

Electrochemical Impedance Spectroscopy." *Electrochimica Acta* 55 (21): 6195–6203. 10.1016/j.electacta.2010.01.027.

Kaghazchi, Laleh, Reza Naderi, and Bahram Ramezanzadeh. 2021. "Synergistic Mild Steel Corrosion Mitigation in Sodium Chloride-Containing Solution Utilizing Various Mixtures of Phytic Acid Molecules and Zn2+ Ions." *Journal of Molecular Liquids* 323: 114589. 10.1016/j.molliq.2020.114589.

Kaur, Ranjeet, Abshar Hasan, Nusrat Iqbal, Samsul Alam, Mahesh Kr Saini, and Syed Kalbe Raza. 2014. "Synthesis and Surface Engineering of Magnetic Nanoparticles for Environmental Cleanup and Pesticide Residue Analysis: A Review." *Journal of Separation Science* 37 (14): 1805–1825. 10.1002/jssc.201400256.

Kazemi, Fatemeh, and Seyed Morteza Naghib. 2020. "Smart Controlled Release of Corrosion Inhibitor from Normal and Stimuli-Responsive Micro/Nanocarriers." In *Corrosion Protection at the Nanoscale*, 161–179. Elsevier. 10.1016/B978-0-12-819359-4.00010-6.

Keller, Michael W., Scott R. White, and Nancy R. Sottos. 2007. "A Self-Healing Poly(Dimethyl Siloxane) Elastomer." *Advanced Functional Materials* 17 (14): 2399–2404. 10.1002/adfm.200700086.

Kowalski, Damian, Mikito Ueda, and Toshiaki Ohtsuka. 2010. "Self-Healing Ion-Permselective Conducting Polymer Coating †" *Journal of Materials Chemistry* 3: 7630–7633. 10.1039/c0jm00866d.

Kulbaba, Kevin, Alison Cheng, Alexandra Bartole, Sharonna Greenberg, Rui Resendes, Neil Coombs, Athena Safa-Sefat, et al. 2002. "Polyferrocenylsilane Microspheres: Synthesis, Mechanism of Formation, Size and Charge Tunability, Electrostatic Self-Assembly, and Pyrolysis to Spherical Magnetic Ceramic Particles." *Journal of the American Chemical Society* 124 (42): 12522–12534. 10.1021/ja0202053.

Kulčar, Rahela, Mojca Friškovec, Nina Hauptman, Alenka Vesel, and Marta Klanjšek Gunde. 2010. "Colorimetric Properties of Reversible Thermochromic Printing Inks." *Dyes and Pigments* 86 (3): 271–277. 10.1016/j.dyepig.2010.01.014.

Kumar, Ashok, L. D. Stephenson, and J. N. Murray. 2006. "Self-Healing Coatings for Steel." *Progress in Organic Coatings* 55 (3): 244–253. 10.1016/j.porgcoat.2005.11.010.

Kuroki, Hidenori, Ihor Tokarev, Dmytro Nykypanchuk, Ekaterina Zhulina, and Sergiy Minko. 2013. "Stimuli-Responsive Materials with Self-Healing Antifouling Surface via 3D Polymer Grafting." *Advanced Functional Materials* 23 (36): 4593–4600. 10.1002/adfm.201300363.

Leal, Débora Abrantes, Izabel Cristina Riegel-Vidotti, Mário Guerreiro Silva Ferreira, and Cláudia Eliana Bruno Marino. 2018. "Smart Coating Based on Double Stimuli-Responsive Microcapsules Containing Linseed Oil and Benzotriazole for Active Corrosion Protection." *Corrosion Science* 130 (April): 56–63. 10.1016/j.corsci.2017.10.009.

Lee, Hyoryong, Hyunchul Choi, Mikyoung Lee, and Sukho Park. 2018. "Preliminary Study on Alginate/NIPAM Hydrogel-Based Soft Microrobot for Controlled Drug Delivery Using Electromagnetic Actuation and near-Infrared Stimulus." *Biomedical Microdevices* 20 (4). 10.1007/s10544-018-0344-y.

Lee, Sue Hyun, Mukesh K. Gupta, Jae Beum Bang, Hojae Bae, and Hak Joon Sung. 2013. "Current Progress in Reactive Oxygen Species (ROS)-Responsive Materials for Biomedical Applications." *Advanced Healthcare Materials* 2 (6): 908–915. 10.1002/adhm.201200423.

Li, Guo Liang, Zhaoliang Zheng, Helmuth Möhwald, and Dmitry G. Shchukin. 2013. "Silica/Polymer Double-Walled Hybrid Nanotubes: Synthesis and Application as Stimuli-Responsive Nanocontainers in Self-Healing Coatings." *ACS Nano* 7 (3): 2470–2478. 10.1021/nn305814q.

Li, Guoqiang, and Damon Nettles. 2010. "Thermomechanical Characterization of a Shape Memory Polymer Based Self-Repairing Syntactic Foam." *Polymer* 51 (3): 755–762. 10.1016/j.polymer.2009.12.002.

Li, Li, Chaofang Dong, Liqin Liu, Jiankuan Li, Kui Xiao, Dawei Zhang, and Xiaogang Li. 2014. "Preparation and Characterization of PH-Controlled-Release Intelligent Corrosion Inhibitor." *Materials Letters* 116: 318–321. 10.1016/j.matlet.2013.11.003.

Li, Lin Xin, Zhi Hui Xie, Carlos Fernandez, Liang Wu, Daojian Cheng, Xiao Hui Jiang, and Chuan Jian Zhong. 2020. "Development of a Thiophene Derivative Modified LDH Coating for Mg Alloy Corrosion Protection." *Electrochimica Acta* 330: 135186. 10.1016/j.electacta.2019.135186.

Li, Zhen, Benye Qin, Xiaoyong Zhang, Ke Wang, Yen Wei, and Yan Ji. 2015. "Self-Healing Anti-Corrosion Coatings Based on Polymers of Intrinsic Microporosity for the Protection of Aluminum Alloy." *RSC Advances* 5 (126): 104451–104457. 10.1039/c5ra19614k.

Liu, Jian, Shi Zhang Qiao, Qiu Hong Hu, and Gao Qing Lu. 2011. "Magnetic Nanocomposites with Mesoporous Structures: Synthesis and Applications." *Small* 7 (4): 425–443. 10.1002/smll.201001402.

Luo, Xiaofan, and Patrick T. Mather. 2010. "Shape Memory Assisted Self-Healing (SMASH) Coating." *Macromolecules* 6 (19): 2146. 10.1021/mz400017x%5Cnhttp://pubs.acs.org/doi/full/10.1021/mz400017x%5Cnhttp://pubs.acs.org/doi/pdfplus/10.1021/mz400017x%5Cnhttp://xlink.rsc.org/?DOI=c001295e%5Cnhttp://pubs.acs.org/doi/abs/10.1021/ma9015888.

Lutz, Alexander, Otto Van Den Berg, Jonas Van Damme, Karen Verheyen, Erwin Bauters, Iris De Graeve, Filip E. Du Prez, and Herman Terryn. 2015. "A Shape-Recovery Polymer Coating for the Corrosion Protection of Metallic Surfaces." *ACS Applied Materials and Interfaces* 7 (1): 175–183. 10.1021/am505621x.

Lv, Li-Ping, Katharina Landfester, and Daniel Crespy. 2014. "Stimuli-Selective Delivery of Two Payloads from Dual Responsive Nanocontainers." *Chemistry of Materials* 26 (11): 3351–3353. 10.1021/cm500923d.

Mahdavian, M., and M. M. Attar. 2009. "The Effect of Benzimidazole and Zinc Acetylacetonate Mixture on Cathodic Disbonding of Epoxy Coated Mild Steel." *Progress in Organic Coatings* 66 (2): 137–140. 10.1016/j.porgcoat.2009.06.013.

Mahmoudian, Mehdi, Ehsan Nozad, Mahmoud Ghasemi Kochameshki, and Mojtaba Enayati. 2018. "Preparation and Investigation of Hybrid Self-Healing Coatings Containing Linseed Oil Loaded Nanocapsules, Potassium Ethyl Xanthate and Benzotriazole on Copper Surface." *Progress in Organic Coatings* 120 (February): 167–178. 10.1016/j.porgcoat.2018.03.014.

McMurray, H. N., D. Williams, G. Williams, and D. Worsley. 2003. "Inhibitor Pretreatment Synergies Demonstrated Using a Scanning Kelvin Probe Technique." *Corrosion Engineering Science and Technology* 38 (2): 112–118. 10.1179/147842203767789186.

Meng, Yuan, Jeh-Chang Yang, Christopher L. Lewis, Jisu Jiang, and Mitchell Anthamatten. 2016. "Photoinscription of Chain Anisotropy into Polymer Networks." *Macromolecules* 49 (23): 9100–9107. 10.1021/acs.macromol.6b01990.

Mercato, Loretta Laureana Del, Marzia Maria Ferraro, Francesca Baldassarre, Serena Mancarella, Valentina Greco, Ross Rinaldi, and Stefano Leporatti. 2014. "Biological Applications of LbL Multilayer Capsules: From Drug Delivery to Sensing." *Advances in Colloid and Interface Science*. Elsevier B.V. 10.1016/j.cis.2014.02.014.

Michal, Brian T., Colin A. Jaye, Emily J. Spencer, and Stuart J. Rowan. 2013. "Inherently Photohealable and Thermal Shape-Memory Polydisulfide Networks." *ACS Macro Letters* 2 (8): 694–699. 10.1021/mz400318m.

Mohammadi, Iman, Taghi Shahrabi, Mohammad Mahdavian, and Mazdak Izadi. 2021a. "Chemical Modification of LDH Conversion Coating with Diethyldithiocarbamate as a Novel Anti-Corrosive Film for AA2024-T3." *Journal of Industrial and Engineering Chemistry* 95: 134–147. 10.1016/j.jiec.2020.12.016.

Mohammadi, Iman, Taghi Shahrabi, Mohammad Mahdavian, and Mazdak Izadi. 2021b. "Zn-Al Layered Double Hydroxide as an Inhibitive Conversion Coating Developed on AA2024-T3 by One-Step Hydrothermal Crystallization: Crystal Structure Evolution and Corrosion Protection Performance." *Surface and Coatings Technology* 409 (August 2020): 126882. 10.1016/j.surfcoat.2021.126882.

Montemor, M. F., D. V. Snihirova, M. G. Taryba, S. V. Lamaka, I. A. Kartsonakis, A. C. Balaskas, G. C. Kordas, et al. 2012. "Evaluation of Self-Healing Ability in Protective Coatings Modified with Combinations of Layered Double Hydroxides and Cerium Molibdate Nanocontainers Filled with Corrosion Inhibitors." *Electrochimica Acta* 60: 31–40. 10.1016/j.electacta.2011.10.078.

Nejman, Alicja, Małgorzata Cieślak, Bogumił Gajdzicki, Bogna Goetzendorf-Grabowska, and Agnieszka Karaszewska. 2014. "Methods of PCM Microcapsules Application and the Thermal Properties of Modified Knitted Fabric." *Thermochimica Acta* 589: 158–163. 10.1016/j.tca.2014.05.037.

Nji, Jones, and Guoqiang Li. 2012. "Damage Healing Ability of a Shape-Memory-Polymer-Based Particulate Composite with Small Thermoplastic Contents." *Smart Materials and Structures* 21 (2). 10.1088/0964-1726/21/2/025011.

Park, Jeong Ho, and Paul V. Braun. 2010. "Coaxial Electrospinning of Self-Healing Coatings." *Advanced Materials* 22 (4): 496–499. 10.1002/adma.200902465.

Patel, Kaushik, Sarah Angelos, William R. Dichtel, Ali Coskun, Ying Wei Yang, Jeffrey I. Zink, and J. Fraser Stoddart. 2008. "Enzyme-Responsive Snap-Top Covered Silica Nanocontainers." *Journal of the American Chemical Society* 130 (8): 2382–2383. 10.1021/ja0772086.

Pirhady Tavandashti, N., M. Ghorbani, A. Shojaei, Y. Gonzalez-Garcia, H. Terryn, and J. M. C. Mol. 2016. "PH Responsive Ce(III) Loaded Polyaniline Nanofibers for Self-Healing Corrosion Protection of AA2024-T3." *Progress in Organic Coatings* 99: 197–209. 10.1016/j.porgcoat.2016.04.046.

Pirhady Tavandashti, N., M. Ghorbani, A. Shojaei, J. M. C. Mol, H. Terryn, K. Baert, and Y. Gonzalez-Garcia. 2016. "Inhibitor-Loaded Conducting Polymer Capsules for Active Corrosion Protection of Coating Defects." *Corrosion Science* 112: 138–149. 10.1016/j.corsci.2016.07.003.

Qian, Bei, Marios Michailidis, Matt Bilton, Theo Hobson, Zhaoliang Zheng, and Dmitry Shchukin. 2019. "Tannic Complexes Coated Nanocontainers for Controlled Release of Corrosion Inhibitors in Self-Healing Coatings." *Electrochimica Acta* 297: 1035–1041. 10.1016/j.electacta.2018.12.062.

Qian, Bei, Zhaoliang Zheng, Marios Michailids, Nicole Fleck, Matthew Bilton, Yan Song, Guoliang Li, and Dmitry Shchukin. 2019. "Mussel-Inspired Self-Healing Coatings Based on Polydopamine-Coated Nanocontainers for Corrosion Protection." *ACS Applied Materials and Interfaces* 11 (10): 10283–10291. 10.1021/acsami.8b21197.

Qian, Hongchang, Dake Xu, Cuiwei Du, Dawei Zhang, Xiaogang Li, Luyao Huang, Leping Deng, Yunchao Tu, Johannes M. C. Mol, and Herman A. Terryn. 2017. "Dual-Action Smart Coatings with a Self-Healing Superhydrophobic Surface and Anti-Corrosion Properties." *Journal of Materials Chemistry A* 5 (5): 2355–2364. 10.1039/C6TA10903A.

Saravanakumar, Gurusamy, Jihoon Kim, and Won Jong Kim. 2017. "Reactive-Oxygen-Species-Responsive Drug Delivery Systems: Promises and Challenges." *Advanced Science* 4 (1). 10.1002/advs.201600124.

Sciurti, E., R. Primavera, D. Di Mascolo, A. Rizzo, A. Balena, S. K. Padmanabhan, F. Rizzi, P. Decuzzi, and M. De Vittorio. 2020. "Ultrasound-Induced Deformation of PLGA-MicroPlates for on-Command Drug Release." *Microelectronic Engineering* 229 (November 2019): 111360. 10.1016/j.mee.2020.111360.

Shchukin, Dmitry G., Mikhail Zheludkevich, and Helmuth Möhwald. 2006. "Feedback Active Coatings Based on Incorporated Nanocontainers." *Journal of Materials Chemistry* 16 (47): 4561–4566. 10.1039/b612547f.

Shchukin, Dmitry G., Mikhail Zheludkevich, Kiryl Yasakau, Sviatlana Lamaka, Mario G. S. Ferreira, and Helmuth Möhwald. 2006. "Layer-by-Layer Assembled Nanocontainers for Self-Healing Corrosion Protection." *Advanced Materials* 18 (13): 1672–1678. 10.1002/adma.200502053.

Shchukin, Dmitry G., Ekaterina V. Skorb, Dmitry V. Sviridov, and Helmuth Mo. 2009. "Light Responsive Protective Coatings W," 6041–6043. 10.1039/b914257f.

Shkirskiy, V., P. Keil, H. Hintze-Bruening, F. Leroux, P. Vialat, G. Lefèvre, K. Ogle, and P. Volovitch. 2015. "Factors Affecting MoO42- Inhibitor Release from Zn2Al Based Layered Double Hydroxide and Their Implication in Protecting Hot Dip Galvanized Steel by Means of Organic Coatings." *ACS Applied Materials and Interfaces* 7 (45): 25180–25192. 10.1021/acsami.5b06702.

Singh, H. K., K. V. Yeole, and S. T. Mhaske. 2016. "Synthesis and Characterization of Layer-by-Layer Assembled Magnesium Zinc Molybdate Nanocontainer for Anticorrosive Application." *Chemical Engineering Journal* 295: 414–426. 10.1016/j.cej.2016.03.055.

Skorb, Ekaterina V., Dmitri Fix, Daria V. Andreeva, Helmuth Möhwald, and Dmitry G. Shchukin. 2009. "Surface-Modified Mesoporous SiO2 Containers for Corrosion Protection." *Advanced Functional Materials* 19 (15): 2373–2379. 10.1002/adfm.200801804.

Song, Young-Kyu, Ye-Hyun Jo, Ye-Ji Lim, Sung-Youl Cho, Hwan-Chul Yu, Byung-Cheol Ryu, Sang-In Lee, and Chan-Moon Chung. 2013. "Sunlight-Induced Self-Healing of a Microcapsule-Type Protective Coating." *ACS Applied Materials & Interfaces* 5 (4): 1378–1384. 10.1021/am302728m.

Stimpfling, Thomas, Pierre Vialat, Horst Hintze-Bruening, Patrick Keil, Viacheslav Shkirskiy, Polina Volovitch, Kevin Ogle, and Fabrice Leroux. 2016. "Amino Acid Interleaved Layered Double Hydroxides as Promising Hybrid Materials for AA2024 Corrosion Inhibition." *European Journal of Inorganic Chemistry* 2016 (13–14): 2006–2016. 10.1002/ejic.201501161.

Sun, Yu Long, Yue Zhou, Qing Lan Li, and Ying Wei Yang. 2013. "Enzyme-Responsive Supramolecular Nanovalves Crafted by Mesoporous Silica Nanoparticles and Choline-Sulfonatocalix[4]Arene [2]Pseudorotaxanes for Controlled Cargo Release." *Chemical Communications* 49 (79): 9033–9035. 10.1039/c3cc45216f.

Szabó, Tamás, Judit Telegdi, and Lajos Nyikos. 2015. "Linseed Oil-Filled Microcapsules Containing Drier and Corrosion Inhibitor - Their Effects on Self-Healing Capability of Paints." *Progress in Organic Coatings* 84: 136–142. 10.1016/j.porgcoat.2015.02.020.

Tedim, J., A. Kuznetsova, A. N. Salak, F. Montemor, D. Snihirova, M. Pilz, M. L. Zheludkevich, and M. G. S. Ferreira. 2012. "Zn-Al Layered Double Hydroxides as Chloride Nanotraps in Active Protective Coatings." *Corrosion Science* 55: 1–4. 10.1016/j.corsci.2011.10.003.

Tedim, J., S. K. Poznyak, A. Kuznetsova, D. Raps, T. Hack, M. L. Zheludkevich, and M. G. S. Ferreira. 2010. "Enhancement of Active Corrosion Protection via Combination of Inhibitor-Loaded Nanocontainers." *ACS Applied Materials and Interfaces* 2 (5): 1528–1535. 10.1021/am100174t.

Telitel, Siham, Yoshifumi Amamoto, Julien Poly, Fabrice Morlet-Savary, Olivier Soppera, Jacques Lalevée, and Krzysztof Matyjaszewski. 2014. "Introduction of Self-Healing Properties into Covalent Polymer Networks via the Photodissociation of Alkoxyamine Junctions." *Polymer Chemistry* 5 (3): 921–930. 10.1039/c3py01162c.

Toorani, M., M. Aliofkhazraei, M. Mahdavian, and R. Naderi. 2021. "Superior Corrosion Protection and Adhesion Strength of Epoxy Coating Applied on AZ31 Magnesium Alloy Pre-Treated by PEO/Silane with Inorganic and Organic Corrosion Inhibitors." *Corrosion Science* 178 (January): 109065. 10.1016/j.corsci.2020.109065.

Tran, The Hai, Ashokanand Vimalanandan, Georgi Genchev, Johannes Fickert, Katharina Landfester, Daniel Crespy, and Michael Rohwerder. 2015. "Regenerative Nano-Hybrid Coating Tailored for Autonomous Corrosion Protection." *Advanced Materials* 27 (25): 3825–3830. 10.1002/adma.201501044.

Vega, J. M., N. Granizo, D. De La Fuente, J. Simancas, and M. Morcillo. 2011. "Corrosion Inhibition of Aluminum by Coatings Formulated with Al-Zn-Vanadate Hydrotalcite." *Progress in Organic Coatings* 70 (4): 213–219. 10.1016/j.porgcoat.2010.08.014.

Vimalanandan, Ashokanand, Li Ping Lv, The Hai Tran, Katharina Landfester, Daniel Crespy, and Michael Rohwerder. 2013. "Redox-Responsive Self-Healing for Corrosion Protection." *Advanced Materials* 25 (48): 6980–6984. 10.1002/adma.201302989.

Wang, Luntao, Leping Deng, Dawei Zhang, Hongchang Qian, Cuiwei Du, Xiaogang Li, Johannes M. C. Mol, and Herman A. Terryn. 2016. "Shape Memory Composite (SMC) Self-Healing Coatings for Corrosion Protection." *Progress in Organic Coatings* 97: 261–268. 10.1016/j.porgcoat.2016.04.041.

Wang, Mingdong, Guangcai Gong, Jing Feng, Ting Wang, Chendi Ding, Baojing Zhou, Wei Jiang, and Jiajun Fu. 2016. "Dual PH-Mediated Mechanized Hollow Zirconia Nanospheres." *ACS Applied Materials and Interfaces* 8 (35): 23289–23301. 10.1021/acsami.6b07603.

Wang, Mingdong, Mengyang Liu, and Jiajun Fu. 2015. "An Intelligent Anticorrosion Coating Based on PH-Responsive Smart Nanocontainers Fabricated via a Facile Method for Protection of Carbon Steel." *Journal of Materials Chemistry A* 3 (12): 6423–6431. 10.1039/c5ta00417a.

Wang, Ting, Juan Du, Sheng Ye, Linghua Tan, and Jiajun Fu. 2019. "Triple-Stimuli-Responsive Smart Nanocontainers Enhanced Self-Healing Anticorrosion Coatings for Protection of Aluminum Alloy." *ACS Applied Materials and Interfaces* 11 (4): 4425–4438. 10.1021/acsami.8b19950.

Wang, Ting, Linghua Tan, Chendi Ding, Mingdong Wang, Jianhua Xu, and Jiajun Fu. 2017. "Redox-Triggered Controlled Release Systems-Based Bi-Layered Nanocomposite Coating with Synergistic Self-Healing Property." *Journal of Materials Chemistry A* 5 (4): 1756–1768. 10.1039/C6TA08547D.

Wang, Wei, Weihua Li, Weijie Fan, Xiaoying Zhang, Liying Song, Chuansheng Xiong, Xiang Gao, and Xiaojie Liu. 2018. "Accelerated Self-Healing Performance of Magnetic Gradient Coating." *Chemical Engineering Journal* 332 (January): 658–670. 10.1016/j.cej.2017.09.112.

Wang, Xian, Linxin Li, Zhi Hui Xie, and Gang Yu. 2018. "Duplex Coating Combining Layered Double Hydroxide and 8-Quinolinol Layers on Mg Alloy for Corrosion Protection." *Electrochimica Acta* 283: 1845–1857. 10.1016/j.electacta.2018.07.113.

Wei, Jie, Xiao Jie Ju, Xiao Yi Zou, Rui Xie, Wei Wang, Ying Mei Liu, and Liang Yin Chu. 2014. "Multi-Stimuli-Responsive Microcapsules for Adjustable Controlled-Release." *Advanced Functional Materials* 24 (22): 3312–3323. 10.1002/adfm.201303844.

Wei, Ting, Zengchao Tang, Qian Yu, and Hong Chen. 2017. "Smart Antibacterial Surfaces with Switchable Bacteria-Killing and Bacteria-Releasing Capabilities." *ACS Applied Materials and Interfaces* 9 (43): 37511–37523. 10.1021/acsami.7b13565.

White, S. R., N. R. Sottos, P. H. Geubelle, J. S. Moore, M. R. Kessler, S. R. Sriram, E. N. Brown, and S. Viswanathan. 2002. "Erratum: Autonomic Healing of Polymer Composites (Nature (2001) 409 (794–797))." *Nature* 415 (6873): 817. 10.1038/415817a.

Xie, Tao. 2011. "Recent Advances in Polymer Shape Memory." *Polymer* 52 (22): 4985–5000. 10.1016/j.polymer.2011.08.003.

Yeganeh, Mahdi, M. Omidi, S. H. H. Mortazavi, A. Etemad, M. H. Nazari, and S. M. Marashi. 2020. "Application of Mesoporous Silica as the Nanocontainer of Corrosion Inhibitor." In *Corrosion Protection at the Nanoscale*, 275–294. Elsevier. 10.1016/B978-0-12-819359-4.00015-5.

Ying, Hanze, Yanfeng Zhang, and Jianjun Cheng. 2014. "Dynamic Urea Bond for the Design of Reversible and Self-Healing Polymers." *Nature Communications* 5: 1–9. 10.1038/ncomms4218.

Zeng, Xiaowei, Wei Tao, Gan Liu, and Lin Mei. 2017. "Polydopamine-Based Surface Modification of Copolymeric Nanoparticles as a Targeted Drug Delivery System for Cancer Therapy." *Journal of Controlled Release* 259 (2017): e150–e151. 10.1016/j.jconrel.2017.03.303.

Zhao, Yi, Rüdiger Berger, Katharina Landfester, and Daniel Crespy. 2015. "Double Redox-Responsive Release of Encoded and Encapsulated Molecules from Patchy Nanocapsules." *Small* 11 (25): 2995–2999. 10.1002/smll.201402521.

Zhao, Yu Hui, Daniela Vuluga, Laurence Lecamp, and Fabrice Burel. 2016. "Photoinitiated Thiol-Epoxy Addition for the Preparation of Photoinduced Self-Healing Fatty Coatings." *RSC Advances* 6 (38): 32098–32105. 10.1039/c6ra03693g.

Zheludkevich, M. L., S. K. Poznyak, L. M. Rodrigues, D. Raps, T. Hack, L. F. Dick, T. Nunes, and M. G. S. Ferreira. 2010. “Active Protection Coatings with Layered Double Hydroxide Nanocontainers of Corrosion Inhibitor.” *Corrosion Science* 52 (2): 602–611. 10.1016/j.corsci.2009.10.020.

Zheludkevich, M. L., R. Serra, M. F. Montemor, I. M. Miranda Salvado, and M. G. S. Ferreira. 2006. “Corrosion Protective Properties of Nanostructured Sol-Gel Hybrid Coatings to AA2024-T3.” *Surface and Coatings Technology* 200 (9): 3084–3094. 10.1016/j.surfcoat.2004.09.007.

Zheludkevich, M. L., R. Serra, M. F. Montemor, K. A. Yasakau, I. M..Miranda Salvado, and M. G. S. Ferreira. 2005. “Nanostructured Sol-Gel Coatings Doped with Cerium Nitrate as Pre-Treatments for AA2024-T3 Corrosion Protection Performance.” *Electrochimica Acta* 51 (2): 208–217. 10.1016/j.electacta.2005.04.021.

Zheludkevich, Mikhail L., Dmitry G. Shchukin, Kiryl A. Yasakau, Helmut Möhwald, and Mario G. S. Ferreira. 2007. “Anticorrosion Coatings with Self-Healing Effect Based on Nanocontainers Impregnated with Corrosion Inhibitor.” *Chemistry of Materials* 19 (3): 402–411. 10.1021/cm062066k.

Zhou, Chaopei, Xiangyang Xie, Hong Yang, Shasha Zhang, Yinke Li, Changchun Kuang, Shiyao Fu, et al. 2019. “Novel Class of Ultrasound-Triggerable Drug Delivery Systems for the Improved Treatment of Tumors.” Research-article. *Molecular Pharmaceutics* 16 (7): 2956–2965. 10.1021/acs.molpharmaceut.9b00194.

10 Carbon Allotropes in Corrosion Prevention
Computational-Based Advancements

Madhulata Shukla
Department of Chemistry, G.B. College, Ramgarh, Kaimur, Veer Kunwar Singh University, Bihar, India

Sumit Kumar
PG Department of Chemistry, Magadh University, Bodh Gaya, Bihar, India

CONTENTS

10.1 INTRODUCTION

Corrosion is a central problem for industry and the environment in terms of austerity and budget [1–3]. Corrosion is a hazardous phenomenon with a devastating effect on technological and industrial applications, particularly in the oil and gas industries [1,3,4]. It is a huge concern for most industrialised countries. Natural gas and crude oil usually contain several highly contaminated products, which are naturally corrosive [5]. Around the world, oil companies are spending a huge proportion of money on tackling corrosion [5,6]. Therefore, controlling the corrosion of metals is an important action to be taken for environmental, technical, and economical aspects to save huge expenses on materials and equipment [3,4]. The proposal of advanced composites as effective anticorrosive materials to avoid corrosion of metals has emerged as a demanding need.

Several industrial processes regarding the efficient execution of highly acidic solutions comprise the economic burden due to corrosion [3,4]. Some of the frequently used industrial methods are known as acid cleaning as well as pickling and oil-well acidification [4,5]. Acidic solutions having high concentration are employed in these processes, to reduce the surface impurities including rusts and scales [7–9]. However, a solution with high concentrations can also cause a massive degradation of metallic components due to corrosion [10–12]. Therefore, some external

DOI: 10.1201/9781003323976-12

additives, mostly chemical species, are being added to the cleaning solutions. These additives are inhibitors for the corrosion process [1,7]. Corrosion inhibitor formulation and anticorrosive coatings control the metallic degradation. This idea has drawn extreme interest in academia as well as in industries [13–15]. Interestingly, not only corrosion inhibitors but also several other procedures such as painting, galvanisation, as well as coating, etc. are also used to moderate the corrosion-related losses [3,4,13]. Organic compounds are also reported as potential inhibitors for metallic corrosion [7,16,17].

Science facilitates the development of a new generation of effective materials for application as corrosion inhibitors. Among others, carbon nanoallotropes are considered one of the most promising materials in the field of corrosion protection due to their exceptional characteristics such as having a large surface area, having excellent mechanical properties, being lightweight, and having easy synthesis. Carbon allotrope-based materials can also be used both as coatings and inhibitors. In the case of coating, it is directly applied over metal surfaces, whereas inhibitors are added to the corrosive medium followed by adsorption on the metal surface. Studies depict that functionalised carbon allotropes stick to the metallic surface and reduce the interaction pursued by aggressive solutions with substrates. Figure 10.1 shows the key characteristics that inhibitor molecules should possess.

This chapter highlights the use of the different categories of carbon nanoallotropes for corrosion protection of metals in the last few decades. These carbon nanoallotropes are graphene, carbon nanotubes (CNTs), and carbon dots (CDs). It also demonstrates their various modes of action/activities as anticorrosive coatings or inhibitors in several corrosive environments [9,18–27]. Carbon allotropes have also been extensively used in many sensing applications for targets such as temperature sensing, pressure detection, magnetics elucidation and measurement of environmental pollutants, and biomolecules [28–31].

Recently, carbon materials, including activated carbon, fullerenes, CNTs, graphene, and other carbon allotropes, have been widely investigated as cocatalysts for semiconductors in photocatalysis [32–35]. Advanced carbon nanomaterials, such as fullerene, graphene, and CNTs are considered the backbone of engineering and scientific research, as advanced carbon materials have

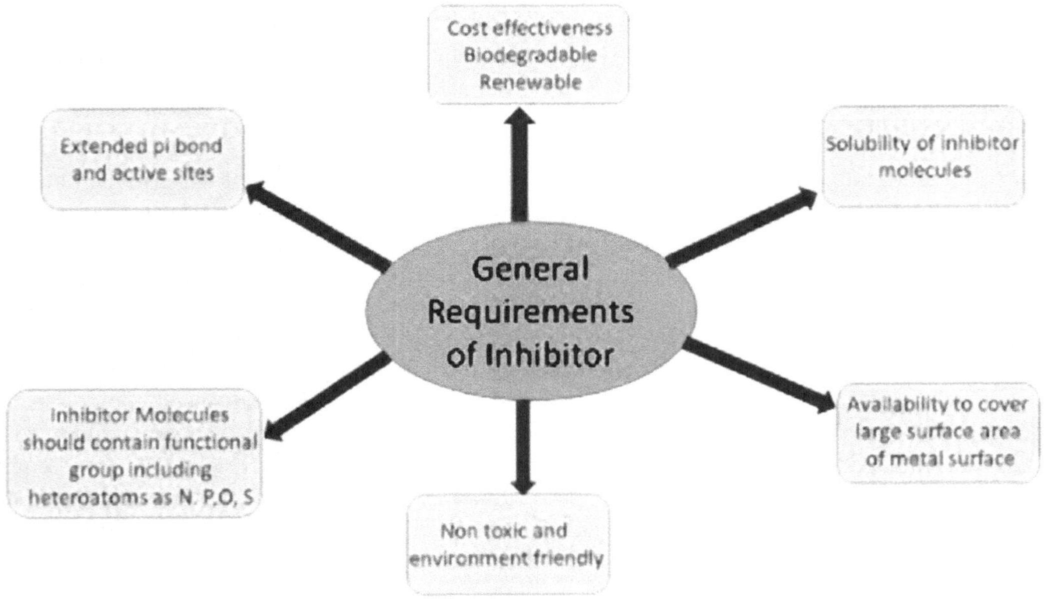

FIGURE 10.1 Key characteristics of inhibitor molecules.

a controllable porous structure, high specific surface area, high electric conductivity, high thermal and chemical stability, high anticorrosion property, as well as compatibility with the composite materials, making them an ideal substitute for noble metal cocatalysts [36–40]. Hence, carbon materials can be used in energy storage as electrocatalysts, intercalation hosts, and ideal substrates for active materials. Sustainable carbon materials are formed using various physical as well as chemical methods. The fabrication methods are used to obtain advanced carbon monoliths with desirable properties. Keeping these overall applications in mind, several monoliths along with allotropes of advanced carbon nanomaterials are used in renewable energy areas, organic photovoltaic technology, and energy conservation procedures. Carbon nanomaterials show a magnificent ability to be used in enhancing biology as well as drug delivery. Furthermore, graphene, CNT, and fullerene are recognised as not only fast but also effective nanocarriers in the case of therapeutic delivery. The chapter conceives the recent demands of advanced carbon materials in the field of technology and innovation as well as scientific research. This chapter focuses on the recent theoretical investigations in light of experimental findings of advanced carbon allotropes as efficient corrosion inhibitors and coatings for the protection of metals/alloys.

10.2 THEORETICAL CALCULATIONS

The experimental process shows the adsorption properties of inhibitors, and sometimes it is challenging to analyse which part of the inhibitor molecule is responsible for the interaction on metal surfaces. Molecular modelling and dynamics of the metal surface play a pivotal role in understanding the mechanism involved during corrosion. Quantum mechanical simulations were performed to get a deeper understanding of the corrosion inhibition mechanism at the molecular and atomic levels. In the last few years, quantum-based density function theory (DFT) calculation has been carried out extensively in this field [41–45]. Similarly, molecular simulation-based analyses, such as molecular dynamic (MD) simulations and Monte Carle (MC) simulations, have also been applied to a greater extent. With the help of DFT, the frontier molecular orbitals (FMOs), the highest occupied molecular orbital (HOMO) and lowest unoccupied molecular orbital (LUMO) can be calculated. Other important global parameters, such as electronegativity, dipole moment, global hardness, global softness, and energy gap can also be calculated using these quantum calculations [16,46]. Discovering new materials has been an important area of research in the past few decades. Theoretical calculation helps a lot in inventing many of the new molecular structures, reaction mechanisms and stabilities without any experimental cost. It simplifies the experimental work of researchers by proposing a tentative reaction mechanism. Molecular simulations are environmentally friendly, inexpensive, immerging techniques that help to perform theoretical calculations for the mechanism of corrosion inhibitors in the aqueous phase on the metal surface. They also determine the orientation and adsorption nature of the interactions between the inhibitor and the metal. Band gap determination, binding energy, interaction energy, frontier molecular orbitals, transport properties, etc., can be calculated using theoretical calculations. The study of the mechanism of adsorption of inhibitors on metal surfaces is needed. The orientation of the inhibitor on the metal surface is significantly important. An orientation sharing more surface area adsorbs more, and therefore it is considered a good inhibitor [26]. Without theoretical calculation, it is difficult to propose, experimentally, the orientation of inhibitors on the metal surface. Hence, before any experimental work, a theoretical study of the effect of the inhibitor on the metal surface is always a clear decision. This may prevent the wastage of chemicals in laboratories, and computational studies can be carried out multiple times to find the proper orientation and the best inhibitor for a particular metal/alloy.

Recently, the determination of molecular interaction energies and polarisation densities in various adsorption studies of aromatic molecules on the different carbon allotropes surfaces have been analysed by Berdullas et. al. using explicit dispersion through Grimme's empirical corrections [44]. Gupta et al. reported the electrochemical and surface studies supported by DFT calculations and MD simulation methods. A commendable relationship between the computational outcomes and experimental results was established [45]. DFT calculations and MD simulations were used to explain the adsorption mode on the aluminum metal surface and to determine the adsorption centers of the Pimpinella Anisum extract inhibitor. DFT, MC, as well as MD calculations show a strong adsorption interaction by the metal surface with the extract molecules [47]. As the properties of any material depend on the structure of molecules in different phases, it is important to understand the structural features of inhibitors. Keeping the importance of carbon allotropes in mind, for in-depth understanding, the computational-based advancements of various carbon allotropes in corrosion prevention have been reviewed below.

10.3 CLASSIFICATIONS OF CARBON ALLOTROPES

Several allotropes of carbon atoms can be shaped with different bond lengths due to their ability to form new and various hybridisations [48,49]. The three-carbon allotropes of the carbon atom that occur naturally are sp^2, sp^3, and hybrid sp^2/sp^3, which characterise the structures of graphite, diamond, and amorphous carbon, respectively. Over the last two decades, an intense amount of research has been carried out regarding the fabrication of naturally occurring carbon allotropes, and with this, new carbon allotropes have been formed with various hybridisations such as graphene, nanotubes, fullerenes, etc. These new carbon allotropes can be materials with different dimensions; fullerenes are a zero-dimensional material (0D), CNTs are one-dimensional (1D), graphene is a two-dimensional material (2D), and diamond is three-dimensional material (3D) [50–54]. The use of carbon allotropes in the field of corrosion prevention has already been a demanding field for the last few decades due to its advanced property and easy abundance in nature and has already been highlighted in a few reviews, but in a discrete manner [2,7,13–15,17,26,55–60]. This review provides a cumulative report of the properties and usage of major carbon allotropes in corrosion prevention (Figure 10.2).

10.3.1 Graphene-Based Materials

Graphene is a two-dimensional monolayer of carbon atoms bonded covalently and arranged in a hexagonal network. It had been accepted as a theoretical structure until confirmed experimentally by the revolutionary work of Novoselov et al. [62]. For this, the Nobel Prize was awarded in 2010. Since then, there has been exponential growth in the field of research related to graphene. It is used for energy storage [63], energy conversion [64], and biomedical applications [65,66] due to its numerous exceptional behaviours. Graphene is also an exciting material for sensing applications due to its high sensitivity towards external stimuli, as every carbon atom is a surface atom, and hence has an enormously high surface-to-volume ratio [67]. The physical properties of graphene can also be altered and improved to permit specific interaction with certain molecules due to the delocalised π-electrons on its surface.

The electrical properties of graphene also play an important role in sensing applications. The electrical conductivity of graphene exhibits a tremendous change in its electrical behaviour after the absorption of molecules on its surface because the molecules may act as electron donors or acceptors, thus affecting the carrier concentration [68,69]. Graphene's mechanical properties are outstanding. Graphene also shows surface-enhanced Raman scattering (SERS) property, permitting trace-level of target molecules to be spotted by intensifying the characteristic Raman signals [70,71]. Other than the electronic properties, graphene also shows characteristic optical properties that are widely used for sensing applications [67]. The preparation of thin graphene flakes is done by mechanical exfoliation [62]. Hernandez et al. introduced a liquid exfoliation technique to obtain a single to a few layers of graphene sheets by dispersion and exfoliation in solvents [72]. Another chemical technique to obtain graphene is the reduction of graphene oxide (GO). GO can be obtained by oxidising graphite using strong oxidising chemicals, as reported

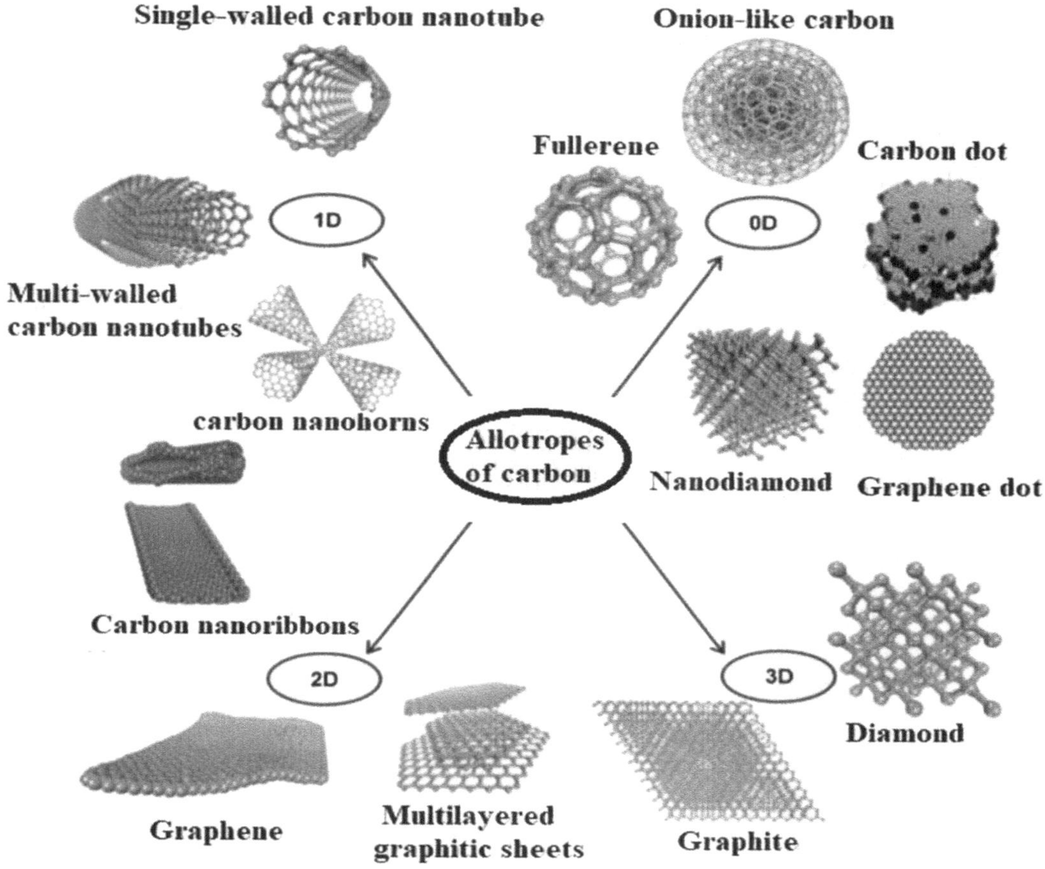

FIGURE 10.2 Allotropes of carbon. Reused with permission from Ref. [61].

by Hummers et al. [73], among others. GO is easily exfoliated in water and can be easily reduced and transformed back into graphene. The modified Hummer method is used to synthesise GO [74]. Other methods of preparing graphene include heat treatment of silicon carbide (SiC) [75,76], CVD [77,78], the intercalative expansion of graphite [79], and epitaxial growth technique [80,81].

Graphene, being a monolayer of graphite, has applications in corrosion prevention. This is due to its interesting properties, which are chemical inertness up to 400°C, and it can be tailored to maintain desired thickness in a meter scale. Furthermore, its deposition can be made mechanically over any surface [82]. In-depth study reveals that the graphene coating does not hamper the optical properties of metal once applied up to four layers [83,84]. Chen et al. have shown from SEM, XPS, and Raman spectroscopic measurements that the graphene layer shows the ability to prevent the oxidation of Cu metal [85].

Prasai et al. have used graphene layers as a protective shield to inhibit the process of corrosion of the metals such as copper. For this, multiple layers of graphene have been loaded on the metal surface mechanically [86]. The Tafel analysis is an efficient electrochemical technique to monitor the proper pathway of the corrosion process using electrochemical impedance spectroscopy. The experimental studies combined with computational chemistry calculations were used to obtain information regarding the anticorrosion mechanism at the molecular level. The computational analysis digs out microscopic information about the corrosion process without the blindness carried out from the experiment procedure.

Li et al. checked the formation of oxide defects in the layers of graphene, which is responsible for corrosion by the breaking of protective layers over graphene by measuring the kinetic and thermodynamic stability of oxygen atoms over pristine graphene [87]. DFT calculation is performed using Vanderbilt pseudo potential scheme and compared with experimental data to conclude oxidation, fluorination, and hydrogenation of the graphene film to make it eligible for protection against corrosion. The DFT calculations conclude that the functionalised graphene has more absorption stability as well as diffusion activity for oxygen atoms over graphene film. It dictates that the destructive oxidation due to the functionalisation of graphene by using chemical methods reduces the protective power of graphene. Both DFT calculations and electronic structure measurements give a collaborative result that functionalisation reduces the oxidative protection of graphene under the availability of atomic oxygen.

The study of graphene metal interfacial structures provides an understanding about interfacial interaction. The study shows single atom thickness of graphene as corrosive prevention coating is sufficient for the protection of Pt(100) in the presence of an oxygen atmosphere (pressure 10^{-7} bar) whereas, in the presence of CO (pressure 10^{-9} bar), the metal is exposed within a very short period (20 s) [88]. DFT calculation reveals the process of the corrosion during the CO-induced lifting over the metal surface [88]. The study shows that CO molecules get intercalated into the sheets of graphene layers and come into contact with the metal surface and get exposed.

Further, the theoretical study shows that the dense two-dimensional honeycomb structure of graphene is capable of protecting the metal surface from corrosion but structural irregularities, as well as intrinsic atomic defects, provides the pores between the structure of the graphene, which makes the corrosion feasible over the metal surface having graphene protecting layers. These imperfections, such as vacancies, stone walls, cracks, and grain boundaries, make the graphene layers feasible for the intercalation of oxidising elements, i.e., corrosive species.

GO is also an efficient material to use for corrosion prevention in various applications for different surfaces [19–24,45,89–94]. GO, along with its functionalised materials, has been exploited for the hydrophobic coating of different metal alloys. The corrosion inhibition efficiency has been measured to understand the behaviour of inhibitors to control the corrosion.

Magnesium alloy, being a biological important material, has been studied to monitor the effect of the addition of Fe, Ni, and Co as impurities to enhance the corrosion rate. To protect from carrion amino acid, functionalised GO has been used as an inhibition layer for magnesium alloys [21]. Similarly in several places GO and its functionalised derivative, such as aromatic amine functionalised GO, perfluorinated polysiloxane-functionalised GO, silane-incorporated GO, polyvinyl alcohol functionalised GO, and amino acid functionalised GO, have been used to inhibit the corrosion efficiently on the magnesium alloy. The computation studied in multiple places supports experimental results and guides understanding of the intriguing mechanism. Palaniappan et al. have studied the octylamine covalently added functionalised GO to monitor the electrochemical stability as well as corrosion inhibition efficiency on magnesium alloy in a 3.5% NaCl medium and performed computational studies [22]. Thus, they found interesting ways to deal with theoretical calculation during experimental work for graphene-based materials.

10.3.2 Carbon Dots

Carbon nanoparticles were identified accidentally during an electrophoresis purification of single-walled carbon nanotubes (SWCNTs) prepared via arc-discharge process, known as CDs [95]. CDs were found to be fluorescent [55]. CDs commonly comprise amorphous or crystalline centers with carbon configuration having sp^2-hybridisation. Research on CDs is also described to exhibit diamond-like sp^3 carbon configuration [96]. Usually two routes are used to synthesise CDs, which are either top-down methods or bottom-up methods. Optical, chemical or thermal processes are also used for the synthesis. In the optical synthesis technique, a laser beam is usually used to remove a carbon object either in water or other solvents [97]. The chemical synthesis method normally uses strong oxidative chemicals such as conc. H_2SO_4 or conc. H_3PO_4 oxidises the carbon precursors to produce fluorescent CDs [98]. Thermal synthesis methods such as direct hydrothermal, pyrolysis, solvothermal, and microwave-assisted pyrolysis are also used for the preparation of CDs [99]. CDs are successful corrosion inhibitors for aluminum, copper or steel. They prevent the corrosion of metals in acidic as well as brine solution. They are also used as a bio-corrosion inhibitor [26]. CDs display outstanding shielding capabilities extending from 84% to 99%. CNTs are normally used in coating systems, whereas graphene-based materials can be used as inhibitors or coating materials in diverse media [27].

Various experiments performed on CDs review the properties showing the corrosion inhibition [8,9,16,18,100–102]. The experimental findings never provide a complete understanding of the responsible part for corrosion inhibition for CDs; therefore, the theoretical advancement based on quantum as well as atomised simulation along with DFT calculations is necessary to achieve the dependency of the mechanism for the corrosion inhibition.

Ye et al. have also performed the calculation for the corrosion inhibition for mild steel in the acidic condition by N-doped citric acid-based CDs and calculated the absorption coefficient as well as free energy for absorption for the corrosion test [8]. They showed that the corrosion obeys the absorption model proposed by Langmuir. The photoabsorption spectroscopy analysis decided that the nitrogen atom doped over the CDs is in the form of pyrrole, pyridine, and graphite-like N. Furthermore, the advancement has been done by utilising molecular simulations such as Monte Carlo and molecular dynamics simulations. This simulation provides theoretical data about inhibition of the corrosion, which leads to information about the site of interaction, the orientation of the interaction molecule, and the nature of absorption on the metal by the inhibitor. By using the simulations, which are mainly based on a classical Newtonian mechanism, transport properties of corrosion-inhibiting molecules and the equilibrium maintained between the metal substrate and these molecules have been measured efficiently in environmentally friendly conditions.

Guo et al. have performed calculations based on density functional theory and molecular dynamics for anticorrosion activity of thiourea derivatives to protect a carbon steel Fe(110) surface [100]. The quantum chemical parameters such as Frontier molecular orbitals, the HOMO-LUMO gap and dipole moment along with Fukui index have been measured with the help of molecular dynamics simulations considering the factors related to temperature coverage and solvent. The temperature and coverage dependence on the absorption energy and efficiency for the three urea-based inhibitors has been reported and firmly collaborated with experimental results.

The process of corrosion inhibition is dependent on the orientation of the inhibitor over the metal surface. The orientation manages the nature of the inhibition efficiency of the molecule. A flat orientation is favourable in comparison to the vertical or tilted orientation because the flat orientation acquires more surface area than other orientations. Interestingly, a molecule once absorbed on those metal surfaces releases a certain amount of energy in the form of heat, resulting in an exothermic reaction. MC simulation and DFT calculations are helpful to measure this released amount of energy. As this process is the association of inhibition deformation and adsorption, the corresponding energies provide total energy for the reaction.

Quantum mechanical calculations along with molecular modelling studies are not feasible for the study of tiny CDs due to their zero-dimensional size. In contrast, the use of organic complexes in the preparation of CDs makes possible several calculations along with Monte Carlo and molecular dynamics simulation for CDs. Liu et al. have studied the corrosion inhibition of metal in acidic conditions by using N-doped CDs [9]. They found that N-doping forms the pyridine or pyrrole-like nitrogen moieties. Electrochemical impedance spectroscopy is used for electrochemical analysis of the materials. The result shows that N-CDs are 24.1, 10.7, and 40.4 $\mu A/cm^2$ of current density. The molecular simulation calculation has been used to understand the corrosion process of metal in the presence of N-doped CDs and provide the adsorption-free energy of pyrrole like N, pyridine like N, and graphitic N, which are −26.94, −25.99, and −25.78 kJ/mol, respectively.

Molecular dynamics, along with quantum calculation, have been used to study the inhibition process of the N-doped CDs acidic environment of hydrochloric acid. Citric acid and histidine have been used in the preparation of these CDs. Electrochemical experiments performed using electrochemical workstations and electrochemical impedance spectroscopy along with computational studies show the dependence of orientation of inhibitor on the metal surface and reported that the used carbon nanodots do extreme protection to the metal surface due to their compact and well-arranged placements.

10.3.3 Nanodiamonds

Diamond is well recognised for its utmost hardness, stability, and higher thermal conductivity properties. The nano-scaled diamonds (NDs) were spotted by Soviet scientists in 1960 [59] in the soot after the eruption in the insufficient supply of oxygen TNT/hexogen composition in an inert environment with no added carbon supply [103]. Although graphene, CNTs, and CDs exhibit sp^2 graphitic carbon, NDs entail sp^3 hybridised carbon atoms only and have a diamond-type morphology. Mostly, the dimension of NDs varies from 2 to 20 nm range, which is significantly smaller than bulk diamond and diamond abrasive powders but are bigger than organic diamond molecules [67]. NDs tend to form aggregates. Although commercial NDs, which comprise big ND clusters, also form aggregates, they cannot be spread via ultrasonication action [104]. Thus, numerous approaches have been established to de-aggregate the NDs. Two procedures explained by Ozawa et al. [105] to breakdown NDs aggregate in several non-aqueous solvents are firstly by applying agitated-medium milling with zirconia microbeads, which can reduce the diameter of the NDs from 200 nm to 4–5 nm within 100 min and secondly by using high-power ultrasonication with the help of zirconia beads. A standard technique of formulating NDs is also by discharging an explosive carbon precursor such as trinitrotoluene and hexogen (1,3,5-triazinane) [106,107].

10.3.4 Carbon Nanotubes

CNTs are the utmost familiar members of the nanocarbon family. Excessive attention from the researcher has been noted in the area of CNTs because of their exceptional mechanical, structural, and electronic behaviour. CNTs can be divided into two types: SWCNTs and multi-walled CNTs (MWCNTs) [108]. SWCNTs consist of a single sheet of carbon atoms exhibiting a sp^2-hybridised structure, which are turned into a cohesive cylinder having diameter and length in the nanometer and micrometer range, respectively. Other than this, MWCNTs comprise numerous coaxial CNTs with an interlayer spacing of 3.4 Å [109]. Substantially, CNTs have excessive length-to-diameter aspect ratios, which make them the most anisotropic nanomaterials. Mechanically, CNTs are among the toughest and strongest fibers known to date, which exhibit strong sp^2 hybrid bonds among the discrete carbon atoms [55]. The foremost reason for the incapability of CNTs to scatter in organic or polar solvents is their surface properties. Although the two ends of the CNTs show oxygen-containing species, which are mostly hydrophilic, the wall that creates a maximum part of the CNT's surface area is hydrophobic [110,111]. Due to the strong van der Waal interaction, CNTs are often held together into bundles. CNTs can be prepared via three main techniques: arc-discharge method [56,112], chemical vapour deposition (CVD), and laser ablation method [113,114].

CNTs are the efficient material to be used as corrosion inhibitors [25,56,58,115–126]. Various physical properties of CNTs measured using DFT calculations, and ab initio quantum mechanical calculations [58,117,119,121,123,125]. The theoretical calculations nicely append to the experimental results and provide depth analysis of different properties depending upon the size of the nanotubes. These calculations are performed using the first principle theory, which leads to solving many-body Schrodinger equations. By restricting the number of molecules, first principle calculations are generally performed to avoid high computational cost usages [115].

The density functional theory calculations have been used to investigate the corrosion inhibition properties of CNTs as the corrosion protection ability of CNTs coating on metals is better than on undoped metal surfaces. This enhancement in the protection against corrosion is due to the slow transportation of water molecules to the metal surface once coated with CNTs. At the same time, it was found that the CNT fills the gap, as well as micro, and holds present on the metal surface, whereas electrochemical hindrance is also an important property to add on. Therefore, the CNTs and their composite are often exploited using theoretical calculations to reveal the inhibition ability on different metal surfaces.

To study the corrosion inhibition process, the absorption of molecular oxygen and hydroxyl ions on CNTs has been studied through theoretical calculations. The study made on CNTs also revealed that the adsorption energy depends upon the various physical parameters such as length, diameter and the end of the nanotube, which can be open or closed. It was also reported that the zigzag tube is more useful for absorption than the armchair type tube.

Butyrskaya et al. studied the mechanism of corrosion protection by using CNTs through computation calculation. The study demonstrates that the use of CNTs significantly enhances the corrosion protection efficiency, which is also supported by hydrophobic behaviour of CNTs against oxygen, hydroxyl ion, and the other ionic group present in the solution.

Interesting studies have been performed by concentrating on the size of CNTs and reported that the size influences the electronic structure of junctions articulated by covalent linking. The DFT study gives an energy gap between HOMO and LUMO, which increases by decreasing the length of the tube. The energy gap is also correlated with the number of atoms involved with CNTs.

Competition modelling of environmental fatigue behaviour has been quantified for hybrid CNTs composite, and the material behaviour is studied using a python best program through both mechanical and environmental loadings. After computational studies, they observed better fatigue performances on applying secondary CNTs as reinforcement.

Ab initio calculation-based study of SWCNTs using various radii has been performed [127]. Various properties of these nanotubes are monitored through density functional theory using a high number of atoms per cell. The result shows that the relaxation effect as well as high-frequency optical phonon modes significantly depend on tube curvature and its radius.

The study of absorption of various gas molecules such as NO_2, O_2 on SWCNTs is performed using first principle calculations [128]. The study also includes the calculation of absorption energy, electronic label calculations, and charge transfer calculations. It was reported that the fluctuation in gas-induced charge at various equilibrium positions significantly depends on its transport properties. At the same time, it was also found that all molecules are adsorbed on CNTs weakly and the molecular absorption depends upon the transport and associated electronic properties of the CNTs.

10.3.5 Fullerene

Fullerene, being a nanoscale carbon-containing soccer-ball shaped material, has significant importance in the field of corrosion inhibition [2,17,57,129–131]. In the defect-free structure, it contains a regular cage-like structure. Fullerene has many applications in various fields, including fuel cells. Here, corrosion prevention is one of the major factors to build an efficient fuel cell. First, the principle DFT calculation of fullerene was performed using Vienna Ab-initio Simulation Package (VASP), and the DFT-D method with Grimme along with plane wave basis function has been employed to estimate the corrosion-related parameters of a pure fullerene and compared with graphite. Interesting, the result shows that the corrosion potential of fullerene (0.086 V vs RHE) is lower than that of graphite (0.207 V vs RHE) [130].

Fullerene also has optical, mechanical, and electronic properties, which pushes it to be used in various materials. Due to the nanosized shape of this compound, it is used as nanofillers on multiple surfaces, which enhances the corrosion inhibition process. Interestingly tribological as well as corrosion prevention performance have been investigated for fullerene C60 filled epoxy

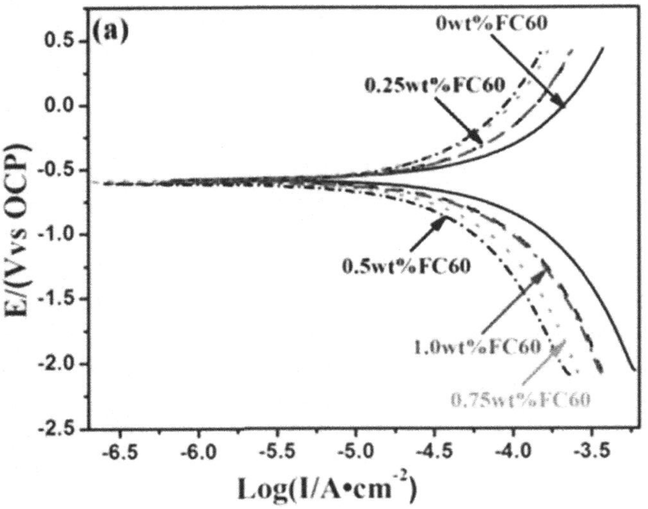

FIGURE 10.3 Tafel plot for the fullerene/epoxy coatings. Reused with permission from Ref. [132].

composite coated surface. The study is performed on the polymeric matrix. Various spectroscopy techniques such as FTIR spectroscopy, Raman spectroscopy, transmission electron microscopy, and scanning electron microscopy are used to monitor the surface-related properties. The corrosion resistance has been measured using electrochemical experiments. Tafel polarisation plot is a graphical interface to show a relationship built between the current produced and electrode potential on the conducting surface. Tafel plot has been used to study the change occurring for epoxy/fullerene composite coating in comparison to the unfilled epoxy-coated surface. Figure 10.3 shows the Tafel polarisation plot in 3.5 wt% NaCl solution at various wt% of fullerene with epoxy composite coating. The optimum %wt is 0.5 for the fullerene/epoxy nanocomposite, above which the nanocomposite performance is found better than unfilled epoxy coating due to the increased number of cracks over the epoxy-containing surfaces. In addition, the fullerene nanomaterial deposition forms aggregates on the surface increasing tribological and corrosion protection performance. The result dictates that incorporation of fullerene to the epoxy coating significantly improves the anticorrosion as well as tribological properties concerning unfilled epoxy coating (Figure 10.3).

Turan et al. have studied magnesium matrix composite reinforced with fullerene to study the corrosion, tribological, and mechanical properties by using the semi-powder technique of metallurgy [131]. They found that the use of fullerene improves the hardness of magnesium matrix composite in comparison to pure magnesium. It is due to the nanoscale size of fullerene so that it is embedded homogeneously over the surface of the magnesium matrix. Therefore, fullerene reinforced magnesium matrix composite claims a lower friction coefficient as well as wear rate than pure magnesium. The Tafel plot shows that the fullerene reinforcement to the magnesium matrix increases the corrosion current density in comparison to bare magnesium. Eventually, the corrosion resistance decreases for the unreinforced magnesium on incorporation fullerene for the matrix. Another study also concluded that magnesium composite reinforced with fullerene provides the highest hardness performance among MWCNT, fullerene, and graphene sheets.

10.4 CONCLUSIONS AND FUTURE DIRECTIONS

Summarising the above discussion, it can be concluded that advanced carbon allotropes have significant use as inhibitors for the corrosion of metals and alloys. In the last few years, these nanomaterials have been getting a lot of attention in the field of scientific research due to their unique chemical, mechanical, thermal, electrical, and optical properties. This includes graphene, fullerene, CDs, CNTs, and other carbon-containing materials. These materials have recently emerged as one of the most efficacious, cost-effective and ecologically friendly materials. These materials are also being used in drug delivery, semiconductor electronics, energy conservation technology, etc. They have been used to develop several materials with different applications depending upon the requirements. Scientific research on advanced carbon allotropes is being carried out on the industrial level to tackle the corrosion occurring in industries, which leads to heavy economic loss. The computation study, as well as experimental measurements, are equally important to understand the corrosion inhibition properties of these materials on metals and alloys. The demand for advanced carbon materials will be further growth for technological innovation.

ACKNOWLEDGEMENT

MLS acknowledges UGC, India (F.30-446/2018 (BSR)) for financial assistance. SK thanks the Department of Science and Technology (DST), India (Grant No. SRG/2019/002284) for financial support.

REFERENCES

1. Gaidis, J. M., *Cem. Concr. Compos.* (2004) **26** (3), 181
2. Kausar, A., *J. Plast. Film Sheeting* (2019) **35** (2), 181
3. McCafferty, E., *Introduction to corrosion science*. Springer Science & Business Media: 2010
4. Fontana, M. G., and Greene, N. D., (1967)
5. Popoola, L. T., *et al., Int. J. Ind. Chem.* (2013) **4** (1), 1
6. Zhao, L., *et al., Mater. Corros.* (2004) **55** (9), 684
7. Goyal, M., *et al., J. Mol. Liq.* (2018) **256**, 565
8. Ye, Y., *et al., J. Hazard. Mater.* (2020) **381**, 121019
9. Liu, Z., *et al., J. Cleaner Product.* (2020) **270**, 122458
10. Haruna, K., *et al., J. Mol. Liq.* (2018) **264**, 515
11. Singh, A., *et al., J. Colloid Interface Sci.* (2020) **560**, 225
12. Ituen, E., *et al., Surf. Interfaces* (2019) **16**, 29
13. Sørensen, P. A., *et al., J. Coat. Technol. Res.* (2009) **6** (2), 135
14. Zhang, Y., *et al., J. Mater. Sci. Technol.* (2022)
15. Finšgar, M., and Jackson, J., *Corros. Sci.* (2014) **86**, 17
16. Verma, D. K., *et al., J. Mol. Struct.* (2021) **1236**, 130294
17. Pourhashem, S., *et al., J. Coat. Technol. Res.* (2020) **17** (1), 19
18. Guo, L., *et al., Res. Chem. Intermed.* (2015) **41** (6), 3729
19. Zhu, Y. X., *et al., J. Phys. D: Appl. Phys.* (2017) **50** (11), 114001
20. Zhang, K., *et al., Vacuum* (2019) **161**, 276
21. Shang, W., *et al., Ind. Eng. Chem. Res.* (2019) **58** (3), 1200
22. Palaniappan, N., *et al., RSC Adv.* (2020) **10** (19), 11426
23. N, P., *et al., RSC Adv.* (2021) **11** (36), 22095
24. Choudhary, S. K., and Das, S., *J. Nanosci. Nanotechnol.* (2017) **17** (3), 2130
25. Dewangan, Y., *et al.*, Carbon Nanotubes as Corrosion Inhibitors. In *Organic corrosion inhibitors*, John Wiley & Sons, Ltd. (2021), pp. 371
26. Berdimurodov, E., *et al., J. Mol. Liq.* (2021), 118124
27. Quadri, T. W., *et al.*, Functionalized carbon allotropes as corrosion inhibitors. In *Functionalized nanomaterials for corrosion mitigation: Synthesis, characterization, and applications*, ACS Publications (2022), pp. 87
28. Zhang, J., *et al., Laser Phys. Lett.* (2014) **11** (3), 035901
29. Wang, Q., and Wang, B.-T., *Sens Actuators, B* (2018) **275**, 332
30. Chan, K. K., *et al., Nano-micro Lett.* (2018) **10** (4), 1
31. Chien, Y. H., *et al., J. Chem. Technol. Biotechnol.* (2018) **93** (6), 1519
32. Ouzzine, M., *et al., Carbon* (2014) **67**, 104

33. Xiang, Q., *et al.*, *Chem. Soc. Rev.* (2012) **41** (2), 782
34. Cao, S., and Yu, J., *J. Photochem. Photobiol. C: Photochem. Rev.* (2016) **27**, 72
35. Yu, H., *et al.*, *Adv. Mater. (Weinheim, Ger.)* (2016) **28** (43), 9454
36. Zhang, N., *et al.*, *Nanoscale* (2012) **4** (19), 5792
37. Wang, C., *et al.*, *Mater. Lett.* (2013) **108**, 336
38. Di, J., *et al.*, *Chem. Eng. J. (Lausanne)* (2015) **281**, 60
39. Wang, J., *et al.*, *Appl. Catalysis B: Environmental* (2017) **204**, 577
40. Han, C., *et al.*, *Nano Today* (2016) **11** (3), 351
41. Popov, Z. I., *et al.*, *Carbon* (2021) **184**, 714
42. Al-Fahdi, M., *et al.*, *Crystals* (2021) **11** (7), 783
43. Oliveira, F. L., *et al.*, *Chem. Phys. Chem.* (2020) **21** (1), 59
44. Ramos-Berdullas, N., *et al.*, *Phys. Chem. Chem. Phys.* (2015) **17** (1), 575
45. Gupta, R. K., *et al.*, *Mater. Chem. Phys.* (2019) **236**, 121727
46. Verma, D. K., *et al.*, *J. Mol. Liq.* (2021) **329**, 115531
47. Akounach, Z., *et al.*, *Int. J. Corros. Scale Inhib.* (2022) **11** (1), 402
48. Miller, E., *et al.*, *Chem. Mater.* (1997) **9** (1), 18
49. Zhao, Z., *et al.*, *ACS Nano* (2011) **5** (9), 7226
50. Zhu, Q., *et al.*, *Phys. Rev. B* (2011) **83** (19), 193410
51. Zhou, X.-F., *et al.*, *Phys. Rev. B* (2010) **82** (13), 134126
52. Ivanovskaya, V., and Ivanovskii, A., *J. Superhard Mater.* (2010) **32** (2), 67
53. Umemoto, K., *et al.*, *Phys. Rev. Lett.* (2010) **104** (12), 125504
54. Wang, Z., *et al.*, *Proc. Natl. Acad. Sci.* (2004) **101** (38), 13699
55. Yap, S. H. K., *et al.*, *Sensors* (2020) **20** (7), 2046
56. Wang, Y., and Yeow, J. T. W., *J. Sens.* (2009) **2009**, e493904
57. P., A. K., *et al.*, *J. Mater. Res.* (2021) **36** (1), 114
58. de Lara-Castells, M. P., and Mitrushchenkov, A. O., *Front. Chem.* (2021) **9**, 796890
59. Zhang, Y., *et al.*, *Compos. Part B: Engineering* (2018) **143**, 19
60. Wang, Y., and Yeow, J. T., *J. Sens.* (2009) **2009**
61. Verma, C., *et al.*, *Nano Select* (2021) **2** (12), 2237
62. Novoselov, K. S., *et al.*, *Science* (2004) **306** (5696), 666
63. Pumera, M., *Energy Environ. Sci.* (2011) **4** (3), 668
64. Bonaccorso, F., *et al.*, *Science* (2015) **347** (6217), 1246501
65. Yang, C., *et al.*, *ACS Appl. Mater. Interfaces* (2018) **11** (3), 2768
66. Shen, H., *et al.*, *Theranostics* (2012) **2** (3), 283
67. Hong, G., *et al.*, *Chem. Rev. (Washington, DC, U. S.)* (2015) **115** (19), 10816
68. Kwak, Y. H., *et al.*, *Biosens. Bioelectron.* (2012) **37** (1), 82
69. Huang, Y., *et al.*, *Nanoscale* (2010) **2** (8), 1485
70. Dresselhaus, M. S., *et al.*, *Nano Lett.* (2010) **10** (3), 751
71. Ling, X., *et al.*, *Nano Lett.* (2010) **10** (2), 553
72. Hernandez, Y., *et al.*, *Nat. Nanotechnol.* (2008) **3** (9), 563
73. Hummers Jr, W. S., and Offeman, R. E., *J. Am. Chem. Soc.* (1958) **80** (6), 1339
74. Zaaba, N., *et al.*, *Procedia Engineering* (2017) **184**, 469
75. Huang, H., *et al.*, *ACS Nano* (2008) **2** (12), 2513
76. Berger, C., *et al.*, *Science* (2006) **312** (5777), 1191
77. Dedkov, Y. S., *et al.*, *Phys. Rev. Lett.* (2008) **100** (10), 107602
78. Sutter, P. W., *et al.*, *Nat. Mater.* (2008) **7** (5), 406
79. Li, X., *et al.*, *Science* (2008) **319** (5867), 1229
80. Zangwill, A., and Vvedensky, D. D., *Nano Lett.* (2011) **11** (5), 2092
81. First, P. N., *et al.*, *MRS Bulletin* (2010) **35** (4), 296
82. Liu, L., *et al.*, *Nano Lett.* (2008) **8** (7), 1965
83. Nair, R. R., *et al.*, *Science* (2008) **320** (5881), 1308
84. Bae, S., *et al.*, *Nat. Nanotechnol.* (2010) **5** (8), 574
85. Chen, S., *et al.*, *ACS Nano* (2011) **5** (2), 1321
86. Prasai, D., *et al.*, *ACS Nano* (2012) **6** (2), 1102
87. Li, Q., *et al.*, *Phys. Chem. Chem. Phys.* (2019) **21** (23), 12121
88. Nilsson, L., *et al.*, *ACS Nano* (2012) **6** (11), 10258
89. Chauhan, D. S., *et al.*, *Prog. Org. Coat.* (2020) **147**, 105741
90. Haghdadeh, P., *et al.*, *J. Taiwan Inst. Chem. Eng.* (2018) **86**, 199

91. Haghdadeh, P., *et al.*, *Prog. Org. Coat.* (2019) **136**, 105243
92. Jiang, F., *et al.*, *Prog. Org. Coat.* (2019) **127**, 70
93. Mehmeti, V., and Podvorica, F. I., *Materials* (2018) **11** (6), 893
94. Punith Kumar, M. K., *et al.*, *Appl. Surf. Sci.* (2020) **533**, 147512
95. Xu, X., *et al.*, *J. Am. Chem. Soc.* (2004) **126** (40), 12736
96. Salinas-Castillo, A., *et al.*, *Chem. Commun. (Cambridge, U. K.)* (2013) **49** (11), 1103
97. Gonçalves, H., *et al.*, *Sens Actuators, B* (2010) **145** (2), 702
98. Chang, M. M. F., *et al.*, *Microchim. Acta* (2016) **183** (6), 1899
99. Li, M., *et al.*, *Chem. Eng. J. (Lausanne)* (2017) **320**, 570
100. Guo, L., *et al.*, *J. Colloid Interface Sci.* (2017) **506**, 478
101. Saha, S. K., *et al.*, *Phys. Chem. Chem. Phys.* (2016) **18** (27), 17898
102. Ye, Y., *et al.*, *J. Cleaner Product.* (2020) **264**, 121682
103. Shenderova, O., *et al.*, *Crit. Rev. Solid State Mater. Sci.* (2002) **27** (3-4), 227
104. Mochalin, V. N., *et al.*, *ACS Nano* (2011) **5** (9), 7494
105. Liang, Y., *et al.*, *ACS Nano* (2009) **3** (8), 2288
106. Danilenko, V. V., *On the history of the discovery of nanodiamond synthesis.* Springer (2004), Vol. 46, pp. 595
107. Shenderova, O. A., and Gruen, D. M., *Ultrananocrystalline diamond: Synthesis, properties and applications.* William Andrew: 2012
108. Iijima, S., and Ichihashi, T., *Nature* (1993) **363** (6430), 603
109. Niyogi, S., *et al.*, *Acc. Chem. Res.* (2002) **35** (12), 1105
110. Bahr, J. L., *et al.*, *J. Am. Chem. Soc.* (2001) **123** (27), 6536
111. Agnihotri, S., *et al.*, *Langmuir* (2005) **21** (3), 896
112. Ando, Y., and Iijima, S., *Jpn. J. Appl. Phys. Part 2 Lett.* (1993) **32**, L107
113. Mubarak, N., *et al.*, *J. Ind. Eng. Chem.* (2014) **20** (4), 1186
114. Thess, A., *et al.*, *Science* (1996) **273** (5274), 483
115. *Carbon nanotubes and nanoparticles: Current and potential applications.* Apple Academic Press: Toronto; New Jersey, 2019
116. *Carbon nanoallotropes-based anticorrosive coatings: Recent advances and future perspectives.* CRC Press: 2021
117. Butyrskaya, E. V., *et al.*, *Comput. Theor. Chem.* (2016) **1090**, 1
118. Frank, S., *et al.*, *Science* (1998) **280** (5370), 1744
119. Gómez, B., *et al.*, *J. Phys. Chem. A* (2005) **109** (39), 8950
120. Jeon, H., *et al.*, *J. Ind. Eng. Chem.* (2013) **19** (3), 849
121. Khorrampour, R., *et al.*, *Physica E: Low-dimensional Syst. Nanostruct.* (2009) **41** (8), 1373
122. Vikramaditya, T., and Sumithra, K., *Comput. Mater. Sci.* (2013) **79**, 656
123. Wu, Y., *et al.*, *Carbon* (2021) **179**, 445
124. Xu, J., *et al.*, *Nanoscale* (2020) **12** (12), 6832
125. Yousefi, A., *et al.*, *J. Bio- Tribo-Corrosion* (2019) **5** (4), 82
126. Zhang, L., *et al.*, *Energy Environ. Sci.* (2012) **5** (3), 6156
127. Sánchez-Portal, D., *et al.*, *Phys. Rev. B* (1999) **59** (19), 12678
128. Zhao, J., *et al.*, *Nanotechnology* (2002) **13** (2), 195
129. Kausar, A., *Mater. Res. Innovations* (2021) **25** (3), 175
130. Noh, S. H., *et al.*, *Nanoscale* (2017) **9** (22), 7373
131. Turan, M. E., *et al.*, *J. Alloys Compd.* (2018) **740**, 1149
132. Liu, D., *et al.*, *Surf. Coat. Technol.* (2016) **286**, 354

Part 3

Anticorrosive Self-Healing Coating Materials and Industrialization

11 Concept of Self-Healing Coatings

Carbon Allotropes as Ideal Substitutes

S.A. Haddadi
Surface Coating and Corrosion Department, Institute for Color Science and Technology, Tehran, Iran

I. Mohammadi
Department of Materials Engineering, Faculty of Engineering, Tarbiat Modares University, Tehran, Iran

M. Izadi
Department of Materials Engineering, Hamedan University of Technology, Hamedan, Iran

M. Mahdavian
Surface Coating and Corrosion Department, Institute for Color Science and Technology, Tehran, Iran

CONTENTS

11.1 INTRODUCTION

Metals and their alloys are key materials that are extensively used in almost all industrial applications. In harsh environments with high levels of humidity, temperature, and corrosive chemicals, the transformation of metallic objects and materials into their oxidised forms has been considered as the main challenge (Fontana and Greene 2018; Talbot and Talbot 2018). Owing to the weak performance of oxidised metallic materials caused by corrosion, they must be replaced with new ones, which compels industries to dedicate a huge budget to fighting corrosion. Based on a

DOI: 10.1201/9781003323976-14

report from *World Corrosion Organization*, the global corrosion expenses per year is equal to about 3% of GDP (Hays and General 2010).

The oxidisation of metals via corrosion reactions can be encompassed by electron transfer and charge transfer reactions (Fontana and Greene 2018; Talbot and Talbot 2018). Between different strategies to limit corrosion reactions, the passivation of metallic surfaces by organic coatings can be introduced as a long-term efficient solution by making a physical barrier between surfaces and harsh environments. For all typical coating and paint systems, their physical barrier performance can be reduced remarkably by applying any mechanical damages. These mechanical failures can provide passages through which humidity and corrosive components can meet the metallic surface and cause corrosion (Hosseini and Makhlouf 2016; Marrion 2004). In this case, corrosion reactions are in advanced stages, while the release of healing agents in the defect can minimise all corrosion activities. As a result, the degradation and deterioration of organic coatings happen in shorter periods compared to intact systems (Hosseini and Makhlouf 2016; Marrion 2004). In recent years, the demands for the active protection of metallic surfaces led to the development of smart coatings. Based on the stimuli applied by intrinsic or extrinsic actions such as pH, UV radiations, mechanical stresses, ion-exchange interactions, and capsule lacerations, smart coatings are capable of altering their properties and characteristics (Hosseini and Makhlouf 2016; Kumar, Stephenson, and Murray 2006). Thus, smart coatings can heal mechanical damages and exhibit self-healing/self-repairing properties. The active healing of damages and cracks in organic coatings recovers their durability and barrier performance (Hosseini and Makhlouf 2016; Stankiewicz, Szczygieł, and Szczygieł 2013).

Because of their high specific surface area, carbon allotropes (CAs) were recently introduced as barrier materials to resist water, oxygen, and ion transport when dispersed in the coating films. The suitable CAs dispersion in the coatings decreases the coating defects and porosities and minimises the electrolyte penetration into the coating structure by zigzagging the diffusion pathway (Samadzadeh et al. 2010). However, in the case of layers with many flaws or mechanical degradation, composites containing these particles may not offer enough protection against the corrosive electrolyte (Urban 2014; Wei et al. 2014). In this circumstance, an active inhibition property based on film-forming agents should be encouraged in the coatings' structure to protect the defected layer. One of the most valuable methods to induce active anticorrosive performance is incorporating the inhibitive agents into the coatings' structure using nanocarriers. They have beneficial characteristics that make them proper candidates for self-healing composite coating design. Decoration of various organic/inorganic inhibitory systems on the surface of the CAs, as the materials with high surface area, is one of the most popular techniques that has been employed to promote the active inhibition of the coatings (Cui, Zhang, et al. 2021). Due to the high specific surface area, the use of the CAs in the self-healing coatings (as the carrier) has various advantages; among them, improving the response to external stimuli and providing a large amount of surface functional groups for forming covalent bonds and intermolecular interactions with the coating matrix can be mentioned (Cui, Zhang, et al. 2021).

11.2 CARBON ALLOTROPES INCORPORATING FILM-FORMING AGENTS

The story of smart coatings began with the incorporation of microcapsules loaded with active agents such as film-forming agents, catalysts, monomers, and other healing agents in polymers (Andreeva and Shchukin 2008). Mechanical stimuli such as scratch, wear, impact, and fatigue cause rupture in embedded capsules, releasing the self-healing agents inside the defect (García, Fischer, and Van Der Zwaag 2011). Possible contacts between released monomers and catalysts initiate the polymerisation reactions, healing the defect. In this approach, the catalyst can be distributed either in the coating matrix or in separate capsules (Kumar, Stephenson, and Murray 2006; Stankiewicz, Szczygieł, and Szczygieł 2013). The direct contact between polymer chains and catalysts may interfere with the coatings' curing reactions or increase the matrix's heterogeneity degree (Hosseini and Makhlouf

2016; Stankiewicz, Szczygieł, and Szczygieł 2013). Consequently, monomers or self-healing agents with self-polymerisation and self-activation capabilities are preferable to be loaded in micro/nanocapsules (Hosseini and Makhlouf 2016; Stankiewicz, Szczygieł, and Szczygieł 2013). For these systems, external factors such as oxygen or moisture initiate the polymerisation or healing reactions of released agents. For instance, the activation of natural linseed and tung oils as self-healing agents occurs by their oxidation in contact with air (Nesterova et al. 2012; Samadzadeh et al. 2011). The performance of capsule-based self-healing coatings depends on various parameters such as the type of capsule wall, wall thickness, capsule diameters, and encapsulation techniques (Parameswaranpillai et al. 2022; Zhao et al. 2012).

As stated in the literature, due to ease in synthesis, tunable size and shell thickness, and variety of functionalities, polymer capsules such as polyurea, poly(urea-formaldehyde), and polyurethane (PU) were developed and incorporated into different organic coatings to induce self-healing properties (Esser-Kahn et al. 2011; Latnikova 2012). Characterisation of such coatings illustrated enhanced self-healing performance after encapsulating corrosion inhibitors, monomers, and other self-healing active agents in polymeric capsules (Hosseini and Makhlouf 2016; Stankiewicz, Szczygieł, and Szczygieł 2013). The weak mechanical strength and dimensional stability of shells in polymeric capsules lead to a challenging dispersion process (Haddadi, Ramazani et al. 2019). The undesired wall fracture of polymeric capsules is caused by shear stresses during the dispersion in a polymer matrix, limiting their applications in designing self-healing materials (Haddadi, Ramazani et al. 2019; Haddadi et al. 2019). Among different categories of inorganic capsules, CAs are good candidates for the fabrication of micro- and nanocapsules as hosts for active self-healing agents (Bekas et al. 2016; Thakur et al. 2022).

Excellent mechanical, chemical, electrical, and thermal characteristics, various morphologies, tunable size, and good dimensional stability are major advantages of CAs for the fabrication of guest-host self-healing systems (Pourhashem et al. 2020; Rafiee et al. 2011). Among different CAs, 0D (e.g., fullerene and carbon-based hollow spheres [CSs]) and 1D (e.g., carbon nanotubes [CNTs]) carbonaceous nanostructures are capable of being utilised as efficient nanocapsules for the encapsulation of self-healing agents due to their unique morphology and/or sufficient specific surface areas and free internal spaces (Thakur et al. 2022; Quadri et al. 2022). Due to the lack of functionalities, some surface pre-treatments, such as surface oxidisation with a mixture of sulphuric acid and nitric acid, have been recommended for CAs before being loaded with active agents and embedded in polymeric coatings (Hosseini and Makhlouf 2016; Hong et al. 2011). Only a few research studies have been conducted on the application of fullerene spherical nanostructures in anticorrosion coatings owing to their challenging synthesis methods and small particle sizes (0.7–1 nm). Nevertheless, the oxidisation and functionalisation of fullerene provided enhanced barrier properties in sol-gel coatings (Pourhashem et al. 2020).

Several studies have been conducted on loading CNTs with active corrosion inhibitors to fabricate CNT-based self-healing coating systems with active inhibition performance (Najmi et al. 2021), whereas self-healing coatings containing CNT-based capsules loaded with film-forming agents have not received enough attention. However, a theoretical molecular dynamic study done by Lanzara et al. exhibited the capability of single-walled CNTs (SWCNTs) for the encapsulation of organic molecules as self-healing agents (Lanzara et al. 2009). The findings of this research showed that it only takes a few picoseconds for organic molecules to be released from ruptured loaded SWCNTs, accelerated by altering the crack size and increasing temperature. Some attempts have been also conducted to intercalate different components with self-healing potentials, such as polymers, surfactants, nanoparticles, and inorganic components, into CNTs. Sinha-Ray et al. loaded CNTs with paraffines and their derivatives mixed with triglycerides using a self-sustained diffusion method (Sinha-Ray, Sahu, and Yarin 2011). Results showed the successful loading of fatty components into CNTs. The melting of loaded components in CNTs above 20°C introduced these CNT capsules as potent carriers for the fabrication of temperature-triggered self-healing coatings. The successful loading of liquid monomers, including dicyclopentadiene and isophorone diisocyanate, has also been performed (Sinha-Ray et al. 2012). Bazilevsky et al. illustrated the intercalation of water, low molecular weight polycaprolactone, surfactant, and polystyrene (PS) nanoparticles in open-ended and noncatalytic CNTs (Bazilevsky et al. 2008). Their studies also proved the efficient loading of low molecular weight polymers such as poly(ethylene oxide) ($MW = \approx 600$ kDa) and poly(caprolactone) ($MW = \approx 80$ kDa) in highly graphitic CNTs (Bazilevsky et al. 2007).

Fabrication of CNT-based hollow capsules is another solution to develop self-healing coatings containing film-forming agents-loaded CNTs. Based on the layer-by-layer (LbL) method for the fabrication of CNT-based hollow capsules, charged CNTs should be deposited on the outer surface of a soft template (e.g., PS and silica particles) (Hong et al. 2011). To induce a uniform opposite surface charge on the surface of core materials, polyelectrolyte components are used. Negatively or positively charged CNTs are synthesised via further modification of oxidised CNTs with carboxylic acid groups (COO^-) or amine groups ($-NH^{3+}$). As reported in (Correa-Duarte et al. 2005), CNT hollow spheres were synthesised using the LbL deposition of multi-walled CNTs (MWCNTs) on the outer layer of negatively charged PS particles in the presence of poly(diallyldimethylammonium chloride) (PDDA) as the positively charged polyelectrolyte. Then, multilayer coatings of CNT/PDDA were deposited on the outer surface of PS particles via electrostatic interactions. PS cores were extracted via washing with tetrahydrofuran (THF). Morphological evaluation of prepared PS/CNT/PDDA core-shells and CNT/PDDA capsules revealed the collapse of CNT/PDDA shells after the core removal.

CSs can be introduced as more desired hosts for self-healing agents (Haddadi et al. 2018). CSs can be fabricated by the deposition of carbonaceous materials onto the outer layers of sacrificial cores via an LbL method (Wang et al. 2016). The characteristics of CSs, such as particle size, functionalities of core and shell, shell thickness, surface charge, porosity, and hybrid carbon structure (sp^2 or sp^3), are controlled and tuned during different synthesis stages (Haddadi et al. 2018; Wang et al. 2016; Liu et al. 2019). Spherical PS, silica and melamine-formaldehyde particles are good candidates for sacrificial cores (Wang et al. 2016). The carbonaceous precursor can be deposited on the surface of core materials via different methods such as in-situ polymerisation, hydrothermal reactions, hydrogen bonding, intermolecular and electrostatic interactions, and covalent bonding (Wang et al. 2016; Hong et al. 2011). The extraction of inorganic or organic core materials can be utilised using solvent extraction, dissolving in an appropriate medium (e.g., acids), and calcination (Hong et al. 2011). The main advantage of templating methods for the fabrication of CSs is that by altering the morphology and particle size of sacrificial cores, a wide range of CSs with various characteristics can be synthesised (Wang et al. 2016; Hong et al. 2011). Haddadi et al. developed self-healing epoxy coatings filled with dual encapsulation of epoxy and polyamine film-forming agents in graphene-based CSs (Haddadi et al. 2018; Haddadi, Ramazani et al. 2019; Haddadi et al. 2019). The carbonisation of polysaccharide shells and then the silica cores extraction by a diluted hydrofluoric (HF) acid led to the formation of CSs. Detailed characterisation of CSs, synthesised by silanised and unsilanised silica cores showed that the amino-functionalisation of silica cores alters the hybrid carbon structure of shells from sp^3 to sp^2. Self-healing epoxy coatings containing different contents of epoxy-doped CSs (Ep-DCS, 2, 5, and 10 wt.%) and polyamine-doped CSs (Am-DCS, 1, 2.5, and 5 wt.%) were fabricated. The schematic presentations of doped CSs, self-healing mechanism and possible healing reactions are shown in Figure 11.1.

After applying scratches, all coatings were preconditioned for 1, 3, and 6 h at room temperature to secure the fulfilment of curing reactions between released film-forming agents. Self-healing investigation of scratched coatings using electrochemical impedance spectroscopy (EIS) illustrated that the healing efficiency increases with an increase in the contents of Ep-DCS and Am-DCS. Also, preconditioned coatings for 6 h presented higher self-healing efficiencies compared to other coating systems.

Regarding 2D CAs (e.g., graphene and graphene oxide [GO]), further modifications via the LbL method and hybridisation with other nanostructures should be performed to alter their morphology

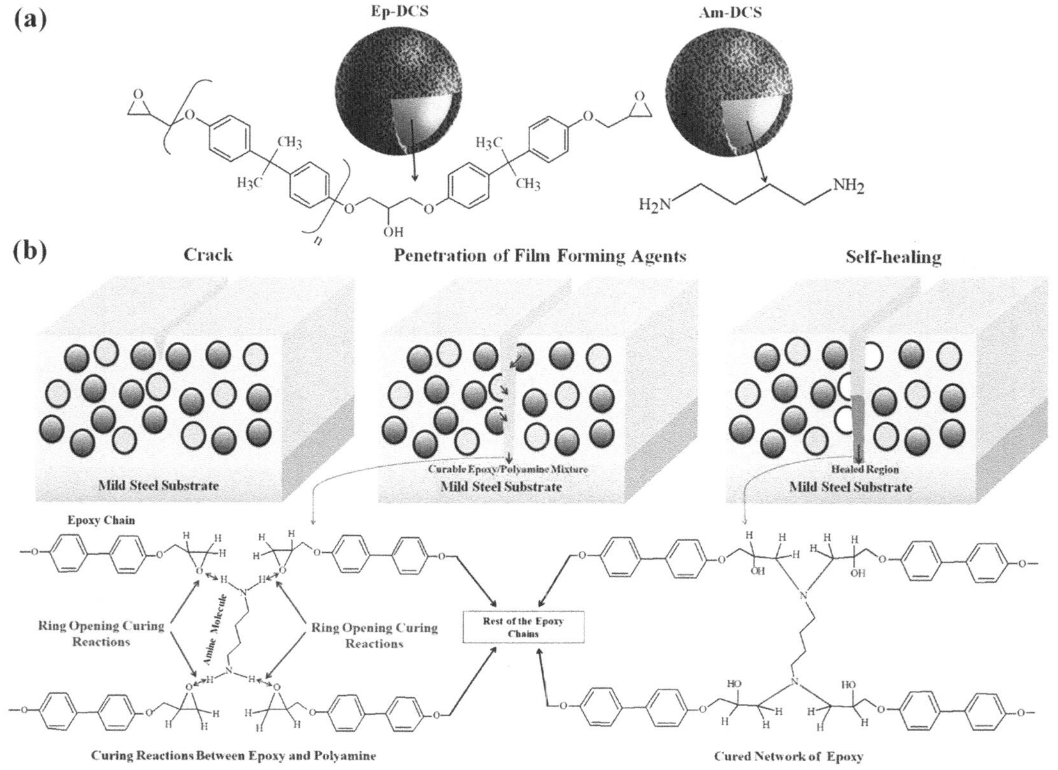

FIGURE 11.1 The schematic presentations of (a) Ep-DCS and Am-DCS and (b) the self-healing mechanism of epoxy coatings and possible healing reactions between released film-forming agents. Reprinted from Haddadi et al. (2019) with permission from Elsevier.

and build free internal spaces (Hong et al. 2011). For these cases, in the first approach, graphene or GO nanosheets are used as nanoplatforms, and secondary nanostructures (e.g., metallic organic frameworks [MOFs]) or polyelectrolyte layers (e.g., polyaniline [PANI] and polydopamine [PDA]) are fabricated on their outer layers (Cui, Zhang, et al. 2021). A wide range of self-healing agents, such as anionic and cationic corrosion inhibitors, can be hosted in such advanced carbon-based nanostructures due to their large specific surface area and limited free spaces (Cui, Zhang, et al. 2021; Haddadi et al. 2022). Tuning the surface charge and functionality of these advanced nanostructures highlights their triggered release potentials (Cui, Zhang, et al. 2021; Ding, Chen, Lv, Zhang, Zhao, Liu, Wang, Gui, Li, and Tang 2019). Film-forming agents and healing monomers are inappropriate guests for such advanced nanostructures due to their high molecular weights, which make their encapsulation process difficult and less efficient. Similar to the CNT-based hollow capsules, graphene-based hollow spheres can be synthesised via the LbL deposition of graphene/GO nanosheets on the outer layer of soft templates (Hong et al. 2011). Owing to the wide lateral size of graphene/GO nanosheets, the repetitive deposition of negatively and positively charged graphene/GO nanosheets on the surface of sacrificial cores and then the removal of cores lead to the fabrication of graphene-based hollow spheres without using polyelectrolyte components (Cui, Zhang, et al. 2021; Hong et al. 2011). Li et al. (2017) developed GO microcapsules using a self-assembly process in Pickering emulsions. The nanoscaled GO shells were formed at the oil/water interfaces in Pickering emulsions loaded with linseed oil as the healing agent (Figure 11.2). GO microcapsules (GOMCs) doped with linseed oil were dispersed uniformly in a waterborne PU resin due to the hydrophilic nature of GO microcapsules. Self-healing PU coatings applied on hot-dip

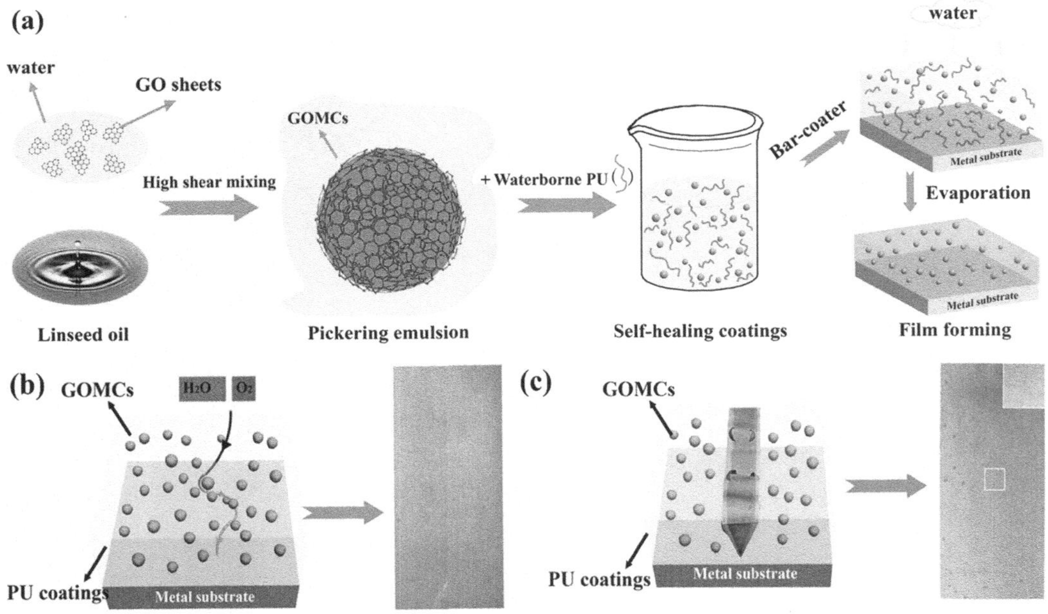

FIGURE 11.2 Schematic illustration of the linseed oil-loaded GO microcapsules and PU self-healing waterborne coatings. Reprinted from Li et al. (2017) with permission from Elsevier.

galvanised steel substrates illustrated remarkable enhancements in both self-healing and physical barrier properties. In another study done by Li et al. (2019), the effects of different parameters such as the concentration of surfactants, pH of the GO suspension, and mass ratio of linseed oil/GO were studied. Results proved that polyetheramine can be used for the GO chemical stitching to improve the chemical stability of GO microcapsules. Significant self-healing performance was observed for self-healing PU coatings filled with 10 wt.% linseed oil loaded GO microcapsules.

Epoxy-silicone self-healing coatings containing chitosan-reduced GO (CH-rGO) microcapsules loaded with polysiloxane as the healing agent was fabricated by Zheng et al. (2021). Acryloyloxy and methoxy silane groups were selected to synthesise the polysiloxane-based healing agent. Different synthesis steps of polysiloxane-loaded CH-rGO microcapsules are presented schematically in Figure 11.3. The characterisation of CH-rGO microcapsules showed that by increasing the CH/GO ratio, the morphology of microcapsules reaches closer to spherical and both particle size and reduction degree of GO nanosheets increase. The incorporation of polysiloxane-loaded CH-rGO microcapsules in the epoxy-silicone coating enhanced both the tensile strength (≈92%) and corrosion mitigation performance of coatings. Assessment of the self-healing capability of coatings showed that superior self-healing properties were obtained in the presence of dual moisture/ultraviolet light stimuli.

11.3 INHIBITOR-LOADED CARBON ALLOTROPES

Corrosion inhibitors are one of the best additives for improving the anticorrosion and self-healing performance of the coatings. When coated metals and alloys are exposed to the corrosive electrolyte, corrosion inhibitors are dissolved and transported to the metal surface. So, the interface corrosion rate is reduced due to different inhibition mechanisms. Despite the essential role of corrosion inhibitors, their introduction into the coatings is limited due to the destructive inhibitor-coating interactions, especially during the curing and application steps. Inhibitor containers are the most effective approach to overcome this problem. Inhibitors are loaded in the containers or on

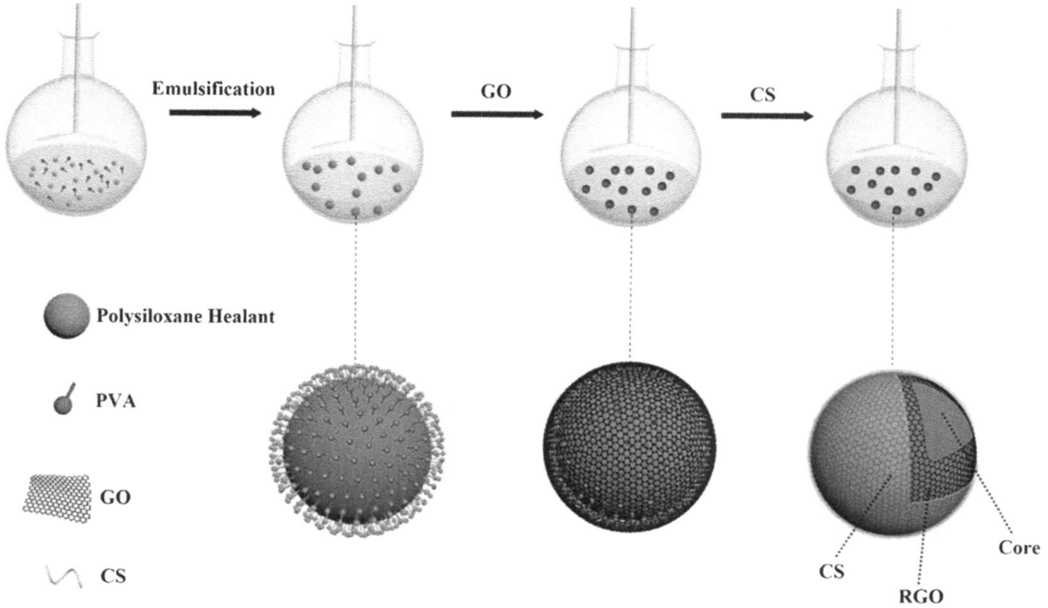

FIGURE 11.3 Different synthesis steps of polysiloxane-loaded CH-rGO microcapsules. Reprinted from Zheng et al. (2021) with permission from Elsevier.

their surface and stabilised until activated by the corrosion initiation. In addition to inhibitor stabilisation, containers have other positive effects such as on-demand release, long-term protection, and barrier action. CAs have attracted attention among different container families due to their special structures (Fayyad et al. 2016). CAs such as CNTs and CSs could be filled with corrosion inhibitors, and others such as graphene and GO flakes could be functionalised with inhibitor molecules (Gupta et al. 2017). In the novel applications, the loading of corrosion inhibitors in the CAs is performed to carry inhibitors and better dispersion of CAs in the coating matrix. This part discusses different CAs loaded/filled with organic/inorganic corrosion inhibitors.

11.3.1 Graphene and Graphene Oxides

Graphene and GO change anticorrosion coatings' physical, mechanical and chemical characteristics. The loading of graphene layers with corrosion inhibitors has several benefits for corrosion protection of coatings (Cui, Xiong et al. 2021; Ramezanzadeh et al. 2016; Nikpour et al. 2017; Ding, Chen, Lv, Zhang, Zhao, Liu, Wang, Gui, Li, Tang et al. 2019; Yu et al. 2016):

- **Improving the flakes–matrix interactions:** surface functional groups could be used to improve the interaction of graphene layers with the coating matrix, leading to fewer defects and better coating integrity against corrosive electrolyte diffusion.
- **Uniform dispersion of layers:** well-dispersed graphene flakes could be obtained using surface modification of layers with organic corrosion inhibitors. The preparation of uniform composite coating is vital regarding self-healing and anticorrosion coatings.
- **Barrier effect against electrolyte diffusion:** electrolyte and corrosive moieties' diffusion paths could be lengthy through parallel orientation of graphene and GO concerning the coating-substrate interface.
- **Enhancing the coating-metal adhesion strength:** anticorrosion coating service life could be improved by strong interaction between adsorbed inhibitor molecules on the

graphene/GO flakes and the oxide layer formed on the metal surface. In many cases, coating with stronger adhesion shows better corrosion protection.

- **Increasing the coating hydrophobicity:** the hydrophobic behaviour of coatings could be improved using functionalised graphene/GO layers. The presence of hydrophobic moieties on the surface of graphene/GO sheets reduces the aqueous electrolyte adsorption and consequent corrosion control on the metal surface.
- **Active protection:** corrosion inhibitors could be released from the graphene/GO surfaces and interact with the metal surface to control the interfacial corrosion rate. Chelates or other compounds could be formed through the interaction of inhibitors with the corrosion products such as OH^- anions or metal cations and precipitated on the metal surface. Also, these chelates could block the coating defects as the main electrolyte paths into the metal-coating interface. Another active protection mechanism is the direct adsorption of released corrosion inhibitors on the substrate's anodic/cathodic areas.

GO layers have been loaded with hydroxyethylidene-1 and 1-diphosphonic acid by Cui et al. and introduced into the epoxy coating (Cui, Xiong et al. 2021). They concluded that adding loaded GO layers improved the coating corrosion protection through chelate formation between Fe^{2+} cations and corrosion inhibitors. The obtained chelates block the diffusion of the electrolyte through the coating defects. In another study, GO was functionalised with an aromatic diamine to participate directly in the curing process of an anticorrosion epoxy coating (Ramezanzadeh et al. 2016). In this situation, the diffusion pathways are reduced and the barrier performance of a coating is enhanced against corrosion. The active self-healing behaviour is another protection mechanism in the presence of inhibitor-loaded GO nanosheets. The GO sheets could be loaded with organic corrosion inhibitors such as plant extract to induce active performance in the coatings (Nikpour et al. 2017; Ding, Chen, Lv, Zhang, Zhao, Liu, Wang, Gui, Li, Tang et al. 2019). Also, inorganic inhibitors, i.e., transition metal cations, have been used to reinforce graphene/GO sheets against corrosion. Sometimes, combining two or more corrosion inhibitors shows a special behaviour called the synergistic effect (Izadi et al. 2020). Figure 11.4 shows the synergistic effect between green corrosion inhibitor-loaded GO nanosheets, and zinc cations in a nanocomposite self-healing epoxy coating applied on the mild steel substrate. The deposited chelates in the artificial defect are the main active protection mechanism.

Nanocontainers are one of the most feasible and effective corrosion protection agents in self-healing coatings (Javidparvar, Naderi, and Ramezanzadeh 2020). Containers based on the CAs are good candidates for reinforcing the coating's corrosion protection. Different organic and inorganic corrosion inhibitors have been loaded on the graphene/GO nanosheets to prepare nanocontainers with on-demand release capability (Liu et al. 2018; Javidparvar, Naderi, and Ramezanzadeh 2020). A pH-sensitive nanocarrier based on GO nanoplates loaded with Ce^{3+} cations has been used to induce a self-healing effect into the epoxy coating structure (Javidparvar et al. 2019). The Ce^{3+} cations could be exchanged with Na^+ cations in the diffused electrolyte and fed into the electrolyte. The released Ce^{3+} cations could interact with hydroxyl anions in the vicinity of the metal-coating interface and form hydroxide/oxide precipitates that control the corrosion rate. The controlled and continual release of corrosion inhibitors from the CA-based nanocontainers guarantees the long-term performance of anticorrosion coating and its self-healing behaviour.

11.3.2 Carbon Nanotubes

CNTs with special structures have been used as anticorrosion additives in protective coatings. The addition of CNTs to the coatings enhances the mechanical strength and improves the corrosion resistance by reducing the coating defects and improving the barrier action of the protective system (Butyrskaya, Nechaeva, and Zapryagaev 2016). Also, the hydrophilicity of the coatings could be engineered and reduced with the addition of CNTs resulting in less electrolyte diffusion into the

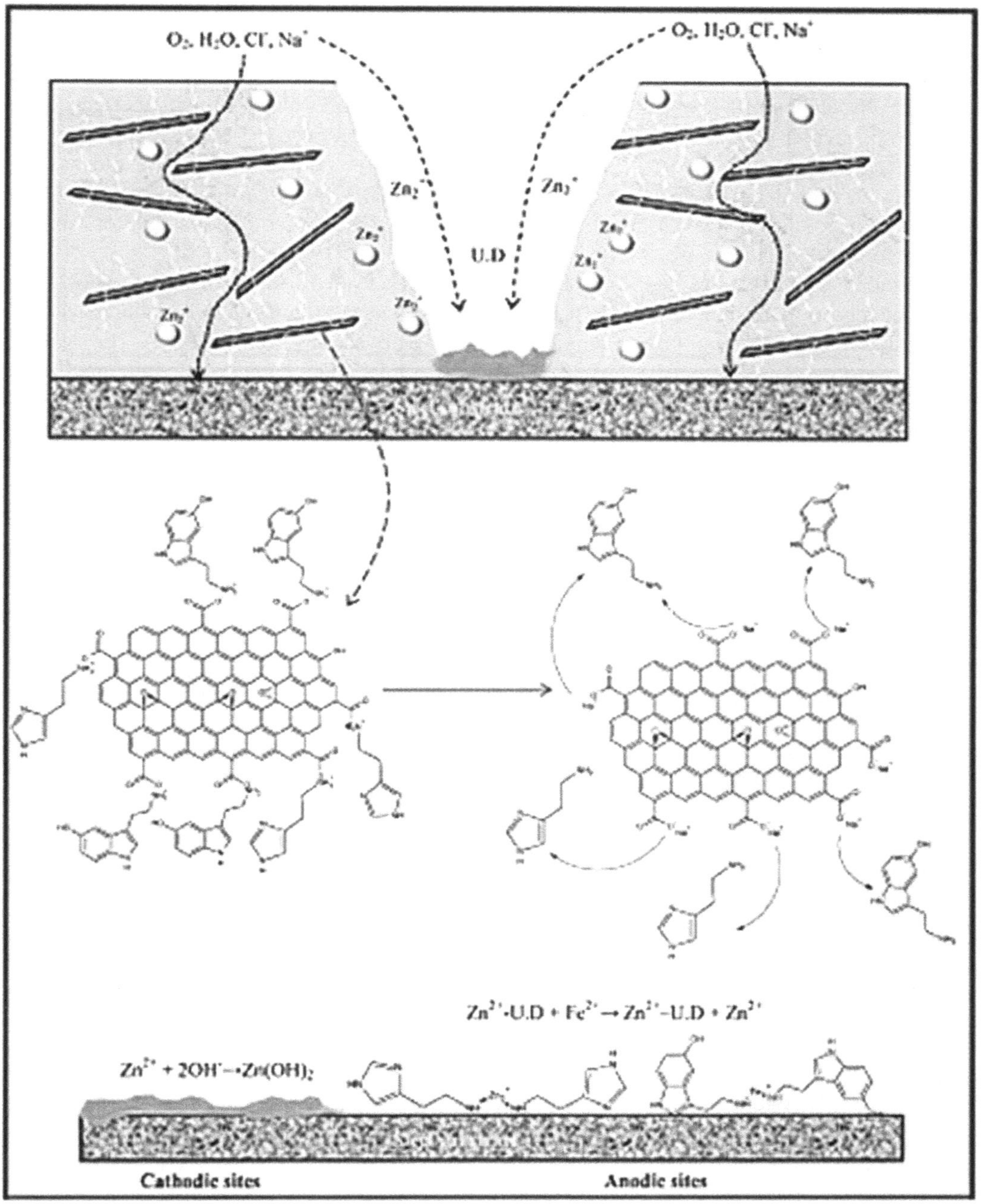

FIGURE 11.4 The common corrosion protection mechanism in the presence of green inhibitor-loaded GO nanosheets and zinc cations in the epoxy-coated mild steel substrate. Reprinted from Nikpour et al. (2017) with permission from Elsevier.

coatings. CNTs could be reinforced with corrosion inhibitors to provide active functionality in the anticorrosion coatings. Surface modification and inhibitor entrapment are the two common methods to prepare inhibitor-loaded CNTs (Yeole, Agarwal, and Mhaske 2016; Cen, Cao, and Chen 2020). The surface functionalising of the CNTs could be performed using covalent and noncovalent methods. Covalent methods could destroy the structural integrity of the CNTs. On the other hand, surface modification of CNTs through noncovalent methods using corrosion inhibitors is an established procedure for anticorrosion composite coatings (Cui et al. 2018). Different organic corrosion inhibitors could be adsorbed on the CNTs' outer surface through π-π interactions and other physical bonds. The noncovalent functionalising of CNTs has two positive effects on the corrosion behaviour of protective coating. First, the main problem of CNT-containing composite coatings (nanotubes aggregation) will be resolved. Second, the corrosion process could be interrupted due to the release of anchored inhibitors on the CNTs (Davoodi, Honarbakhsh, and Farzi 2015). The nanotubes could also be used as capsules and feed the corrosion inhibitors into the coatings for on-demand release.

Although it seems that the inner lumen of CNTs could be loaded with corrosion inhibitors, there are limited reports in the literature. An LBL method has been used to entrap 2-mercaptobenzothiazole between two polypyrrole and polyacrylic acid (Yeole, Agarwal, and Mhaske 2016). The designed containers are swallowed depending on pH as smart reservoirs and fed corrosion inhibitors into the damaged areas of the coating-metal interface. The released organic inhibitors could be adsorbed on the cathodic or anodic regions of the substrate and reduce the destructive interfacial reactions. Also, chelate formed between inhibitor molecules and metal cations or compounds formed due to the other interactions of inhibitors and corrosion products blocks the damaged surficial area and electrolyte approaching the interface. However, other inhibition mechanisms have been reported for CNTs containers. Some corrosion inhibitors, i.e., poly(2-butylaniline), are interacted with corrosive moieties such as Cl^- anions (Cui et al. 2018). So, Cl^- anions are prohibited from reaching the metal surface, and the occurrence of uniform and localised corrosion will be postponed. On the other hand, metal salts are well-known examples of inorganic corrosion inhibitors that could be loaded on the CNTs to improve the corrosion protection of coatings. The inner and outer surface of CNTs has been loaded with nitrates of lanthanum and cerium and used to protect magnesium alloys coated with silane compounds (Montemor and Ferreira 2008).

11.3.3 Carbon Hollow Spheres

CSs with unique characteristics is another form of CAs used to feed corrosion inhibitors into the protective coatings. The different characteristics of CS's shell structure could be adjusted to deliver corrosion inhibitors with different sizes and affinities (Li et al. 2016). CSs could be filled with corrosion inhibitors and used as pH-responsive containers. After successful loading, pH variations can lead to inhibitor release from the CS structure. Any changes in shell chemical structure such as surface group protonation or dissolution of surficial moieties alter the surface charge of CSs.

Conversely, inhibitor molecules could also be ionised or protonated according to the electrolyte pH (Haddadi et al. 2021). Thus, an inhibitor-loaded container based on CS could be acquired, capable of releasing the corrosion inhibitor into the coatings in the designed situation of the coating-electrolyte-metal system (Liu et al. 2021). A schematic illustration of the inhibitive action of CS nanocontainer loaded with 2-mercaptobenzimidazole embedded in an epoxy coating is provided in Figure 11.5. In addition to the aforementioned inhibitive behaviour of CSs in protective coatings, improving the coating integrity and reducing structural defects as the main electrolyte diffusion paths are other benefits of composite coatings reinforced with CS nanocontainers (Haddadi et al. 2021). CSs with inorganic inhibitors have also been used to protect metals in the form of metal salt-loaded carriers (Behgam, Mahdavian, and Ramazani 2020). The released metal cations, i.e., transition metals, form hydroxide/oxide surface products at the metal-coating interface reducing the rate of corrosion at the defect areas.

11.3.4 Other Carbon Allotropies

Carbon dots are environmentally friendly nanomaterials with dimensions of less than 10 nm. Different top-down and bottom-up methods could synthesise carbon dots from carbon-rich sources

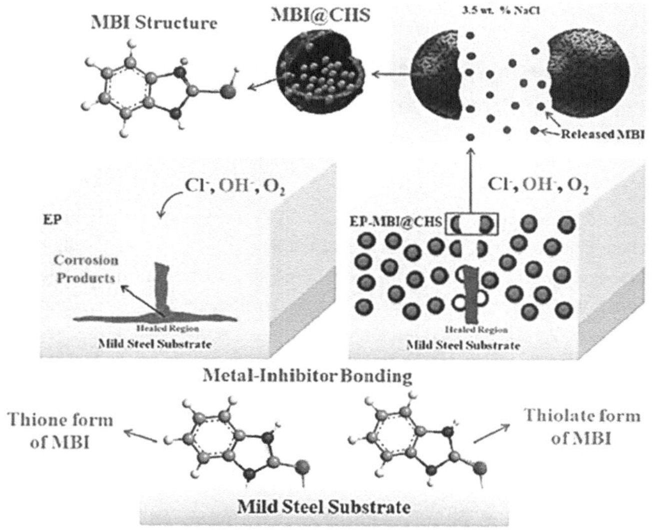

FIGURE 11.5 The CS nanocontainer loaded with 2- mercaptobenzimidazole and its inhibitive action against saline solution for coated mild steel. Reprinted from Haddadi et al. (2021) with permission from Elsevier.

such as CNTs, graphite, GO, starch, and others (Al Jahdaly et al. 2021). One of the most attractive characteristics of the carbon dots is their good functionalisation, which makes them good candidates for drug delivery, photoluminescence/electrochemical sensing, and design of corrosion inhibitors (Yang et al. 2019; Al Jahdaly et al. 2021). Like the other CAs, carbon dots could be loaded with corrosion inhibitors such as imidazole ionic liquid to improve the corrosion resistance of metals. Different surface modifications could be used for the consequent grafting of corrosion inhibitors on the surface of carbon dots. During pre-modification steps, the created surface functional groups act as a bridge between carbon dots and corrosion inhibitors (Yang et al. 2019). Modified carbon dots also improve the barrier characteristics of the coatings.

Fullerene, carbon nanofibers, and nanodiamonds (NDs) are other CAs could be used as corrosion inhibitor reservoirs. The reported results about incorporating inhibitor carriers based on the above-mentioned CAs in the protective coatings are rare. NDs with high surface-to-volume ratio and chemical inertness have been loaded with dodecylamine and incorporated into the epoxy matrix to protect mild steel plates (Rahmani, Shojaei, and Pirhady Tavandashti 2020). Coating deficiencies such as porosities are considerably reduced in the presence of loaded NDs. Also, the obtained composite coating's mechanical and wear resistance properties are improved due to the high hardness of NDs. However, the main protective effect of ND containers is due to the on-demand release of corrosion inhibitors. The inhibitor deprotonation-ionisation due to the local pH changes is the main reason for the release of inhibitive molecules from the NDs surface (Rahmani, Shojaei, and Pirhady Tavandashti 2020). Released inhibitor molecules could disrupt the interfacial corrosion progress.

Among the CAs, GO is the most frequently used one in industrial applications. Table 11.1 provides a quick review of the inhibitor-loaded CAs.

11.4 CONDUCTIVE POLYMER-GRAFTED CARBON ALLOTROPES

Attributed to their excellent environmental stability, simple processability, and significant redox properties, conductive polymers (CPs) such as PANI, polypyrrole (PPy), and others have been recently defined as the most valuable materials to induce the self-healing performance in coatings'

TABLE 11.1
A Quick Review of the Investigated CAs Loaded With Corrosion Inhibitors and Their Anti-Corrosion Effect on the Protective Coatings

Carbon Allotropes	Corrosion Inhibitor	Probable Protective Mechanism
Graphene/GO	hydroxyethylidene-1, 1-diphosphonic acid, P-Phenylenediamine, Urtica Dioica leaf extract, Aminoazobenzene, diaminoazobenzene, L-cysteine, Cerium (III) cations, benzimidazole, metronidazole, Cyclodextrin, octylamine, Zn (II) cations, *p*-aminophenol, 3-aminopropyl triethoxysilane, 3-glysidyloxypropyl trimethoxysilane, Cyclodextrin, Pr (III) cations,	1. On-demand release of corrosion inhibitor 2. Defect reduction 3. Barrier effect 4. Localised corrosion inhibition 5. Adhesion promotion (Dry & Wet) 6. elimination of corrosive moieties 7. Water adsorption reduction (improving hydrophobicity) 8. Coating compactness
CNTs	2-aminopyridine, 2-mercaptobenzothiazole, poly(2-butylaniline), Bis-[triethoxy amino] silane, ,5-diphenyl-imidazole, polybenzoxazine, Polyindole, 3-aminophenoxyphthalonitrile	
Carbon hollow spheres	2-mercaptobenzimidazole, corrosion inhibitor M16, Zinc Cations, walnut extract	
Other CAs	dodecylamine, imidazole ionic liquid, N-doped, polydopamine, 3-aminopropyltriethoxysine, sodium dodecyl sulphate	

structure (Tallman et al. 2002; Armelin et al. 2009; Fazli-Shokouhi, Nasirpouri, and Khatamian 2019, 2021). However, PANI has been more favoured for fabricating self-healing anticorrosive coatings due to its doping/de-doping chemistry structure, various oxidation states, and the ability to act as a corrosion mitigation system (Chandrasekhar 2018). The CPs may be directly polymerised on the surface of metal substrates with dopant molecules or grafted on the surface of the carriers and then dispersed into the coatings' structure. In this condition, a conductive and electroactive film with great redox ability can be formed on the underlying substrate. The improvement in the anticorrosive properties of the coatings by using CPs has been the subject of various research works (Zhang et al. 2018; Wei et al. 2015; Yan et al. 2021; Fazli-Shokouhi, Nasirpouri, and Khatamian 2021). It was illustrated that CPs could act as the inhibitive agent to induce the self-healing property in the coating structure and electrochemically protect exposed metallic substrates. It has also been shown that functionalising CAs with CPs changes their surface reactivity from hydrophilic to slightly hydrophobic, improving dispersibility and compatibility with a polymer matrix (Hayatgheib et al. 2018).

The CPs can be decorated on the surface of the CAs through various methods, such as electrochemical polymerisation, in situ chemical polymerisation, solution mixing, interfacial polymerisation, and electrophoretic routs (Kumar, Kumar, and Awasthi 2018; Yazdi et al. 2019). For the polymer that does not melt through the heat treatment (e.g., PANI), the melt mixing procedure is not applicable for synthesising the PANI/CAs nanocomposites (Yazdi et al. 2019). For the PANI/CNT nanocomposite, it was reported that before in situ polymerisation, through π-π interactions, aniline molecules can be adsorbed on the surface of the CNT (Amrollahi et al. 2019). In this condition, the polymerisation process for forming the CP layer begins from the adsorbed molecules acting as nucleation sites, which leads to the wrapping of CPs on the CNTs' entire surface (Khodadadi Yazdi et al. 2018). Based on the literature, the functionalisation of CAs may increase their interaction with PANI chains and their dispersibility in organic solvents. For instance, amino-functionalised CNT serves as reactive sites in aniline polymerisation where PANI chains may be chemically grafted to the surface of CNTs (Jeon et al. 2010).

PANI molecules, a well-known conducting polymer, were used to induce the self-healing characteristics of different coatings when grafted on the surface of the CAs. The three types of PANI oxidation states are pernigraniline base (PB), emeraldine base (EB), and leucoemeraldine base (LB), which are assigned to complete oxidation, semi-complete oxidation, and total reduction of PANI, respectively (Akbarzadeh et al. 2019). However, among these form of PANI, only protonated emeraldine salt (ES) behave as a conductor and can be used as the active agent to improve the self-healing properties of the coatings (Bhadra et al. 2009). For the layer containing PANI on steel substrate under conditions of immersion in the electrolyte, it was illustrated that the electrochemical potential of the metallic surface is increased upon transforming the PANI structure by consuming the released electron through the corrosion process (Wessling 1994; Gupta et al. 2013). The transformation of PANI into different states and the change of surface electrochemical potential encourage the formation of the γ-Fe_2O_3 layer instead of the α-Fe_2O_3 layer at the coating/metallic substrate interface (Wessling 1994; Gupta et al. 2013). The creation of the γ-Fe_2O_3 layer prevents the formation of iron hydroxides. Iron hydroxides cause the coating delamination due to their large volume. Research studies illustrated that direct doping of functionalised CAs into coatings structure is the most straightforward route for fabricating self-healing coatings. It was demonstrated that anilines' redox catalytic characteristics facilitate the production of passivation coatings on metal surfaces. Ye et al. synthesised aniline trimer (AT) functionalised graphene sheets (SAT-G) by the silanisation and intercalation of trianiline precursors (Ye et al. 2019). Results illustrated that the epoxy coating containing 0.5 wt.% SAT-G has excellent long-term corrosion resistance and self-healing properties. The good anticorrosive performance of the coatings after adding SAT-G to the coating matrix can be due to the well-distribution of the graphene sheets into the epoxy coatings structure, which may result in maximising their shielding efficacy. Also, by capturing electrons, a silanised aniline trimer with intact electroactivity induces the formation of the passivation layer consisting mainly of Fe_2O_3 and Fe_3O_4 (Figure 11.6). In the presence of the CPs in the structure of the coatings, they act as the catalyst for the formation of the passive film and

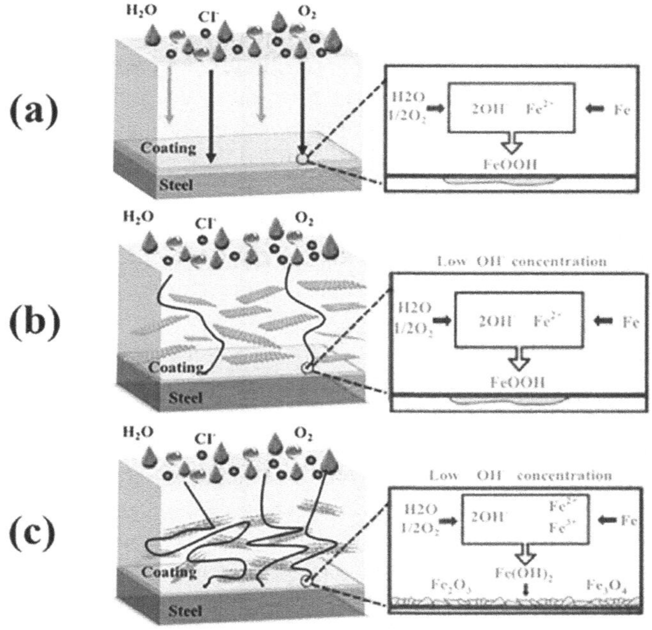

FIGURE 11.6 The protection mechanism: (a) neat epoxy coating, (b) epoxy coating by adding 0.05 wt.% GO, and (c) epoxy coating after adding 0.05 wt.% PPy/GO. Reprinted from Zhu et al. (2020) with permission from Elsevier.

then improve the self-healing protection performance of the coating. While in the absence of the CPs in the coating structure, the corrosion process occurs at the interface of the coating/metallic substrate, leading to coating delamination.

The oxidising of the CAs is the first process for wrapping most of the CPs on their surface. Because the CPs have a positive charge and the oxidised CAs have a negative charge, research results illustrated that the adsorption of the CPs on the oxidised CAs occurs via both covalent and electrostatic interactions. Akbarzadeh et al. (2019) studied the effect of adding CNT grafting with PANI on the anticorrosion protection performance of the silane coating. Their results illustrated that the self-healing anticorrosive silane coating is produced after incorporating the CNT/PANI in the coating structure. It was also demonstrated that the zinc-rich epoxy coating's barrier performance and cathodic protection properties significantly improved after employing the PANI/GO in the coating matrix as an inhibitive pigment (Ramezanzadeh et al. 2017). The conductive path between the zinc particles is created in the zinc-rich coating, and the cathodic protection is significantly enhanced when the conductive polyaniline fiber is decorated on the surface of the graphene sheets.

Similar results about improving self-healing performance of the coatings after decorating the poly(2-butylaniline) and polypyrrole on the surface of the CAs (as the inhibitive pigment) were illustrated in the literature (Zhu et al. 2020; Chen et al. 2017). It was shown that modifying graphene sheets with poly(2-butylaniline) lead to stable dispersivity. It also encourages the creation of a passive film on the metal surface (Chen et al. 2017). The polypyrrole molecules enhance the self-healing performance of the coating by the formation of Fe_3O_4 and Fe_2O_3 passive films on steel surfaces (Qiu et al. 2017). However, the catalytic passivation of the metallic substrates after using the poly(2-butylaniline) and polypyrrole has only been reported in a few cases (Ding, Chen, Lv, Zhang, Zhao, Liu, Wang, Gui, Li, and Tang 2019). Ioniţă and Prună (2011) fabricated PPy-modified CNT pigments and dispersed them into the coating matrix. Adding polypyrrole/CNT to the coatings matrix was observed to enhance its active anticorrosion resistance. Also, Qiu et al. offered synergistic corrosion protection for the mild steel substrate after employing the graphene sheets decorated with PPy in the coating matrix (Qiu et al. 2017).

Although the previously published works illustrated that the coatings' active/barrier anticorrosive properties were improved after the incorporation of conductive polymers into the structure of the coatings, the utility of CPs for corrosion prevention is controversial. Based on the literature, the coatings' corrosion breakdown is accelerated when the added CPs fail to passivate the surface of the metallic substrate. It can be related to more considerable flaws in its structure, which can increase cation mobility (Rohwerder and Michalik 2007). Furthermore, the molecular composition and the interfacial chemistry of the CAs decorated with CPs can significantly affect the anticorrosive performance of the coatings. The modification of CPs by doping inhibitive agents is an effective technique for encouraging the creation of passive metal surfaces (self-healing performance of the coating) and improving the coatings' anticorrosion functioning (Rohwerder and Michalik 2007; Deshpande et al. 2014).

The research work results illustrated that using the CPs along with corrosion inhibitors can dramatically enhance their redox catalytic passivation performances. The synergistic effect between the PANI molecules and zinc or cerium oxide to improve the self-healing anticorrosion protection performance of the coatings was previously illustrated (Ramezanzadeh, Bahlakeh, and Ramezanzadeh 2018; Najmi et al. 2021). It was shown that the synergistic effect might lead to the formation of a high-quality passivation film on the surface of metallic substrates. Najmi et al. (2021) synthesised grafted multi-walled CNT (MWCNT) with PANI molecules and Zn^{2+} cations as an inhibitive pigment for improving the active protection performance of the epoxy coating. For this purpose, the MWCNTs were first oxidised and then modified with PANI nanofibers and Zn^{2+} cations and dispersed in epoxy coatings. Their results illustrated that the coatings' barrier/active protection performance was significantly enhanced after incorporating MWCNT decorated with PANI-Zn into the coating matrix. In the absence of the PANI-Zn, the MWCNT suppressed the cathodic reaction by absorbing oxygen. Although in the presence of the grafted PANI-Zn system, the adsorbing of the PANI and Zn^{2+} cations

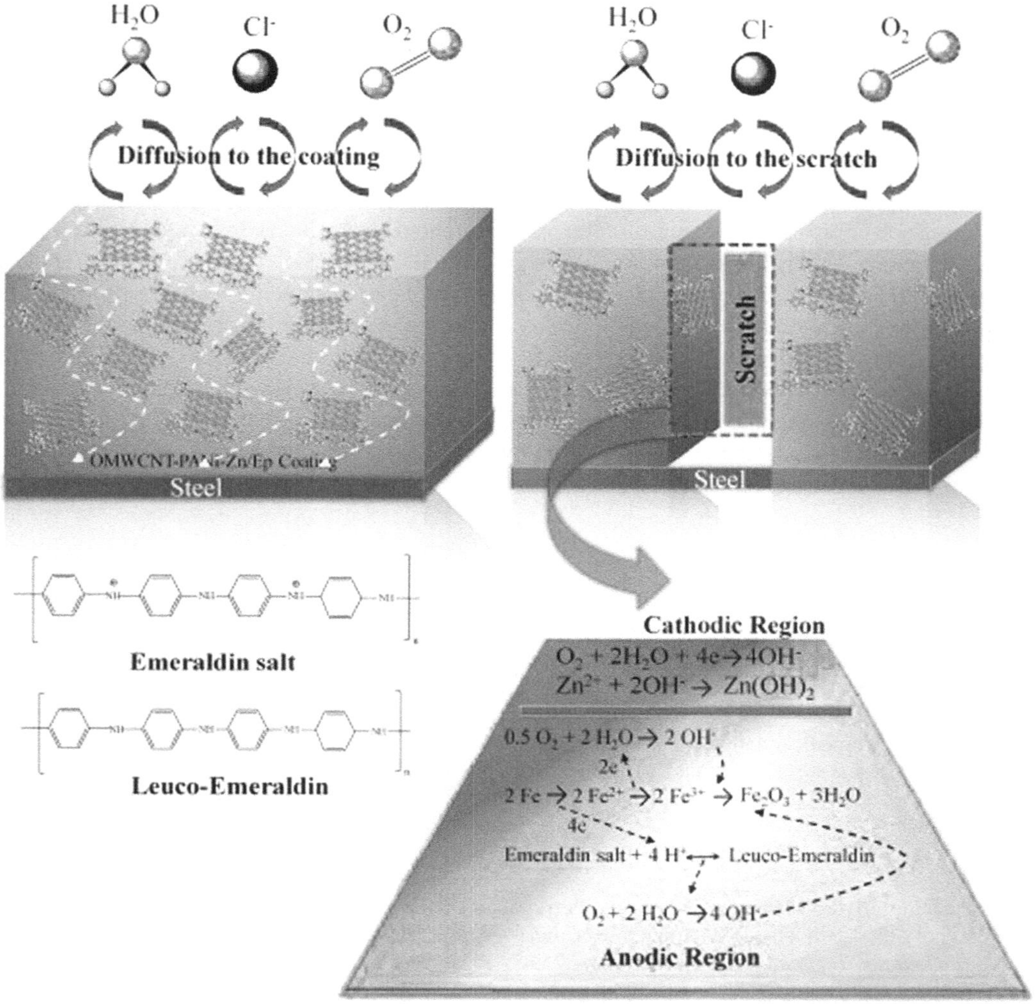

FIGURE 11.7 Schematic representation of the self-healing anticorrosive performance of the epoxy coating after incorporation of MWCNT grafted with PANI molecules and Zn^{2+} cations into the coating structure. Reprinted from Najmi et al. (2021) with permission from Elsevier.

at the anodic and cathodic sites is the primary inhibition mechanism. Their results illustrated that releasing the inhibitive species from the O-MWCNT-PANI-Zn structure may lead to the fabrication of stable zinc hydroxides and Zn/PANI complex compound at the coating/metallic substrate interface (Najmi et al. 2021). A schematic representation of the self-healing effect of the epoxy coating incorporating MWCNT grafted with PANI molecules and Zn^{2+} cations is provided in Figure 11.7. This mechanism enhances the PANI's catalytic role in fabricating the passive film on the metal surface. Similar results were reported for GO-PPy-Zn structure synthesising to improve the self-healing/barrier anticorrosion performance of epoxy coating (Mohammadkhani et al. 2020).

To fabricate an inhibitive pigment for the epoxy coating, the PANI-CeO_2 was decorated on the GO sheets (Ramezanzadeh, Bahlakeh, and Ramezanzadeh 2018). The results illustrated that GO nanosheets' active and barrier corrosion inhibition properties were dramatically enhanced by decorating them with the CeO_2 and PANI compounds. A synergistic corrosion inhibition can be achieved by doping GO with the PANI with CeO_2. The main inhibition mechanism is forming a self-healing passive film containing PANI and CeO_2 on the active sites of the metallic substrate.

11.5 SUMMARY

Self-healing/self-repairing coatings have found great potential for corrosion protection. Thanks to their ability to entrap corrosion inhibitors and film formers and their ability to be grafted by conductive polymers, CAa are a great candidate for application in self-healing/self-repairing coatings. A variety of coatings with self-healing/self-repairing have been developed based on carbon allotropes. The compatibility of CAs in the coating matrix greatly impacts the coating properties. Such compatibility depends on their surface chemistry, which can be tuned with surface functionalisation. Among different CAs, 0D (e.g., fullerene and CSs), 1D (e.g., CNTs), and 2D (e.g., graphene and GO) carbonaceous nanostructures are capable of being loaded with corrosion inhibitors or healing agents for subsequent release to the damaged area of the coatings. They can be also grafted with conductive polymers, which in turn facilitate the production of passivating coatings on metal surfaces due to their redox catalytic characteristics. The 2D structures are most frequently CAs used for corrosion inhibitor loading and conductive polymer grafting, while CSs with 0D structures have been frequently employed for encapsulating healing agents generally based on resins or air-drying oils.

REFERENCES

Akbarzadeh, Sajjad, Mohammad Ramezanzadeh, Bahram Ramezanzadeh, Mohammad Mahdavian, and Reza Naderi. 2019. "Fabrication of highly effective polyaniline grafted carbon nanotubes to induce active protective functioning in a silane coating." *Industrial & Engineering Chemistry Research* 58 (44):20309–20322.

Al Jahdaly, Badreah Ali, Mohamed Farouk Elsadek, Badreldin Mohamed Ahmed, Mohamed Fawzy Farahat, Mohamed M. Taher, and Ahmed M. Khalil. 2021. "Outstanding graphene quantum dots from carbon source for biomedical and corrosion inhibition applications: A review." *Sustainability* 13 (4):2127.

Amrollahi, S., B. Ramezanzadeh, H. Yari, M. Ramezanzadeh, and M. Mahdavian. 2019. "Synthesis of polyaniline-modified graphene oxide for obtaining a high performance epoxy nanocomposite film with excellent UV blocking/anti-oxidant/anti-corrosion capabilities." *Composites Part B: Engineering* 173:106804.

Andreeva, Daria V., and Dmitry G. Shchukin. 2008. "Smart self-repairing protective coatings." *Materials Today* 11 (10):24–30.

Armelin, Elaine, Álvaro Meneguzzi, Carlos A. Ferreira, and Carlos Alemán. 2009. "Polyaniline, polypyrrole and poly (3, 4-ethylenedioxythiophene) as additives of organic coatings to prevent corrosion." *Surface and Coatings Technology* 203 (24):3763–3769.

Bazilevsky, Alexander V., Kexia Sun, Alexander L. Yarin, and Constantine M. Megaridis. 2007. "Selective intercalation of polymers in carbon nanotubes." *Langmuir* 23 (14):7451–7455.

Bazilevsky, Alexander V., Kexia Sun, Alexander L. Yarin, and Constantine M. Megaridis. 2008. "Room-temperature, open-air, wet intercalation of liquids, surfactants, polymers and nanoparticles within nanotubes and microchannels." *Journal of Materials Chemistry* 18 (6):696–702.

Behgam, Rahemeh, Mohammad Mahdavian, and Ahmad Ramazani. 2020. "Fabrication of hollow carbon spheres doped with zinc cations to enhance corrosion protection of organosilane coatings." *Surfaces and Interfaces* 21:100696. doi: 10.1016/j.surfin.2020.100696.

Bekas, D.G., Kyriaki Tsirka, Dimitris Baltzis, and Alkiviadis S. Paipetis. 2016. "Self-healing materials: A review of advances in materials, evaluation, characterization and monitoring techniques." *Composites Part B: Engineering* 87:92–119.

Bhadra, Sambhu, Dipak Khastgir, Nikhil K. Singha, and Joong Hee Lee. 2009. "Progress in preparation, processing and applications of polyaniline." *Progress in polymer science* 34 (8):783–810.

Butyrskaya, E.V., L.S. Nechaeva, and S.A. Zapryagaev. 2016. "Theoretical study of the corrosion protection mechanism by carbon nanotubes." *Computational and Theoretical Chemistry* 1090:1–5. doi: 10.1016/j.comptc.2016.05.012.

Cen, Hongyu, JiaoJiao Cao, and Zhenyu Chen. 2020. "Functionalized carbon nanotubes as a novel inhibitor to enhance the anticorrosion performance of carbon steel in CO_2-saturated NaCl solution." *Corrosion Science* 177:109011. doi: 10.1016/j.corsci.2020.109011.

Chandrasekhar, Prasanna. 2018. *Conducting polymers, fundamentals and applications*. Springer.

Chen, Cheng, Shihui Qiu, Mingjun Cui, Songlv Qin, Guoping Yan, Haichao Zhao, Liping Wang, and Qunji Xue. 2017. "Achieving high performance corrosion and wear resistant epoxy coatings via incorporation of noncovalent functionalized graphene." *Carbon* 114:356–366.

Correa-Duarte, Miguel A., Adam Kosiorek, Witold Kandulski, Michael Giersig, and Luis M. Liz-Marzan. 2005. "Layer-by-layer assembly of multiwall carbon nanotubes on spherical colloids." *Chemistry of Materials* 17 (12):3268–3272.

Cui, Gan, Chuchu Zhang, Ailing Wang, Xin Zhou, Xiao Xing, Jianguo Liu, Zili Li, Qiqi Chen, and Qinglong Lu. 2021. "Research progress on self-healing polymer/graphene anticorrosion coatings." *Progress in Organic Coatings* 155:106231.

Cui, Jincan, Zhi Xiong, Hanxun Qiu, Jing Li, and Junhe Yang. 2021. "Functionalized graphene oxide: Carrier for corrosion inhibitor and barrier in waterborne epoxy coatings." *Composites Part A: Applied Science and Manufacturing* 144:106354. doi: 10.1016/j.compositesa.2021.106354.

Cui, Mingjun, Siming Ren, Shihui Qiu, Haichao Zhao, Liping Wang, and Qunji Xue. 2018. "Non-covalent functionalized multi-wall carbon nanotubes filled epoxy composites: Effect on corrosion protection and tribological performance." *Surface and Coatings Technology* 340:74–85.

Davoodi, Ali, Saleheh Honarbakhsh, and Gholam Ali Farzi. 2015. "Evaluation of corrosion resistance of polypyrrole/functionalized multi-walled carbon nanotubes composite coatings on 60Cu–40Zn brass alloy." *Progress in Organic Coatings* 88:106–115. doi: 10.1016/j.porgcoat.2015.06.018.

Deshpande, Pravin P., Niteen G. Jadhav, Victoria J. Gelling, and Dimitra Sazou. 2014. "Conducting polymers for corrosion protection: A review." *Journal of Coatings Technology and Research* 11 (4):473–494.

Ding, Rui, Si Chen, Jing Lv, Wei Zhang, Xiao-dong Zhao, Jie Liu, Xiao Wang, Tai-jiang Gui, Bing-jun Li, and Yong-zheng Tang. 2019. "Study on graphene modified organic anti-corrosion coatings: A comprehensive review." *Journal of Alloys and Compounds* 806:611–635.

Ding, Rui, Si Chen, Jing Lv, Wei Zhang, Xiao-dong Zhao, Jie Liu, Xiao Wang, Tai-jiang Gui, Bing-jun Li, Yong-zheng Tang, and Wei-hua Li. 2019. "Study on graphene modified organic anti-corrosion coatings: A comprehensive review." *Journal of Alloys and Compounds* 806:611–635. doi: 10.1016/j.jallcom.2019.07.256.

Esser-Kahn, Aaron P., Susan A. Odom, Nancy R. Sottos, Scott R. White, and Jeffrey S. Moore. 2011. "Triggered release from polymer capsules." *Macromolecules* 44 (14):5539–5553.

Fayyad, Eman M., Kishor Kumar Sadasivuni, Deepalekshmi Ponnamma, and Mariam Al Ali Al-Maadeed. 2016. "Oleic acid-grafted chitosan/graphene oxide composite coating for corrosion protection of carbon steel." *Carbohydrate Polymers* 151:871–878. doi: 10.1016/j.carbpol.2016.06.001.

Fazli-Shokouhi, Sara, Farzad Nasirpouri, and Maasoumeh Khatamian. 2019. "Polyaniline-modified graphene oxide nanocomposites in epoxy coatings for enhancing the anticorrosion and antifouling properties." *Journal of Coatings Technology and Research* 16 (4):983–997.

Fazli-Shokouhi, Sara, Farzad Nasirpouri, and Maasoumeh Khatamian. 2021. "Epoxy-matrix polyaniline/p-phenylenediamine-functionalised graphene oxide coatings with dual anti-corrosion and anti-fouling performance." *RSC Advances* 11 (19):11627–11641.

Fontana, Mars Guy, and Norbert D. Greene. 2018. *Corrosion engineering*. McGraw-hill.

García, S.J., H.R. Fischer, and S. Van Der Zwaag. 2011. "A critical appraisal of the potential of self healing polymeric coatings." *Progress in Organic Coatings* 72 (3):211–221.

Gupta, Gunjan, Nick Birbilis, Anthony B. Cook, and Anand Sawroop Khanna. 2013. "Polyaniline-lignosulfonate/epoxy coating for corrosion protection of AA2024-T3." *Corrosion Science* 67:256–267.

Gupta, Rajeev Kumar, Manisha Malviya, Chandrabhan Verma, and M.A. Quraishi. 2017. "Aminoazobenzene and diaminoazobenzene functionalized graphene oxides as novel class of corrosion inhibitors for mild steel: Experimental and DFT studies." *Materials Chemistry and Physics* 198:360–373. doi: 10.1016/j.matchemphys.2017.06.030.

Haddadi, S.A., A. Ramazani SA, M. Mahdavian, P. Taheri, and J.M.C. Mol. 2019. "Mechanical and corrosion protection properties of a smart composite epoxy coating with dual-encapsulated epoxy/polyamine in carbon nanospheres." *Industrial & Engineering Chemistry Research* 58 (8):3033–3046.

Haddadi, S.A., S.A.A. Ramazani, M. Mahdavian, P. Taheri, and J.M.C. Mol. 2018. "Fabrication and characterization of graphene-based carbon hollow spheres for encapsulation of organic corrosion inhibitors." *Chemical Engineering Journal* 352:909–922.

Haddadi, S.A., A. Ramazani SA, M. Mahdavian, P. Taheri, J.M.C. Mol, and Y. Gonzalez-Garcia. 2019. "Self-healing epoxy nanocomposite coatings based on dual-encapsulation of nano-carbon hollow spheres with film-forming resin and curing agent." *Composites Part B: Engineering* 175:107087.

Haddadi, Seyyed Arash, Parisa Najmi, Navid Keshmiri, Nicolas Tanguy, Colin van der Kuur, Ning Yan, Tizazu Mekonnen, and Mohammad Arjmand. 2022. "Cerium-doped tannic acid-reduced graphene oxide nanoplatform/epoxy nanocomposite coatings with enhanced mechanical and Bi-functional corrosion protection properties." *Composites Part B: Engineering* 109969.

Haddadi, Seyyed Arash, Ahmad Ramazani SA, Mohammad Mahdavian, and Mohammad Arjmand. 2021. "Epoxy nanocomposite coatings with enhanced dual active/barrier behavior containing graphene-based carbon hollow spheres as corrosion inhibitor nanoreservoirs." *Corrosion Science* 185:109428. doi: 10.1016/j.corsci.2021.109428.

Hayatgheib, Y., B. Ramezanzadeh, P. Kardar, and M. Mahdavian. 2018. "A comparative study on fabrication of a highly effective corrosion protective system based on graphene oxide-polyaniline nanofibers/epoxy composite." *Corrosion Science* 133:358–373.

Hays, George F., and PE Director General. 2010. "World corrosion organization." *Corrodia. NACE International, Houston.*

Hong, Jinkee, Jung Yeon Han, Hyunsik Yoon, Piljae Joo, Taemin Lee, Eunyong Seo, Kookheon Char, and Byeong-Su Kim. 2011. "Carbon-based layer-by-layer nanostructures: From films to hollow capsules." *Nanoscale* 3 (11):4515–4531.

Hosseini, Majid, and Abdel Salam Hamdy Makhlouf. 2016. *Industrial applications for intelligent polymers and coatings*. Vol. 1. Springer.

Ioniţă, Mariana, and Alina Pruná. 2011. "Polypyrrole/carbon nanotube composites: Molecular modeling and experimental investigation as anti-corrosive coating." *Progress in Organic Coatings* 72 (4):647–652.

Izadi, Mazdak, Ali Rajaei Rad, Taghi Shahrabi, and Iman Mohammadi. 2020. "The combined action of l-cysteine and l-histidine as a significant eco-friendly protective system to enhance the corrosion protection performance of AA2024-T3 alloy in 0.1 M NaCl solution: Electrochemical and surface studies." *Materials Chemistry and Physics* 250:122997. doi: 10.1016/j.matchemphys.2020.122997.

Javidparvar, Ali Asghar, Reza Naderi, and Bahram Ramezanzadeh. 2020. "L-cysteine reduced/functionalized graphene oxide application as a smart/control release nanocarrier of sustainable cerium ions for epoxy coating anti-corrosion properties improvement." *Journal of Hazardous Materials* 389:122135. doi: 10.1016/j.jhazmat.2020.122135.

Javidparvar, Ali Asghar, Reza Naderi, Bahram Ramezanzadeh, and Ghasem Bahlakeh. 2019. "Graphene oxide as a pH-sensitive carrier for targeted delivery of eco-friendly corrosion inhibitors in chloride solution: Experimental and theroretical investigations." *Journal of Industrial and Engineering Chemistry* 72:196–213. doi: 10.1016/j.jiec.2018.12.019.

Jeon, In-Yup, Sang-Wook Kang, Loon-Seng Tan, and Jong-Beom Baek. 2010. "Grafting of polyaniline onto the surface of 4-aminobenzoyl-functionalized multiwalled carbon nanotube and its electrochemical properties." *Journal of Polymer Science Part A: Polymer Chemistry* 48 (14):3103–3112.

Khodadadi Yazdi, Mohsen, Ghodratollah Hashemi Motlagh, Sadaf Saeedi Garakani, and Ali Boroomand. 2018. "Effects of multiwall carbon nanotubes on the polymerization model of aniline." *Journal of Polymer Research* 25 (12):1–15.

Kumar, Anil, Vinod Kumar, and Kamlendra Awasthi. 2018. "Polyaniline–carbon nanotube composites: Preparation methods, properties, and applications." *Polymer-Plastics Technology and Engineering* 57 (2):70–97.

Kumar, Ashok, L.D. Stephenson, and John Nelson Murray. 2006. "Self-healing coatings for steel." *Progress in Organic Coatings* 55 (3):244–253.

Lanzara, G., Y. Yoon, H. Liu, S. Peng, and W.I. Lee. 2009. "Carbon nanotube reservoirs for self-healing materials." *Nanotechnology* 20 (33):335704.

Latnikova, Alexandra. 2012. "Polymeric capsules for self-healing anticorrosion coatings." *Universität Potsdam.*

Li, Jing, Qingkang Feng, Jincan Cui, Qianqian Yuan, Hanxun Qiu, Shanglin Gao, and Junhe Yang. 2017. "Self-assembled graphene oxide microcapsules in Pickering emulsions for self-healing waterborne polyurethane coatings." *Composites Science and Technology* 151:282–290.

Li, Jing, Zhenwei Li, Qingkang Feng, Hanxun Qiu, Guangzhi Yang, Shiyou Zheng, and Junhe Yang. 2019. "Encapsulation of linseed oil in graphene oxide shells for preparation of self-healing composite coatings." *Progress in Organic Coatings* 129:285–291.

Li, Peng, Jingyan Liu, Yuwei Wang, Yang Liu, Xiuna Wang, Kyung-Wan Nam, Yong-Mook Kang, Mingbo Wu, and Jieshan Qiu. 2016. "Synthesis of ultrathin hollow carbon shell from petroleum asphalt for high-performance anode material in lithium-ion batteries." *Chemical Engineering Journal* 286:632–639. doi: 10.1016/j.cej.2015.10.102.

Liu, Chengbao, Haichao Zhao, Peimin Hou, Bei Qian, Xiao Wang, Chunyan Guo, and Liping Wang. 2018. "Efficient graphene/cyclodextrin-based nanocontainer: Synthesis and Host–Guest inclusion for self-healing anticorrosion application." *ACS Applied Materials & Interfaces* 10 (42):36229–36239. doi: 10.1021/acsami.8b11108.

Liu, Sha, Yao Jing, Tingting Zhang, Jin Zhang, Fei Xu, Qiang Song, Qian Ye, Shujuan Liu, and Weimin Liu. 2021. "Excellent tribological and anti-corrosion performances enabled by novel hollow graphite carbon nanosphere with controlled release of corrosion inhibitor." *Chemical Engineering Journal* 412:128648. doi: 10.1016/j.cej.2021.128648.

Liu, Tao, Liuyang Zhang, Bei Cheng, and Jiaguo Yu. 2019. "Hollow carbon spheres and their hybrid nanomaterials in electrochemical energy storage." *Advanced Energy Materials* 9 (17):1803900.

Marrion, Alastair. 2004. *The chemistry and physics of coatings*. Royal Society of Chemistry.

Mohammadkhani, Rahman, Mohammad Ramezanzadeh, Samaneh Saadatmandi, and Bahram Ramezanzadeh. 2020. "Designing a dual-functional epoxy composite system with self-healing/barrier anti-corrosion performance using graphene oxide nano-scale platforms decorated with zinc doped-conductive polypyrrole nanoparticles with great environmental stability and non-toxicity." *Chemical Engineering Journal* 382:122819.

Montemor, M.F., and M.G.S. Ferreira. 2008. "Analytical characterisation and corrosion behaviour of bis-aminosilane coatings modified with carbon nanotubes activated with rare-earth salts applied on AZ31 Magnesium alloy." *Surface and Coatings Technology* 202 (19):4766–4774. doi: 10.1016/j.surfcoat.2008.04.071.

Najmi, Parisa, Navid Keshmiri, Mohammad Ramezanzadeh, and Bahram Ramezanzadeh. 2021. "Synthesis and application of Zn-doped polyaniline modified multi-walled carbon nanotubes as stimuli-responsive nanocarrier in the epoxy matrix for achieving excellent barrier-self-healing corrosion protection potency." *Chemical Engineering Journal* 412:128637.

Nesterova, Tatyana, Kim Dam-Johansen, Lars Thorslund Pedersen, and Søren Kiil. 2012. "Microcapsule-based self-healing anticorrosive coatings: Capsule size, coating formulation, and exposure testing." *Progress in Organic Coatings* 75 (4):309–318.

Nikpour, Bahar, Bahram Ramezanzadeh, Ghasem Bahlakeh, and Mohammad Mahdavian. 2017. "Synthesis of graphene oxide nanosheets functionalized by green corrosion inhibitive compounds to fabricate a protective system." *Corrosion Science* 127:240–259. doi: 10.1016/j.corsci.2017.08.029.

Parameswaranpillai, Jyotishkumar, Nisa V. Salim, Harikrishnan Pulikkalparambil, and Sanjay Mavinkere Rangappa. 2022. *Micro-and nano-containers for smart applications*. Springer.

Pourhashem, Sepideh, Ebrahim Ghasemy, Alimorad Rashidi, and Mohammad Reza Vaezi. 2020. "A review on application of carbon nanostructures as nanofiller in corrosion-resistant organic coatings." *Journal of Coatings Technology and Research* 17 (1):19–55.

Qiu, Shihui, Wei Li, Wenru Zheng, Haichao Zhao, and Liping Wang. 2017. "Synergistic effect of polypyrrole-intercalated graphene for enhanced corrosion protection of aqueous coating in 3.5% NaCl solution." *ACS Applied Materials & Interfaces* 9 (39):34294–34304.

Quadri, Taiwo W., Lukman O. Olasunkanmi, Omolola E. Fayemi, and Eno E. Ebenso. 2022. "Functionalized carbon allotropes as corrosion inhibitors." In *Functionalized nanomaterials for corrosion mitigation: Synthesis, characterization, and applications*, 87–114. ACS Publications.

Rafiee, Mohammad A., Fazel Yavari, Javad Rafiee, and Nikhil Koratkar. 2011. "Fullerene–epoxy nanocomposites-enhanced mechanical properties at low nanofiller loading." *Journal of Nanoparticle Research* 13 (2):733–737.

Rahmani, Pooria, Akbar Shojaei, and Nahid Pirhady Tavandashti. 2020. "Nanodiamond loaded with corrosion inhibitor as efficient nanocarrier to improve anticorrosion behavior of epoxy coating." *Journal of Industrial and Engineering Chemistry* 83:153–163. doi: 10.1016/j.jiec.2019.11.023.

Ramezanzadeh, B., M.H. Mohamadzadeh Moghadam, N. Shohani, and M. Mahdavian. 2017. "Effects of highly crystalline and conductive polyaniline/graphene oxide composites on the corrosion protection performance of a zinc-rich epoxy coating." *Chemical Engineering Journal* 320:363–375.

Ramezanzadeh, B., S. Niroumandrad, A. Ahmadi, M. Mahdavian, and M.H. Mohamadzadeh Moghadam. 2016. "Enhancement of barrier and corrosion protection performance of an epoxy coating through wet transfer of amino functionalized graphene oxide." *Corrosion Science* 103:283–304. doi: 10.1016/j.corsci.2015.11.033.

Ramezanzadeh, Bahram, Ghasem Bahlakeh, and Mohammad Ramezanzadeh. 2018. "Polyaniline-cerium oxide (PAni-CeO2) coated graphene oxide for enhancement of epoxy coating corrosion protection performance on mild steel." *Corrosion Science* 137:111–126.

Rohwerder, Michael, and Adam Michalik. 2007. "Conducting polymers for corrosion protection: What makes the difference between failure and success?" *Electrochimica Acta* 53 (3):1300–1313.

Samadzadeh, M., S. Hatami Boura, M. Peikari, A. Ashrafi, and M. Kasiriha. 2011. "Tung oil: An autonomous repairing agent for self-healing epoxy coatings." *Progress in Organic Coatings* 70 (4):383–387.

Samadzadeh, M., S. Hatami Boura, M. Peikari, S.M. Kasiriha, and A. Ashrafi. 2010. "A review on self-healing coatings based on micro/nanocapsules." *Progress in Organic Coatings* 68 (3):159–164.

Sinha-Ray, S., D.D. Pelot, Z.P. Zhou, A. Rahman, X-F Wu, and A.L. Yarin. 2012. "Encapsulation of self-healing materials by coelectrospinning, emulsion electrospinning, solution blowing and intercalation." *Journal of Materials Chemistry* 22 (18):9138–9146.

Sinha-Ray, S., R.P. Sahu, and A.L. Yarin. 2011. "Nano-encapsulated smart tunable phase change materials." *Soft Matter* 7 (19):8823–8827.

Stankiewicz, Alicja, Irena Szczygieł, and Bogdan Szczygieł. 2013. "Self-healing coatings in anti-corrosion applications." *Journal of Materials Science* 48 (23):8041–8051.

Talbot, David E.J., and James D.R. Talbot. 2018. *Corrosion science and technology*. CRC press.

Tallman, Dennis E., Geoff Spinks, Anton Dominis, and Gordon G. Wallace. 2002. "Electroactive conducting polymers for corrosion control." *Journal of Solid State Electrochemistry* 6 (2):73–84.

Thakur, Abhinay, Ashish Kumar, Praveen Kumar Sharma, Shveta Sharma, Richika Ganjoo, and Humira Assad. 2022. "Carbon allotropes as corrosion inhibitors." In *Functionalized nanomaterials for corrosion mitigation: Synthesis, characterization, and applications*, 115–140. ACS Publications.

Urban, Marek W. 2014. *Self-Healing Polymers. From Principles to Applications*. Edited by Wolfgang H. Binder. Wiley Online Library.

Wang, Xiaojing, J.I. Feng, Yaocai Bai, Qiao Zhang, and Yadong Yin. 2016. "Synthesis, properties, and applications of hollow micro-/nanostructures." *Chemical Reviews* 116 (18):10983–11060.

Wei, Huige, Yiran Wang, Jiang Guo, Nancy Z. Shen, Dawei Jiang, Xi Zhang, Xingru Yan, Jiahua Zhu, Qiang Wang, and Lu Shao. 2015. "Advanced micro/nanocapsules for self-healing smart anticorrosion coatings." *Journal of Materials Chemistry A* 3 (2):469–480.

Wei, Zhao, Jian Hai Yang, Jinxiong Zhou, Feng Xu, Miklós Zrínyi, Patrick H. Dussault, Yoshihito Osada, and Yong Mei Chen. 2014. "Self-healing gels based on constitutional dynamic chemistry and their potential applications." *Chemical Society Reviews* 43 (23):8114–8131.

Wessling, Berhard. 1994. "Passivation of metals by coating with polyaniline: Corrosion potential shift and morphological changes." *Advanced Materials* 6 (3):226–228.

Yan, Dashuai, Jialiang Liu, Zhenhua Zhang, Yanli Wang, Meng Zhang, Dalei Song, Tao Zhang, Jingyuan Liu, Fei He, and Jun Wang. 2021. "Dual-functional graphene oxide-based nanomaterial for enhancing the passive and active corrosion protection of epoxy coating." *Composites Part B: Engineering* 222:109075.

Yang, Dongping, Yuwei Ye, Yue Su, Shuang Liu, Dirong Gong, and Haichao Zhao. 2019. "Functionalization of citric acid-based carbon dots by imidazole toward novel green corrosion inhibitor for carbon steel." *Journal of Cleaner Production* 229:180–192. doi: 10.1016/j.jclepro.2019.05.030.

Yazdi, Mohsen Khodadadi, Hoda Saeidi, Payam Zarrintaj, Mohammad Reza Saeb, and Masoud Mozafari. 2019. "PANI-CNT nanocomposites." In *Fundamentals and emerging applications of polyaniline*, 143–163. Elsevier.

Ye, Yuwei, Dawei Zhang, Tong Liu, Zhiyong Liu, Jibin Pu, Wei Liu, Haichao Zhao, Xiaogang Li, and Liping Wang. 2019. "Superior corrosion resistance and self-healable epoxy coating pigmented with silanzied trianiline-intercalated graphene." *Carbon* 142:164–176.

Yeole, K.V., I.P. Agarwal, and S.T. Mhaske. 2016. "The effect of carbon nanotubes loaded with 2-mercaptobenzothiazole in epoxy-based coatings." *Journal of Coatings Technology and Research* 13 (1):31–40. doi: 10.1007/s11998-015-9730-z.

Yu, Zongxue, Liang Lv, Yu Ma, Haihui Di, and Yi He. 2016. "Covalent modification of graphene oxide by metronidazole for reinforced anti-corrosion properties of epoxy coatings." *RSC Advances* 6 (22):18217–18226. doi: 10.1039/C5RA23595B.

Zhang, Qiang, Libin Liu, Chenguang Pan, and Dong Li. 2018. "Review of recent achievements in self-healing conductive materials and their applications." *Journal of Materials Science* 53 (1):27–46.

Zhao, Yang, Wei Zhang, Le-ping Liao, Si-jie Wang, and Wu-jun Li. 2012. "Self-healing coatings containing microcapsule." *Applied Surface Science* 258 (6):1915–1918.

Zheng, Nan, Jie Liu, Yuhang Wang, Chunmei Li, and Qiuyu Zhang. 2021. "Preparation of chitosan-reduced graphene oxide (CS-RGO) microcapsules and its application in UV/moisture-induced self-healing coatings." *Progress in Organic Coatings* 151:106055. doi: 10.1016/j.porgcoat.2020.106055.

Zhu, Qingsong, E.N. Li, Xianhu Liu, Weiqiang Song, Yilong Li, Xinchao Wang, and Chuntai Liu. 2020. "Epoxy coating with in-situ synthesis of polypyrrole functionalized graphene oxide for enhanced anticorrosive performance." *Progress in Organic Coatings* 140:105488.

12 Graphene Composite-Based Anticorrosion Coatings

Mohammad Mobin, Saman Zehra, and Ruby Aslam
Corrosion Research Laboratory, Department of Applied Chemistry, Faculty of Engineering and Technology, Aligarh Muslim University, Aligarh, India

CONTENTS

12.1 INTRODUCTION

The scientific community has long been interested in graphene, whether it is in its pure or composite form. Most of the study, however, was conducted within the previous 10 years or fewer, and its use in the anticorrosion coating has only just been well known (Fontana 1986). The distinct chemical, structural, electrical, and mechanical characteristics of graphene indicate its potential use in a wide range of fields. Graphene and derivatives of graphene have extensively been utilised in various anticorrosion coating applications (Cui et al. 2019, Mohammadi et al. 2018).

The advantages of graphene as a coating include its high mechanical and electrical properties, impermeability to gases and liquids, minimal chemical reactivity, and the ability to have a coating that is much thinner than conventional coatings while still performing better. In one investigation, stainless steel was coated with a 200 nm thick hybrid graphene oxide layer. The findings suggested that the GO coating alone did not offer significant corrosion protection. Still, when combined with other materials, it produced remarkable results (Pourhashem et al. 2017). In a second investigation, cold-rolled steel was coated with an oil-based graphene oxide base in a saltwater environment (Singhbabu et al. 2015). The study found that their graphene oxide coating was extremely successful in a high-salt environment and decreased corrosion by over 10,000 times compared to the steel in the noncoated control group.

Dumee group (Dumée et al. 2015) in a work grew flakes of graphene directly on the stainless-steel substrate. The team was able to effectively add between 3 and 15 layers to the steel, improving its electrical and corrosion-resistant characteristics without degrading the characteristics of the stainless-steel substrate.

Industrial applications are still constrained significantly despite the exceptional anticorrosion performance of pure and graphene composite anticorrosion coatings. Corrosion of the metal happens more quickly when the pure graphene coating is damaged. Given the available technology, it is also challenging to extend and industrialise this process. In the field of anticorrosion, graphene can also be used to disperse filler particles into coating matrixes to create composite anticorrosive coatings. By

DOI: 10.1201/9781003323976-15

combining graphene's adhesive properties with the matrix's ability to form films, the graphene composite coating enhances the overall performance of the coating. In industrial syntheses and applications, the conventional coating production method can also be used for the preparation and coating of graphene composite anticorrosive coatings. This chapter will discuss anticorrosion coating materials based on graphene composites. The anticorrosion performance, the dispersibility of graphene in the coating matrix, and the anticorrosion mechanism are all covered in detail in this chapter.

12.2 APPLICATION OF GRAPHENE AND ITS ANALOGUE IN ANTICORROSION COATINGS

"Bottom-up" methods of creating graphene coatings include chemical vapour deposition (CVD), thermal breakdown, or organic synthesis, which combine molecular building blocks into monolayers or multilayers of graphene (Kong et al. 2019, Vashist et al. 2015). The CVD process is the most promising method for producing excellent, scalable graphene sheets and films on various metal substrates (aluminium [Zhang et al. 2017], copper [Zhang 2020], nickel [Sun et al. 2019], etc.). This technology has been sufficiently refined to produce thin films with a 30-inch lateral dimension (Bae et al. 2010) and can mechanically transfer the graphene layer to any surface. The ground-breaking pure coatings for preventing metal corrosion are CVD-grown graphene sheets. The corrosion resistance of copper in chloride-containing conditions may be increased by one to two orders of magnitude using just one or two graphene layers (Raman et al. 2012). Due to the high cost of precursor gases, delicate machinery, high processing temperatures necessary for high-quality graphene deposition, and other circumstances that arise during the deposition of multi-component materials, CVD-grown graphene films remain constrained for use (Kakaei et al. 2019). It is also essential to ensure that these ultra-thin films are wear resistant since the integrity of anticorrosive coatings is crucial for their efficacy and longevity.

Graphene/polymer nanocomposite coatings are frequently created by using "top-down" methods such as in-situ polymerisation, solution mixing, melt compounding, and others (Umoren et al. 2019). Graphene is first created as a composite enhancer and then incorporated into the coating. Because the effectiveness of the coating and reinforcing phases work together to create the overall performance for composites, the qualities of graphene nanosheets may be "scaled up" to the macro-characteristics. The requirements for graphene in composite coatings are frequently more lenient, and this material generally comprises 0.05% to 1.0% of the mass (Jiang et al. 2016, Yang et al. 2016, Ding et al. 2019, Cui et al. 2019, Mohammadi et al. 2018). As a result, the price of reinforced organic coatings might be greatly reduced. Graphene-based materials are increasingly used in polycrystalline graphene (PG), graphene oxide (GO), reduced graphene oxide (rGO), and graphene quantum dots (GQDs) in addition to pure graphene (Morales-Narváez et al. 2017).

The essential graphene family member GO is routinely offered for sale at a reasonable price. GO is viewed as potential for preventing metal corrosion because of its similarity to graphene in terms of chemical inertia, thermal and chemical stability, and ion impermeability (Ahmad et al. 2018). The oxygen-containing functional groups can function as active sites for further modification or functionalisation since they are simple to process (Zhang et al. 2017). These groups are also required for the liquid dispersion of graphene. Other graphene derivatives, including those that have been modified using covalent and noncovalent grafting methods (Thanh et al. 2018), are also made using GO as a starting point. Along with the intrinsic features, these modifications have the potential to drastically modify graphene's properties and offer variation to its surface chemistry.

12.3 GRAPHENE-BASED ANTICORROSION COATING: A LITERATURE SURVEY

Numerous research studies (Dittanet et al. 2017, Ammar et al. 2017, Ramezanzadeh et al. 2016, Mochalin et al. 2015, Xie et al. 2005, Shadlou et al. 2013, Miller et al. 2010, Coleman et al. 2006,

Georgakilas et al. 2015, Singh et al. 2011) have examined the usage of graphene in multifunctional coatings. Liu et al. (2016a) created graphene/epoxy coatings with two distinct compositions: 0.5 and 1.0 wt.%. Their findings demonstrated that the inclusion of GNPs improved the epoxy coating's resistance to corrosion and that coatings containing 0.5 wt.% graphene performed better. Chang et al. covered the effects of graphene dispersion on coating performance in their work (Chang et al. 2014). Liu et al. (2016b) examined the tribological and anticorrosion properties of epoxy coating improved by graphene nanoparticles. They found that when the amount of graphene was more than 0.5 wt.%, the nanoparticle aggregation resulted in significant material degradation.

Contrary to past studies, Yu et al. (2014) found that materials with higher graphene nanoparticle concentrations nevertheless exhibited outstanding corrosion and gas barrier resistance. They looked at nanocomposites with graphene levels ranging from 0.5 to 2.0 wt.%, and they came to the conclusion that adding 2.0 wt.% of GNP improved the performance of the materials' anticorrosion and gas barrier properties. Experimental evaluation of the efficacy of a graphene/epoxy coating containing 1.0 wt.% of graphene nanoparticles was done by Monetta et al. (2015). They discovered that graphene improved the corrosion resistance of polymers. They stated that the final coating's adhesion and water resistance remained mostly unchanged, and the addition of graphene had little effect on epoxy's ability to cure. Finally, the future of coating systems seems bright thanks to graphene. Some of the results might not agree since the research's foci were different. For instance, it is still up for debate how much graphene should be included in the nanocomposite coatings in order to enhance the reinforcement. While a few research (Chang et al. 2014, Liu et al. 2016, Monetta et al., 2015) also offered corrosion resistance, which is often required in metallic constructions, the bulk of studies (Singh et al. 2013, Chang et al. 2012, Singh et al. 2013, Ahmadi-Moghadam et al. 2015, Liang et al. 2009) focused on the mechanical, electrical, or tribological characteristics of the nanocomposites. This study aims to investigate the potential impact of GNPs on epoxy glue on the production of high-performance coatings for use on oil and gas pipelines. Ultrasonication and high-speed disc introduction were part of the dispersion strategy. The impact of various nanocarbon additions on the electrical and mechanical properties were examined. The effects of nanofillers on polymer matrix were evaluated using electrochemical impedance spectroscopy (EIS), Taber abraser, pull-off strength, and water contact angle tests. Using atomic force microscopy and scanning electron microscopy, assess the polymer matrix's nanofiller reinforcement and surface roughness (SEM).

The performance of the graphene composite's corrosion resistance in the preceding section was classified into covalent and noncovalent alterations depending on the type of chemical interaction that took place between graphene and the other components. The corrosion behaviour of coatings using modified graphene is covered in this section (Imani et al. 2018). Covalent modifications of graphene often include the bonding of polymer chains, small molecules, or nanoparticles, in contrast to noncovalent modifications, which involve the interaction of hydrogen bonds, and ionic bonds (Yu et al. 2013, Liu et al. 2018). While noncovalent alteration gives graphene its exceptional conductivity and extremely large specific surface area, covalent modification preserves graphene's chemical stability and mechanical properties (Imani et al. 2018). Therefore, researchers should choose the optimal modification techniques based on their needs. Because of its qualities as a physical barrier, silane is frequently employed as a metal preparation. The hydroxyl group on the metal surface is joined by the Si-OH produced when silane is hydrolysed, which enhances the adhesion of the coated substrate to the metal substrate. However, when the solution evaporates, many micropores are left, allowing corrosive materials to come into contact with the metal and cause corrosion (Liu et al. 2018). Parhizkar et al. (Parhizkar et al. 2018) covalently functionalised GO using 3-(triethoxysilyl)propyl isocyanate (TEPI, IGO) and 3-aminopropyltriethoxysilane (APTES, AGO). In order to improve the adhesion of the silane coating to the top epoxy coating, AGO and IGO were obtained, which decreased the porosity of the silane structure on the metal surface. Salt spray testing showed that the SC-AGO/EP and SC-IGO/EP coatings scarcely had any corrosion byproducts or blisters around the scribing lines. This results from the close contact between the metal and the graphene composite coating, which limits the production of corrosion products and blisters and hinders the diffusion of the corrosive medium inside the coating.

In order to create modified Graphene IL-GO, Liu et al. (2018) employed a straightforward grafting technique to graft imidazole ions onto a GO surface. This modified Graphene IL-GO was evenly spread throughout the epoxy-based aqueous coating matrix and was also soluble in water. The epoxy-based aqueous coating's ILGO addition of 0.5 wt.% significantly improved the metal's resistance to corrosion. The SVET maps and the local electrochemical behaviour of the graphene composite covering on the steel electrode were analysed using the scanning vibrating electrode method (SVET). In the scanning zone for pure epoxy-coated steel, the average anode current increased from 3.10 Acm2 (1 hour) to 4.75 Acm2 (30 hours), showing anode disintegration. The anticorrosion properties of the IL-GO composite coating were demonstrated by the anode current, which decreased over time and peaked at a low value (2.20 Acm2) at 30 hours. After 300 hours of salt spray testing, the scratches on the metal coated with graphene composite remained largely corrosion-free, but after 100 hours of salt spray testing, scratches on metal coated with pure epoxy and epoxy coatings containing 0.5 wt.% rGO both revealed the presence of numerous corrosion products at the scratch. The IL-GO composite coating exhibited the highest binding strength of the three samples tested for pull-off, which decreased coating peeling and improved anticorrosion performance. Coatings incorporating graphene simply have their anticorrosion properties non-covalently functionalised through intermolecular interactions.

The interactions between 2-aminothiazoles may facilitate the dispersion of graphene in the epoxy matrix (PAT). Qiu et al. treated graphene with 2-aminothiazole to produce a PAT-G-EP graphene composite covering (PAT). Figure 12.1 displays the schematic diagram. Water absorption tests and electrochemical tests in a 3.5 wt.% NaCl solution evaluated the barrier properties of epoxy and graphene composite coatings. In EEC, Rp and Rct are widely employed to denote a coating's barrier capabilities. According to Figure 12.1, a coating loaded with 1 wt.% of graphene had poor corrosion resistance, whereas a coating loaded with 0.5 wt.% of graphene always had better Rp values than other coatings. The PAT-G-EP graphene composite coating absorbed less than 0.5% of the solution after soaking for 240 hours, which was 89% less than pure epoxy coatings. After being submerged in a 3.5 wt.% NaCl solution for 80 days, the composite coating still demonstrated good corrosion resistance. The composite coating containing 0.5 wt.% graphene showed the lowest wear rate at the same period, which was 69.48% lower than that of pure EP (Qiu et al. 2018).

Ding et al. (2018) used Raman, UV–vis, and X-ray photoelectron spectroscopy (XPS) to investigate interactions between graphene and the hydroxyl epoxy phosphate monomer (PGHEP). The water and coating matrix are equally distributed with the noncovalent graphene coating that PGHEP modified. The corrosion resistance of the coatings was evaluated using the EIS and salt spray tests. Aqueous epoxy coatings incorporating graphene at 0.5 wt.% (PGHEP-G0.5%/WEP) or 1 wt.% (PGHEP-G1%/WEP) had greater Rc and Rct values. PGHEP made it easier for a passivation phosphate coating to grow on the substrate metal, and modified graphene made it harder for corrosive fluids to enter it, increasing corrosion resistance. All coated samples reached consistent water absorption after 7 days of immersion, as seen in Figure 12.2. WEP and PGHEP / WEP both absorbed 0.83 wt.% of the water after 15 days of immersion, whereas PGHEP-G0.5%/WEP and PGHEP-G1%/WEP only absorbed 0.52 wt.% (0.55 wt.%). Additionally, the water contact angle for PGHEP-G0.5 %/WEP was 83.5° as opposed to 50.6° for WEP. In order to lessen the coatings' hydrophobicity and water absorption, graphene was added, and a hydrophobic phosphate layer was created.

Comparatively to covalent surface modification, noncovalent surface modification of graphene is simple and does not require complex chemical processes. It is fascinating that the multiple forms of graphene may work together to promote corrosion resistance and that graphene coatings may simultaneously contain covalent and noncovalent surface modifications. GO/Fe_3O_4 hybrid NPs were altered by Zhan et al. (2018) utilising the biomimetic polymers silane and dopamine (KH550). Through interactions, dopamine became bonded to the surface of graphene, and KH5560 covalently altered the material. The micro-hardness and corrosion resistance of the composite

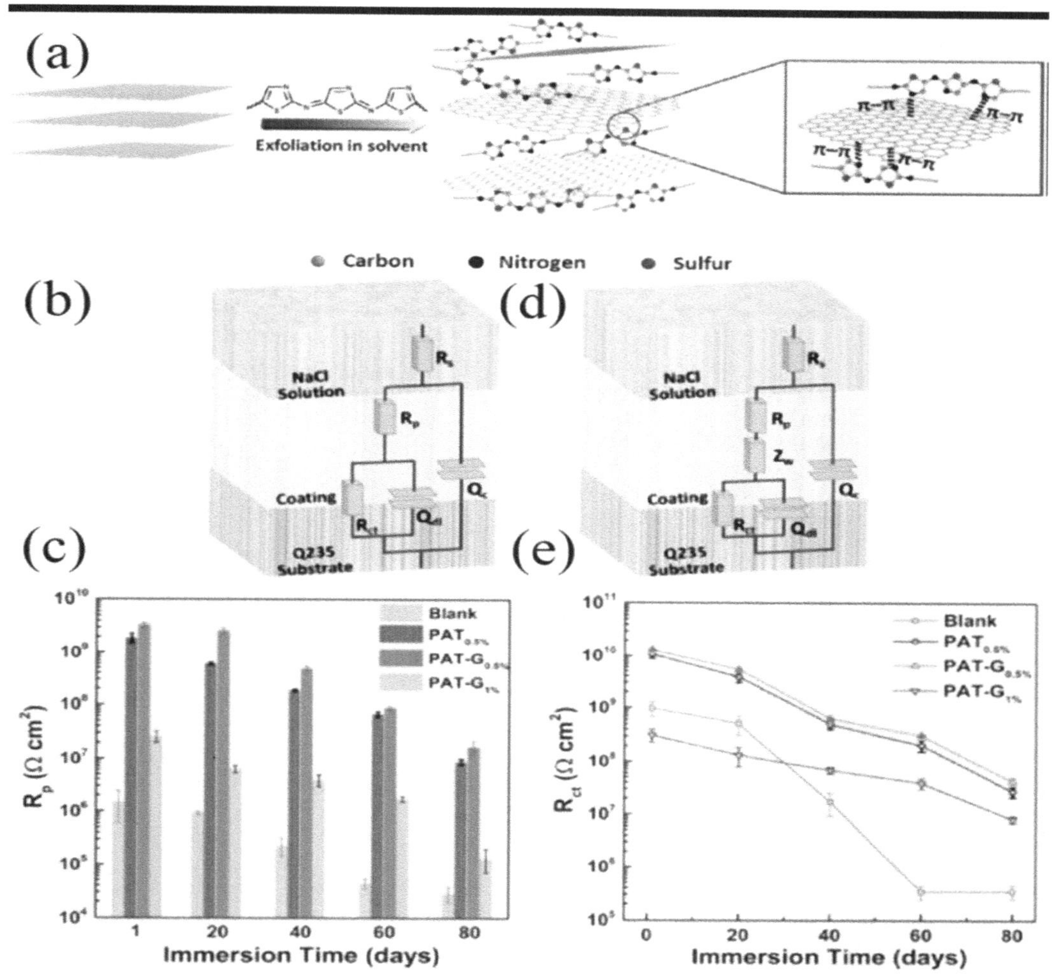

FIGURE 12.1 (a) Schematic representation of the preparation of PAT-G hybrids; (b), (c) Electrical equivalent circuit models and the evolution of the (d) Rp and (e) Rct values with immersion in the 3.5 wt % NaCl solution (Qiu et al. 2018).

coating peaked at a combined NP concentration of 0.5 wt.%. The major culprits behind steel corrosion are H_2O and O_2. Pure epoxy coatings include small channels that permit easy passage of water and oxygen. With the addition of the graphene derivative, the H_2O and O_2 permeability channels are closed, and the graphene composite coating successfully blocks the penetration of the corrosive fluids. The coating is protected from corrosive media by a physical barrier that is formed by the extremely dispersible graphene. A more compact link between the coated substrate and the shielding metal is also produced by the modified graphene. Because the metal and coating are in closer proximity to one another, the coating peels off substantially more slowly.

12.4 ANTICORROSION MECHANISM OF THE GRAPHENE COMPOSITE COATING

The two aspects listed below are the main causes of the graphene composite coating's effective anticorrosion properties. First, graphene acts as a "barrier" in the coating, perhaps preventing the penetration of corrosive substances and the propagation of fractures. Second, graphene improves

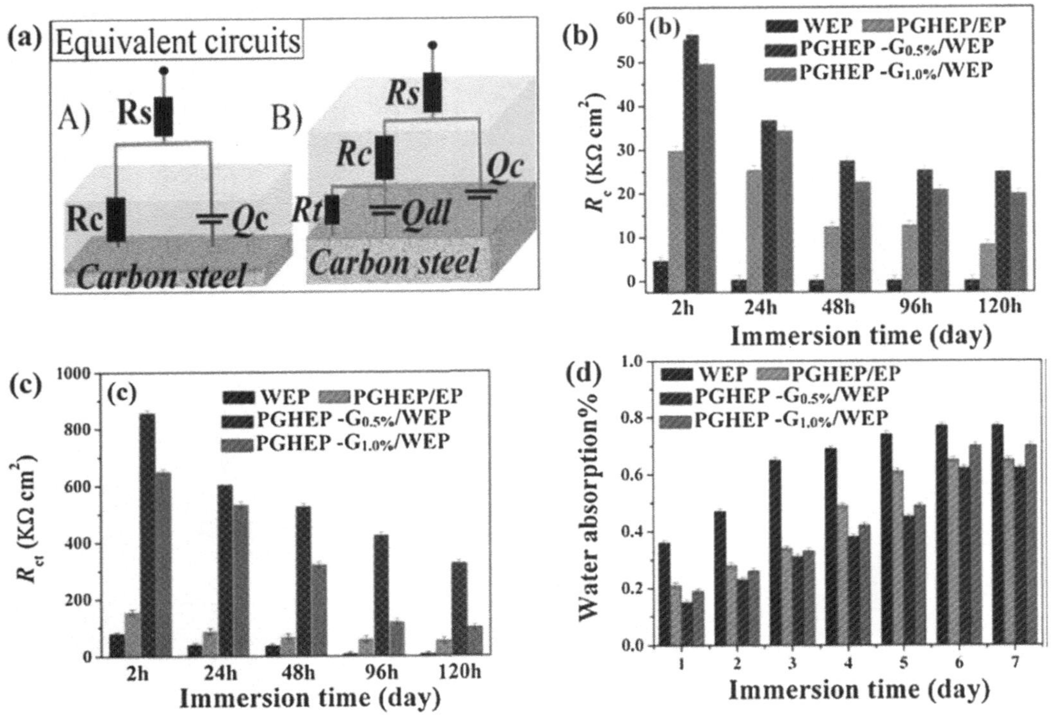

FIGURE 12.2 (a) The equivalent circuits; (b) the fitting results of Rc; (c) the fitting results of Rct; and (d) the water absorption of WEP, PGHEP/WEP, PGHEP-G0.5% and PGHEP-G1% coatings as a function of time in water at room temperature (Ding et al. 2018).

the protective qualities of the sacrificial anode by serving as a pathway for electron flow between anode particles and the substrate in the coating.

12.4.1 The "Barrier" Effect of Graphene

a. ***Corrosive substances cannot pass through graphene.*** Pure polymer coatings may allow corrosive media to penetrate and enter the substrate along the coating's thickness. As soon as the corrosive media encounter the coated metal interface, the corrosion process begins. As shown in Figure 12.3, the aromatic ring of graphene, which is formed of sp2-hybridised carbon atoms, has a high electron density and may block all molecules (Lee et al. 2008). So, the impermeability of graphene is a characteristic of its unique structural design. Graphene may provide better shielding protection during the infiltration process of coatings with included graphene by lengthening the infiltration route. The coating's life and corrosion resistance can therefore be greatly extended and increased (Papageorgiou et al. 2017, Pierleoni et al. 2016, Tan and Thomas 2016).
b. ***Graphene hinders crack propagation in the coating.*** Cracks will occur during the coating's preparation and application. Often, these tiny fractures serve as the ideal point of entry for corrosive media to infiltrate metals. These minute fracture defects decrease the coating's ability to resist corrosion. The fracture's surface area grows as it becomes broader and deeper. As more corrosive media are added to the coating, severe localised corrosion might occur. Graphene, which prevents fractures from propagating and restricts the infiltration of corrosive fluids, may greatly increase the coating's hardness. The energy needed for crack formation is absorbed by graphene, which also affects how polymers crystallise in

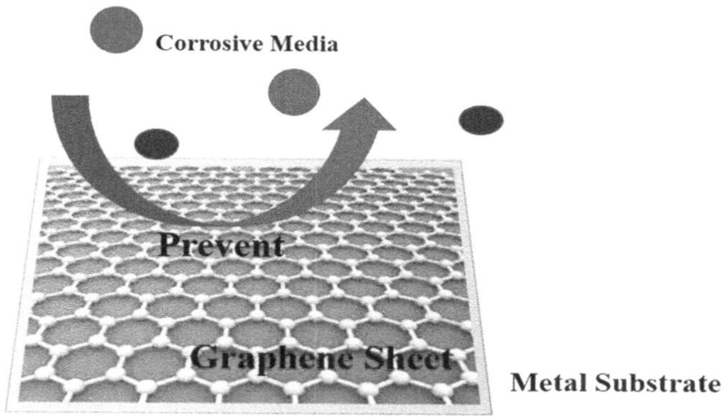

FIGURE 12.3 Graphene acts as an anticorrosion barrier to prevent oxidation.

composite coatings and thus reduces fracture propagation (Wang et al. 2019). Because the opening displacement of the crack tip is smaller than the graphene size, the coating fracture may be pinned without growing when it reaches the graphene-matrix interface. The debonding of graphene is brought on by the tip stress during fracture propagation. Microcracks develop on the graphene or even inside of it. The fracture energy is mostly used by the generated microcracks. If the external forces continue to increase, crack deflection may occur via the damaged graphene (Park et al. 2015, Guo et al. 2018, Jin et al. 2018). By deviating from the original direction of expansion, the fractures increase the crack's path of propagation. As a result, crack propagation resistance has increased. As well as slowing the spread of fractures in the covering, the graphene fracture also takes part of the energy generated by those cracks (Wang et al. 2017, Bafekrpour et al. 2015). The behaviour of polymer crystallisation can also be influenced by the addition of graphene to crystalline and semi-crystalline polymers. High-toughness crystals of the polymer might form. By increasing the number of crystals and altering their crystal shapes, the polymer matrix's hardness can be increased (Zhang et al. 2015, Jin et al. 2013).

12.4.2 Graphene Enhances the Role of the Sacrificial Anode

Because there must be effective electron pathways between the zinc particles and the metal substrate, zinc-rich coatings function effectively as sacrificial anodes. The coating's binder is frequently nonconductive to electricity. Due to this behaviour, the coating may contain a significant amount of isolated zinc particles that are not in touch with the metal substrate. As a result, it is challenging to protect metal substrates well (Marchebois et al. 2004, Zhou et al. 2019, Teng et al. 2018, Shirehjini et al. 2016). Two objectives are achieved by adding graphene to the coating: it transmits the zinc powder particles and serves as a cathode.

Graphene serves as the zinc powder particles' cathode. The coating is applied with a considerable amount of cathode using a significant amount of graphene. As a result, reactions might happen when the external electrolyte makes contact with the graphene and zinc powder particles. It is not necessary for the electrolyte to keep permeating in search of a cathode for the zinc powder particles. At the same time, the created insoluble corrosion product gradually plugs the permeation channel, preventing additional electrolyte solution penetration and enhancing the coating's corrosion resistance. The coating's polymer also makes it difficult to communicate electronically. More zinc powder particles are separated because the corrosion products are nonconductive, which reduces the amount of zinc powder needed. With its exceptional conductivity, graphene has been used as a connection between zinc powder particles for electronic communication. By establishing a significant conductive network in the coating, numerous isolated zinc particles are in electronic communication with the interface

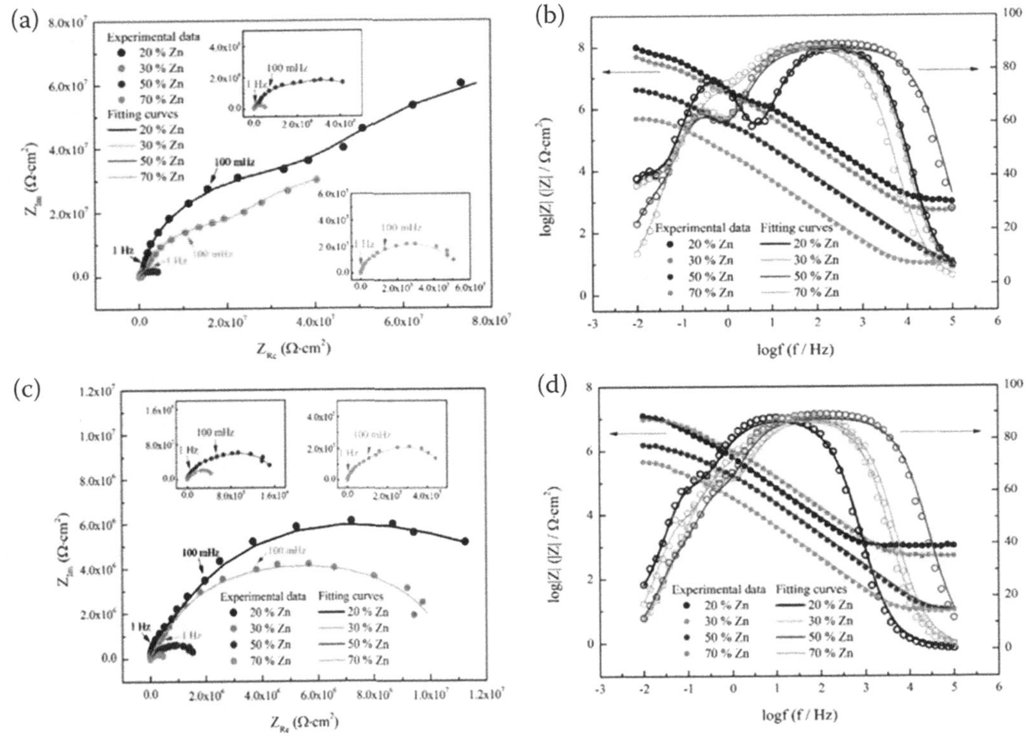

FIGURE 12.4 EIS and fitting curves of dry coatings: (a) No graphene coatings, (b) 0.3 wt.% graphene coatings (Ding et al. 2018a).

zinc or metal substrate. The sacrificial anode reaction then continues (Ding et al. 2018a). The results of anticorrosion performance by utilising EIS are represented in Figure 12.4, (a) for graphene-free coating and (b) for 0.3 wt.%-graphene coating. As the amount of conductive zinc and graphene in the coatings grew, so did their impedance. Each EIS displayed three-time constants, with the middle one being the easiest to see. The high-frequency region's time constant was represented in the Bode chart. The Nyquist chart revealed the time constant in the low-frequency area, but graphene-containing coatings' time constant was less obvious.

To fit the impedance spectrum, zinc distribution in the coatings had to be discussed. As shown in Figure 12.5, zinc was divided into isolated zinc and interfacial zinc depending on whether it came into contact with the inner or outer coating interface.

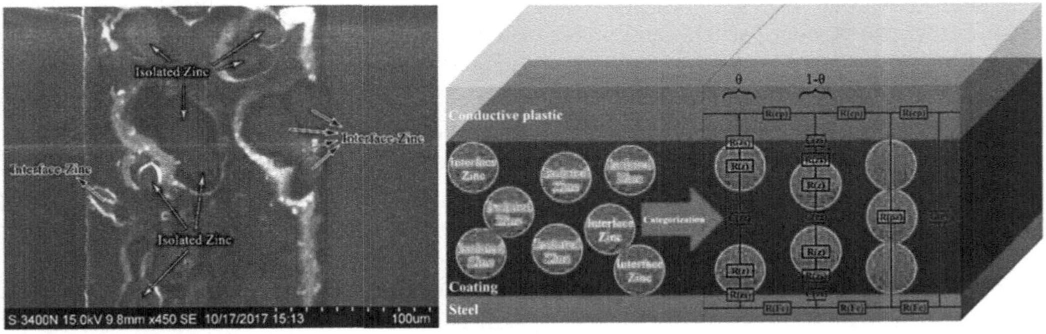

FIGURE 12.5 Classification of zinc particles and transmission line model of dry coatings (Ding et al. 2018a).

12.5 LIMITATIONS ASSOCIATED WITH THE GRAPHENE-BASED ANTICORROSION COATINGS

In composite coatings, graphene serves as a filler and corrosive media barrier. Due to its tendency to promote corrosion, graphene does not offer extended protection. Between the graphene and the metal substrate, a galvanic pair is formed. Unfortunately, Galvanic corrosion is accelerated by the presence of graphene since it serves as a cathode for most metals. In general, polymers are not electrically conductive; therefore, galvanic corrosion between the polymer covering and the substrate is not an issue. However, the polymer coating's conductivity improves by many orders of magnitude with the addition of graphene (Ding et al. 2018b), and galvanic corrosion is still possible (Sun et al. 2015). Regarding the aforementioned issue, there are three methods to increase the corrosion resistance of graphene coatings. One way to avoid galvanic contact between graphene and metal is to chemically lower its conductivity or enclose it in an insulating material. The second technique involves turning the graphene into an anode, for instance by covering it with a more active anode material (Zn). The third tactic is seeking alternatives with dependable electric insulation and other properties like those of graphene.

Ding et al. (2019) created nonconductive graphene by synthesising it with a furan epoxide monomer (FdE) and a Diels-Alder method (FmG). FdE is a highly dispersible insulating polymer that is compatible with the coating used to enclose graphene and is not electrically conductive. The sample was immersed in an aqueous solution for 120 days before the coating was removed from its surface. The sample coated in pure epoxy coating (FmG0) displayed pit corrosion, according to SEM examination, but the other samples coated in the graphene composite coating were free of corrosion-related flaws. Atomic force microscopy (AFM) images show that the pure epoxy coating failed to shield the substrate from corrosion, resulting in the surface developing large holes, as opposed to the substrate covered by the graphene composite coating, which remained flat and uninjured. The sample still exhibited high corrosion resistance after electrochemical testing.

SiO_2 was wrapped around graphene by Sun et al. to create graphene-SiO_2 nanosheets (GSs). To prevent direct contact between the metal and graphene, SiO_2 was utilised as a barrier. SiO_2 evenly coated graphene, and no naked graphene was discovered, according to the SEM and TEM characterisation of the GSs. Graphene's 3D conductive network was broken simultaneously, creating a weakly conductive GS structure that stopped it from encouraging corrosion. Corrosion could not be accelerated by graphene when the coating was compromised [73]. The conductivity of graphene can also be decreased by doping with electron dopants.

B-doped graphene (BG) was created synthetically and combined with an aqueous polyurethane (PU) resin by Ding et al. (2019). These scientists discovered that the coating's corrosion resistance has significantly increased. The excellent insulating qualities of BG prevent the galvanic response, which accounts for this improvement. By way of the sacrificial anode reaction of zinc, the zinc-rich epoxy coating shields the metal (Park and Shon 2015). Polymers do not, however, have excellent electrical conductivity in coatings. Therefore, the Zn concentration is frequently more than 80% to improve the consistency of electron transport between Zn and metal. On the other hand, high Zn levels decrease the coating's hardness and adhesive qualities and increase its viscosity, making it difficult to spray (Marchebois et al. 2002). The majority of the zinc is not utilised until the coating breaks. Thus, there is a significant loss of the zinc-rich epoxy coating. However, only a small quantity of zinc is used as an anode.

To improve the consistency of electron transport between the zinc particles and the metal, Hayatdavoudi et al. (2017) appropriately added graphene to the zinc-rich epoxy coating. In this instance, a greater number of zinc particles are activated, increasing its consumption. The coating's corrosion resistance increases concurrently with the barrier effect of graphene. Ding et al. created a graphene-containing epoxy coating rich in zinc and explained how graphene affects the coating using EIS. There are five steps to the application of the graphene-containing zinc-rich epoxy coating: initial shielding, wavering, cathodic protection, shielding, and failure. The cathodic protection lasts longer thanks to the graphene's shielding effect, which slows down the infiltration of corrosive media into the coating. The electron transport channel is increased, and the use of the zinc particles is improved by the simultaneous connection of the metals and zinc particles.

A new approach to addressing the galvanic corrosion issue related to graphene is made possible by the search for insulating graphene-like materials. The most promising insulating 2D barrier layer to date is hexagonal boron nitride (h-BN), often known as "white graphene." H-BN has exceptional mechanical, thermal, and impermeability qualities in a single layer and is electrically insulated (Camilli et al. 2019). Vacuum filtration was used by Yi et al. (2014) to create 2.5 nm thick boron nitride nanosheets on nylon. The polymer is shielded against corrosion by oxygen atoms by the barrier and bonding properties of h-BN.

Cui et al. (2017) used a carboxylated aniline trimer as a noncovalent dispersion to mix water-soluble h-BN into an epoxy coating. At a concentration of just one wt.%, the uniform distribution of the h-BN layers considerably improved the coating's water barrier performance and corrosion resistance. Their team also used poly (2-butyl aniline, or PBA) as a dispersion to distribute h-BN as a nanofiller in tetrahydrofuran. Epoxy's hardness was increased, and h-"labyrinth BN's effect" extended the corrosive media's penetration by spreading it in epoxy. Furthermore, the coating structure grew denser because PBA passivated the metal surface. As a result, the epoxy coating supplemented with h-BN may offer long-term stability in protection. However, h-BN synthesis is more expensive than producing graphene (Cui et al. 2018); thus, scientists have been looking for new graphene-like 2D barrier materials and low-cost, large-scale h-BN synthesis procedures.

12.6 CONCLUSION AND FUTURE OUTLOOK

Pure graphene coatings may effectively safeguard metals by separating corrosive media from the base. Due to its potential for mass manufacturing, CVD has mostly been utilised to manufacture graphene in laboratories today. On the other hand, long-term graphene flaws could speed up substrate metal degradation. Graphene's service life and mechanical strength will both be impacted by local oxidation. To produce pure graphene coatings with a large area, low defect density, and excellent resistance to mechanical damage, it is thus required to optimise the CVD preparation method. Additionally, it is essential to advance research into the physical vapour deposition (PVD) technique for growing graphene. The PVD technique may be used to regulate the amount of graphene layers, which removes the reliance on the transfer of graphene layers. This technique significantly lowers the cost of production by using amorphous carbon as the carbon source. Because of its "barrier" effect, graphene composite coatings have more satisfactory corrosion resistance than traditional coatings.

Corrosive chemicals cannot penetrate graphene, but on the other side, fissures cannot spread. Graphene could improve the defence of sacrificial anodes in the context of zinc-rich coatings. Graphene improves the corrosion resistance of coatings based on the aforementioned principles. Research is still needed on the defence mechanism of graphene composite coatings.

The current and potential distribution of metals under the protection of graphene composite coatings should be investigated in the future using micro-zone testing techniques such tow electrodes and microelectrodes. Additionally, artificial acceleration experiments should be used to understand the failure derivation process of the graphene composite covering. To offer shielding defence, graphene must be equally dispersed and organised in coatings.

Graphene modification techniques can be applied to improve the dispersion of graphene in coatings and ensure its long-term stability in polymer matrix. The original constants, with the middle one being the easiest structure of graphene, however, is harmed by the alteration of the covalent bonds. These chemical substances are also very poisonous and used in big quantities, which makes them easily able to create pollution and jeopardise worker health. Systematic research is essential to advance environmentally friendly modification technologies for graphene, aiming to preserve its natural structure. This preservation is crucial to enhance dispersion, minimize environmental pollution, and mitigate health risks. Effective non-covalent modification techniques for graphene include pi-pi stacking, ion bonding, and hydrogen bonding. Enhancing the ordered organisation of graphene in coatings is another topical problem. With an ordered rather than a random arrangement of graphene in the coating matrix, the penetration channel may be lengthened more readily, increasing the barrier to the entrance of corrosive substances in the coating. Graphene coatings parallel to metal substrates have been developed in a number of studies, however, the mechanism guiding the material's even distribution has not been fully analysed.

There is currently no documentation on how to create a highly ordered graphene composite coating, and graphene dispersion cannot be effectively controlled. However, rather than a highly ordered distribution in the ideal sense, the ordered distribution of graphene in the current work only manages to acquire a relatively uniform orientation. Ordered dispersion may significantly enhance the performance of graphene composite coatings, which will be a hot research topic in the future. Graphene aqueous anticorrosive coatings contain less hazardous heavy metals and volatile organic compounds than Graphene solvent-borne anticorrosive coatings (VOCs). Consequently, interest in graphene aqueous anticorrosive coatings has grown. As a result, one crucial area of study for heavy anticorrosion coatings is the development of water-based coatings. Graphene aqueous coatings still face a few technological challenges that have yet to be resolved. High-quality graphene aqueous anticorrosive coatings are challenging to create, and they have poor film-forming and abrasion resistance abilities. Due to the persistence of residual water-based groups in it, the aqueous anticorrosion coating has low shielding capacity against corrosive media like water and oxygen. Since the surface tension of water is higher than the crucial wetting surface tension of the pigment, it is difficult for the aqueous coating to achieve the high penetration and dispersion of pigments, such as graphene. It is still highly desired to change graphene correctly and match it with the suitable associated aqueous coating system in order to maximise the dispersibility and compatibility of graphene in coatings. It is crucial to investigate the best graphene concentration for waterborne coatings and how to enhance their ability to guard against corrosive media like water and oxygen. Finding 2D barrier materials that resemble graphene is also important. Due of graphene's high conductivity, galvanic corrosion is unquestionably a problem with coatings that contain graphene.

ACKNOWLEDGEMENTS

One of the author gratefully acknowledge funding from the Department of Science and Technology (DST), New Delhi, India, for the funding under the Women Scientist (WOS-A) program [File number: **SR-WOS-A-ET-54-2019**].

REFERENCES

Ahmad, H., Fan, M., and Hui, D. 2018. Graphene oxide incorporated functional materials: A review. *CompositesPart B-Engineering*, 145:270–280.

Ahmadi-Moghadam, B., Sharafimasooleh, M., Shadlou, S., and Taheri, F. 2015. Effect of functionalization of graphene nanoplatelets on the mechanical response of graphene/epoxy composites. *Material Design*, 66:142–149.

Ammar, S., Ramesh, K., Ma, I., Farah, Z., Vengadaesvaran, B., Ramesh, S., and Arof, A.K. 2017. Studies on SiO_2-hybrid polymeric nanocomposite coatings with superior corrosion protection and hydrophobicity. *Surface Coating and Technology*, 324:536–545.

Bae, S., Kim, H., Lee, Y., Xu, X., Park, J.-S., Zheng, Y.I., Balakrishnan, J., Lei, T., Ri Kim, H., Song, Y.I., Kim, Y.-J., Kim, K.S., Özyilmaz, B., Ahn, J.-H., Hong, B.H., and Iijima, S. 2010. Roll-to-roll production of 30-inch graphene films for transparent electrodes. *Nature Nanotechnology*, 5(8):574–578.

Bafekrpour, E., Ladani, R.B., Wang, C.H., Ghorbani, K., Wu, S., Zhang, J., Mouritz, A.P., and Kinloch, A.J. 2015. Aligning multilayer graphene flakes with an external electric field to improve multifunctional properties of epoxy nanocomposites. *Carbon*, 94:607–618.

Camilli, L., Yu, F., Cassidy, A., Hornekær, L., and Bøggild, P. 2019. Challenges for continuous graphene as a corrosion barrier. *2D Materials*, 6:022002.

Chang, C.-H., Huang, T.-C., Peng, C.-W., Yeh, T.-C., Lu, H.-I., Hung, W.-I., Weng, C.-J., Yang, T.-I., and Yeh, J.-M. 2012. Novel anti-corrosion coatings prepared from polyaniline/graphene composites. *Carbon*, 50:5044–5051.

Chang, K.-C., Hsu, M.-H., Lu, H.-I., Lai, M.-C., Liu, P.-J., Hsu, C.-H., Ji, W.-F., Chuang, T.-L., Wei, Y., Yeh, J.-M. et al. 2014. Room-temperature cured hydrophobic epoxy/graphene composites as corrosion inhibitor for cold-rolled steel. *Carbon*, 66:144–153.

Coleman, J.N., Khan, U., Blau, W.J., and Gun'ko, Y.K. 2006. Small but strong: A review of the mechanical properties of carbon nanotube–polymer composites. *Carbon*, 44:1624–1652.

Cui, G., Bi, Z., Zhang, R., Liu, J., Yu, X., and Li, Z. 2019. A comprehensive review on graphene-based anti-corrosive coatings. *Chemical Engineering Journal*, 373:104–121. doi: 10.1016/j.cej.2019.05.034.

Cui, M., Ren, S., Chen, J. Liu, S., Zhang, G., Zhao, H., Wang, L., and Xue, Q. 2017. Anti-corrosive performance of waterborne epoxy coatings containing water-dispersible hexagonal boron nitride (h-BN) nanosheets. *Applied Surface Science*, 397:77–86.

Cui, M., Ren, S., Qin, S., Xue, Q., Zhao, H., and Wang, L. 2018. Processable poly(2- butylaniline)/hexagonal boron nitride nanohybrids for synergetic anti-corrosive reinforcement of epoxy coating. *Corrosion Science*, 131:187–198.

Cui, M., Ren, S., Pu, J., Wang, Y., Zhao, H., and Wang, L. 2019. Poly(o-phenylenediamine) modified graphene toward the reinforcement in corrosion protection of epoxy coatings. *Corrosion Science*, 159:108131.

Ding, J., ur Rahman, O., Peng, W., Dou, H., and Yu, H. 2018. A novel hydroxyl epoxy phosphate monomer enhancing the anti-corrosive performance of waterborne graphene/epoxy coatings. *Applied Surface Science*, 427:981–991.

Ding, J., Zhao, H., Ji, D., Xu, B., Zhao, X., Wang, Z., Wang, D., Zhou, Q., and Yu, H. 2019. Achieving long-term anticorrosion via the inhibition of graphene's electrical activity. *Journal of Materials Chemistry A*, 7(6):2864–2874.

Ding, R., Zheng, Y., Yu, H., Li, W., Wang, X., and Gui, T. 2018a. Study of water permeation dynamics and anti-corrosion mechanism of graphene/zinc coatings. *Journal of Alloys and Compounds*, 748:481–495.

Ding, J.H., Zhao, H.R., Yan, Z., Zhao, X., and Yu H.B. 2018b. A long-term anticorrsive coating through graphene passivation. *Carbon*, 138:197–206.

Dittanet, P., Pearson, R.A., and Kongkachuichay, P. 2017. Thermo-mechanical behaviors and moisture absorption of silica nanoparticle reinforcement in epoxy resins. *International Journal of Adhesion and Adhesives*, 78:74–82.

Dumée, F.L., He, L., Wang, Z., Sheath, P., Xiong, J., Feng, C., Tan, M.Y., She, F., Duke, Gray, M.S., Pacheco, Hodgson, A.P., Majumder, M., and Kong, L. 2015. Growth of nano-textured graphene coatings across highly porous stainless steel supports towards corrosion resistant coatings. *Carbon*, 87:395–408.

Fontana, M.G. 1986. *Corrosion Engineering* 3rd edn (McGraw-Hill, 1986).

Georgakilas, V., Perman, J.A., Tucek, J., and Zboril, R. 2015. Broad family of carbon nanoallotropes: Classification, chemistry, and applications of fullerenes, carbon dots, nanotubes, Graphene, nanodiamonds, and combined superstructures. *Chemical Reviews*, 115:4744–4822.

Guo, S., Xia, L., Wu, H., Wang, J., Zhang, X., Jin, X., and Li, C. 2018. Achieving high-performance poly (styrene-b-ethylene-ranbutylene-b-styrene) nanocomposites with tannic acid functionalized graphene oxide. *Composites Science and Technology*, 158:137–146.

Hayatdavoudi, H., and Rahsepar, M. 2017. A mechanistic study of the enhanced cathodic protection performance of graphene-reinforced zinc rich nanocomposite coating for corrosion protection of carbon steel substrate. *Journal of Alloys and Compounds*, 727:1148–1156.

Imani, R., Mohabatpour, F., and Mostafavi, F. 2018. Graphene-based nano-Carrier modifications for gene delivery applications. *Carbon*, 140:569–591.

Jiang, S., Liu, Z., Jiang, D., Cheng, H., Han, J., and Han, S. 2016. Graphene as a nanotemplating auxiliary on the polypyrrole pigment for anticorrosion coatings. *High Performance Polymer*, 28(6):747–757.

Jin, J., Rafiq, R., Gill, Y.Q., and Song, M. 2013. Preparation and characterization of high performance of graphene/nylon nanocomposites. *European Polymer Journal*, 49:2617–2626.

Jin, X., Guo, S., Wu, H., Zhang, X., and Wang, J. 2018. Effect of tunable styrene content on achieving high-performance poly(styrene-b-ethylene-ran-butylene-b-styrene)/graphene oxide nanocomposites. *Composites Science and Technology*, 164:229–237.

Kakaei, K., Esrafili, M.D., and Ehsani, A. 2019. Graphene and anticorrosive properties. *Interface Science & Technology*, 27:303–337.

Kong, W., Kum, H., Bae, S.-H., Shim, J., Kim, H., Kong, L., Meng, Y., Wang, K., Kim, C., and Kim, J. 2019. Path towards graphene commercialization from lab to market. *Nature Nanotechnology*, 14 (10):927–938.

Lee, C., Wei, X., Kysar, J.W., and Hone, J. 2008. Measurement of the elastic characteristics and inherent strength of monolayer graphene. *Science*, 321:385–388.

Liang, J., Wang, Y., Huang, Y., Ma, Y., Liu, Z., Cai, J., Zhang, C., Gao, H., and Chen, Y. 2009. Electromagnetic interference shielding of graphene/epoxy composites. *Carbon*, 47:922–925.

Liu, C., Qiu, S., Du, P., Zhao, H., and Wang, L. 2018. An ionic liquid-graphene oxide hybrid nanomaterial: Synthesis and anti-corrosive applications. *Nanoscale*, 10:8115–8124.

Liu, S., Gu, L., Zhao, H., Chen, J., and Yu, H. 2016a. Corrosion resistance of graphene-reinforced waterborne epoxy coatings. *Journal of Material Science and Technology*, 32:425–431.

Liu, D., Zhao, W., Liu, S., Cen, Q., and Xue, Q. 2016b. Comparative tribological and corrosion resistance properties of epoxy composite coatings reinforced with functionalized fullerene C60 and Graphene. *Surface Coating Technology*, 286:354–364.

Marchebois, H., Touzain, S., Joiret, S., Bernard, J., and Savall, C. 2002. Zinc-rich powder coatings corrosion in sea water: Influence of conductive pigments. *Progress in Organic Coatings*, 45:415–421.

Marchebois, H., Savall, C., Bernard, J., and Touzain, S. 2004. Electrochemical behavior of zinc-rich powder coatings in artificial sea water. *Electrochimica Acta*, 49:2945–2954.

Miller, S.G., Bauer, J.L., Maryanski, M.J., Heimann, P.J., Barlow, J.P., Gosau, J.-M., and Allred, R.E. 2010. Characterization of epoxy functionalized graphite nanoparticles and the physical properties of epoxy matrix nanocomposites. *Composites Science and Technology*, 70:1120–1125.

Mochalin, V.N., and Gogotsi, Y. 2015. Nanodiamond–polymer composites. *Diamond and Related Materials*, 58:161–171.

Mohammadi, A., Barikani, M., Doctorsafaei, A.H., Isfahani, A.P., Shams, E., and Ghalei, B. 2018. Aqueous dispersion of polyurethane nanocomposites based on calix [4]arenes modified graphene oxide nanosheets: Preparation, characterization, and anti-corrosion properties. *Chemical Engineering Journal*, 349:466–480.

Mohammadi, S., and Roohi, H. 2018. Influence of functionalized multi-layer Graphene on adhesion improvement and corrosion resistance performance of zinc-rich epoxy primer. *Corrosion Engineering Science & Technology*, 53:422–430.

Monetta, T., Acquesta, A., and Bellucci, F. 2015. Graphene/epoxy coating as multifunctional material for aircraft structures. *Aerospace*, 2:423–434.

Morales-Narváez, E., Sgobbi, L.F., Machado, S.A.S., and Merkoçi, A. 2017. Graphene encapsulated materials: Synthesis, applications, and trends. *Progress in Material Science*, 86:1–24.

Papageorgiou, D.G., Kinloch, I.A., and Young, R.J. 2017. Mechanical properties of graphene and graphene-based nanocomposites. *Progress in Materials Science*, 90:75–127.

Parhizkar, N., Ramezanzadeh, B., and Shahrabi, T. 2018. Corrosion protection and adhesion properties of the epoxy coating applied on the steel substrate pre-treated by a sol-gel based silane coating filled with amino and isocyanate silane functionalized graphene oxide nanosheets. *Applied Surface Science*, 439:45–59.

Park, S.M., and Shon, M.Y. 2015. Effects of multi-walled carbon nano tubes on corrosion protection of zinc rich epoxy resin coating. *Journal of Industrial & Engineering Chemistry*, 21:1258–1264.

Park, Y.T., Qian, Y., Chan, C., Suh, T., Nejhad, M.G., Macosko, C.W., and Stein, A. 2015. Epoxy toughening with low graphene loading. *Advanced Functional Materials*, 25:575–585.

Pierleoni, D., Xia, Z.Y., Christian, M., Ligi, S., Minelli, M., Morandi, V., Doghieri, F., and Palermo, V. 2016. Graphene-based coatings on polymer films for gas barrier applications. *Carbon*, 96:503–512.

Pourhashem, S., Rashidi, A., Vaezi, M.R., and Bagherzadeh, M.R. 2017. Excellent corrosion protection performance of epoxy composite coatings filled with amino-silane functionalized graphene oxide. *Surface and Coatings Technology*, 317:1–9.

Qiu, S., Liu, G., Li, W., Zhao, H., and Wang, L. 2018. Noncovalent exfoliation of graphene and its multifunctional composite coating with enhanced anti-corrosion and tribological performance. *Journal of Alloys & Compounds*, 747:60–70.

Raman, R.K.S., Banerjee, P.C., Lobo, D.E., Gullapalli, H., Sumandasa, M., Kumar, A., Choudhary, L., Tkacz, R., Ajayan, P.M., and Majumder, M. 2012. Protecting copper from electrochemical degradation by graphene coating. *Carbon*, 50(11):4040–4045.

Ramezanzadeh, B., Haeri, Z., and Ramezanzadeh, M.A. 2016. Facile route of making silica nanoparticles-covered graphene oxide nanohybrids (SiO_2-GO); fabrication of SiO_2-GO/epoxy composite coating with superior barrier and corrosion protection performance. *Chemical Engineering Journal*, 303:511–528.

Shadlou, S., Alishahi, E., and Ayatollahi, M. 2013. Fracture behavior of epoxy nanocomposites reinforced with different carbon nano-reinforcements. *Composite Structure*, 95:577–581.

Shirehjini, F.T., Danaee, I., Eskandari, H., and Zarei, D. 2016. Effect of nano clay on corrosion protection of zinc-rich epoxy coatings on steel 37. *Journal of Materials Science & Technology*, 32:1152–1160.

Singh, B.P., Nayak, S., Nanda, K.K., Jena, B.K., Bhattacharjee, S., and Besra, L. 2013. The production of a corrosion resistant graphene reinforced composite coating on copper by electrophoretic deposition. *Carbon*, 61:47–56.

Singh, B.P., Jena, B.K., Bhattacharjee, S., and Besra, L. 2013. Development of oxidation and corrosion resistance hydrophobic graphene oxide-polymer composite coating on copper. *Surface Coating Technology*, 232:475–481.

Singh, V., Joung, D., Zhai, L., Das, S., Khondaker, S.I., and Seal, S. 2011. Graphene based materials: Past, present and future. *Progress in Material Science*, 56:1178–1271.

Singhbabu, Y.N., Sivakumar, B., Singh, J.K., Bapari, H., Pramanick, A.K., and Sahu, R.K. 2015. Efficient anti-corrosive coating of cold-rolled steel in a seawater environment using an oil-based graphene oxide ink. *Nanoscale*, 7(17):8035–8047.

Sun, W., Wang, L., Wu, T., Pan, Y., and Liu, G. 2015. Inhibited corrosion-promotion activity of graphene encapsulated in nanosized silicon oxide. *Journal of Materials Chemistry A*, 3:16843–16848.

Sun, Y., Kandan, K., Shivareddy, S., Farukh, F., and Bailey, R. 2019. Effect of sliding conditions on the macroscale lubricity of multilayer graphene coatings grown on nickel by CVD. *Surface Coating & Technology*, 358:247–255.

Tan, B., and Thomas, N.L. 2016. A review of the water barrier properties of polymer/clay and polymer/graphene nanocomposites. *Journal of Membrane Science*, 514:595–612.

Teng, S., Gao, Y., Cao, F., Kong, D., Zheng, X., Ma, X., and Zhi, L. 2018. Zinc-reduced graphene oxide for enhanced corrosion protection of zinc-rich epoxy coatings. *Progress in Organic Coatings*, 123:185–189.

Thanh, T.D., Chuong, N.D., Hien, H.V., Kshetri, T., Tuan, L.H., Kim, N.H., and Lee, J.H. 2018. Recent advances in two-dimensional transition metal dichalcogenides graphene heterostructured materials for electrochemical applications. *Progress in Material Science*, 96:51–85.

Umoren, S.A., and Solomon, M.M. 2019. Protective polymeric films for industrial substrates: A critical review on past and recent applications with conducting polymers and polymer composites/nanocomposites. *Progress in Material Science*, 104:380–450.

Vashist, S.K., and Luong, J.H.T. 2015. Recent advances in electrochemical biosensing schemes using graphene and graphene-based nanocomposites. *Carbon*, 84:519–550.

Wang, J., Li, C., Zhang, X., Xia, L., Zhang, X., Wu, H., and Guo, S. 2017. Polycarbonate toughening with reduced graphene oxide: Toward high toughness, strength and notch resistance. *Chemical Engineering Journal*, 325:474–484.

Wang, J., Jin, X., Li, C., Wang, W., Wu, H., and Guo, S. 2019. Graphene and graphene derivatives toughening polymers: Toward high toughness and strength. *Chemical Engineering Journal*, 370:831–854.

Xie, X.-L., Mai, Y.-W., and Zhou, X.-P. 2005. Dispersion and alignment of carbon nanotubes in polymer matrix: A review. *Materials Science and Engineering: Report*, 49:89–112.

Yang, Z., Sun, W., Wang, L., Li, S., Zhu, T., and Liu, G. 2016. Liquid-phase exfoliated fluorographene as a two dimensional coating filler for enhanced corrosion protection performance. *Corrosion Science*, 103:312–318.

Yi, M., Shen, Z., Zhao, X., Liang, S., and Liu, L., 2014. Boron nitride nanosheets as oxygen-atom corrosion protective coatings. *Applied Physics Letters*, 14:143101–143101-5.

Yu, Y.-H., Lin, Y.-Y., Lin, C.-H., Chan, C.-C., and Huang, Y.-C. 2014. High-performance polystyrene/graphene-based nanocomposites with excellent anti-corrosion properties. *Polymer Chemistry*, 5:535–550.

Yu, Y.H., Lin, Y.Y., Lin, C.H., Chan, C.C., and Huang, Y.C. 2013. High-performance polystyrene/graphene-based nanocomposites with excellent anti-corrosion properties. *Polymer Chemistry*, 5:535–550.

Zhan, Y., Zhang, J., Wan, X., Long, Z., He, S., and He, Y. 2018. Epoxy composites coating with Fe3O4 decorated graphene oxide: Modified bio-inspired surface chemistry, synergistic effect and improved anti-corrosion performance. *Applied Surface Science*, 436:756–767.

Zhang, H., Wang, S., Lin, Y., Feng, M., and Wu, Q. 2017. Stability, thermal conductivity, and rheological properties of controlled reduced graphene oxide dispersed nanofluids. *Applied Thermal Engineering*, 119:132–139.

Zhang, L., Li, Y., Wang, H., Qiao, Y., Chen, J., and Cao, S. 2015. Strong and ductile poly (lactic acid) nanocomposite films reinforced with alkylated graphene nanosheets. *Chemical Engineering Journal*, 264:538–546.

Zhang, X., Zhou, Q., Yuan, M., Liao, B., Wu, X., and Ying, M. 2020. Etching-controlled preparation of large-area fractal graphene by low-pressure CVD on polycrystalline Cu substrate. *Material Today Communication*, 24:101093.

Zhang, Y., Fu, Y., Edwards, M., Jeppson, K., Ye, L., and Liu, J. 2017. Chemical vapor deposition grown graphene on Cu-Pt alloys. *Materials Letter*, 193:255–258.

Zhou, S., Wu, Y., Zhao W., Yu, J., Jiang, F., Wu, Y., and Ma, L. 2019. Designing reduced graphene oxide/zinc rich epoxy composite coatings for improving the anti-corrosion performance of carbon steel substrate. *Materials & Design*, 169:107694.

13 Industrial and Economic Impact of Carbon Allotropes as Anticorrosive Coatings

A.M. Fadl and M.I. Abdou
Production Department, Egyptian Petroleum Research Institute, Nasr City, Cairo, Egypt

CONTENTS

Engineering for corrosion protection is utilised to prevent and control degradation processes on metallic surfaces, particularly carbon steel substrates. Different application techniques were provided to prolong material life and reduce corrosion. Nanocontainers might be utilised to significantly reduce corrosion on many surfaces. The examined research findings supported the notion that nanostructured materials reduce corrosion rates while improving corrosion inhibitory effectiveness with material loading. Many surfaces have been coated with organic coatings for various roles, such as protection, decoration, or a mix of the two, in addition to other multi-functional qualities. Because steel is one of the most frequently used building elements and is frequently exposed to harsh environments, 80% of steel surfaces worldwide need to be painted to protect them. Coatings are the best corrosion mitigation applicant on steel surfaces for superb protection. In order to formulate the ideal nanomaterial with a variety of vehicles and achieve the preferred corrosion mitigation behaviour on a variety of surfaces, coating producers must choose the correct nanomaterial.

Utilisation of nanotechnology becomes the preferable future technology in different applied fields. The term "nanotechnology" is means the preparation, analysis and use of structures of less than 100 nanometres (nm) in at least one dimension. Applicably, the incorporation of nano-swed materials and systems offers multi-functional properties versus neat epoxy for enhancing their appearance and use in severe conditions.

DOI: 10.1201/9781003323976-16

The principle aim of this chapter is to demonstrate the influence of incorporation of nano-sized materials via organic coatings to display a passive barrier to prohibit the transmission zigzag paths of corrosive species and support corrosion mitigation performance on various surfaces, especially carbon steel.

13.1 GENERAL CORROSION AND CORROSION CONTROL ASPECTS

13.1.1 Definition of Corrosion

The damaging attack on metals or alloys caused by a chemical or electrochemical response with their surroundings is known as corrosion [1,2]. This procedure could produce solid, liquid, or gaseous byproducts. The items' physical and chemical characteristics are crucial since they frequently affect the rate at which corrosion may develop later. Metals are destroyed by corrosion in one of two ways: (a) through direct chemical assault at high temperatures in a dry environment; or (b) through electrochemical reactions at lower temperatures in a water- or moisture-wet environment [3,4].

13.1.2 Why Metals Corrode

Corrosion is a natural process that tends to the lowest possible energy levels (much like how water flows to its lowest level). Iron and steel, for instance, naturally have a propensity to combine with other chemical elements in order to return to their lowest energy levels [1,2]. Iron and steel regularly react with oxygen and water, both of which are present in most natural surroundings, to generate hydrated iron oxides (rust), which are chemically similar to the original iron ore. This process is done in order to return to lower energy levels. However, the interaction of oxygen and water can result in additional compounds with various colours:

- $Fe_2O_3.H_2O$, the main ingredient of reddish-brown rust, is H_2O (hydrous ferrous oxide), which is also known as Fe $(OH)_3$. Hematite, the most typical iron ore, can be formed from it.
- $Fe_3O_4.H_2O$ (hydrated magnetite), also called ferrous ferrite ($Fe_2O_3.FeO$), is most often green but can be deep blue in the presence of organic complexes.
- Fe_3O_4 (magnetite) is black.

The energy needed to transform iron ore into metallic iron is recovered when the iron corrodes to create the original material. The life cycle of corrosion in a steel product is depicted in Figure 13.1.

13.1.3 The Economic Importance of Corrosion

Every sector of the economy is impacted by corrosion, from the integrated circuit to the reinforced concrete bridge. According to estimates, the expense of corrosion accounts for 4% of the gross national product [1]. These numbers include the following:

- **Direct losses:** alteration of equipment and materials damaged by corrosion.
- **Indirect losses:** cost of repair and loss of production.
- **Cost of corrosion protection:** application of surface coatings, cathodic protection systems, and the use of more expensive corrosion-resistant materials.
- **Cost of corrosion prevention:** maintenance, inspections, corrosion prevention by design.

However, there is no denying that the sums of money involved are rather substantial. Safety is also another crucial part of corrosion. Damage from corrosion can make it unsafe for equipment or

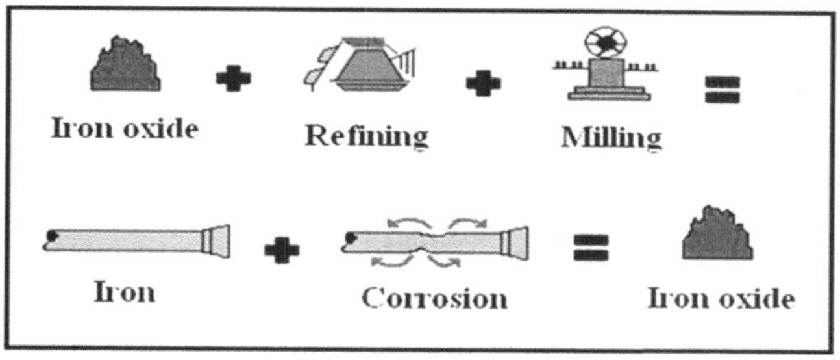

FIGURE 13.1 Refining corrosion cycle [1].

installations to operate, and it can also be the cause of fatal accidents and other serious mishaps. Furthermore, corrosion is a waste of raw materials and, indirectly, of energy [1].

13.2 SURFACE COATINGS FOR CORROSION PROTECTION AND CONTROL

An anode (the positive pole), a cathode (the negative pole), an electrolyte that is continuously conducting, and a conductor that facilitates the flow of electrons from the anode to the cathode make up any corrosion cell [1]. This cell needs to be inhibited in order to stop corrosion. To do this, one of the following options is available: (1) making the electrolyte nonconductive; (2) disconnecting the anode and cathode electrically; (3) isolating the cathode or anode from the electrolyte; or (4) removing the electrical potential from the system. Practically, the most popular ways to counteract corrosion include using corrosion-resistant metals, cathodic protection, anodic protection, corrosion inhibitors, or protective coatings [1,5].

A great film-based substance that can withstand industrial or marine environments and prevent catastrophic structural collapse constitutes is considered an idealistic protective coating. Additionally, a protective coating film needs to be able to withstand ion penetration and keep ions from reaching the metal, withstand moisture osmosis, and maintain its appearance under a variety of weathering circumstances for a fair amount of time [6]. Effective anticorrosive coating design is a difficult skill. In addition to the chemistry involved in the creation and formulation of coatings, it necessitates a thorough understanding of corrosion concepts [7].

13.3 MECHANISMS OF PROTECTION

Why steel corrodes even after painting—especially when the paint is a multi-coat system—is a subject that users of protective coatings frequently ask. Therefore, the user must have a thorough understanding of the mechanism of coating protection. The following three explanations can each be used to describe how the coatings' protective mechanism works.

13.3.1 Barrier Mechanism

The barrier mechanism functions by increasing the resistance against disbondment, decreasing the permeability to moisture, and providing resistance to the flow of corrosive ions [8,9].

13.3.1.1 Working of Barrier Mechanism

The thick undercoat, which has a dry film thickness (DFT) of 100 to 150 microns, provides the primary barrier effect. This undercoat may include two coatings with an even greater DFT in some instances of extremely harsh corrosive environments. This undercoat layer diminishes the amount of moisture that seeps through the coating, and because it acts as a thick barrier, it also prevents the cathodic reaction, which is an interaction between oxygen and the substrate. It resists the flow of corrosive ions like chlorides and sulphates to the substrate since it is a cross-linked polymer substance. The barrier has a low moisture vapour transmission rate (MVTR), which prevents moisture from penetrating the film. This resistance is increased by the presence of the foliar pigment micaceous iron oxide (MIO).

13.3.2 Inhibitive/Passivation Mechanism

Utilising anticorrosive/inhibitive pigments, the inhibitive/passivation mechanism slows down corrosion at the coating-metal contact [10].

13.3.2.1 Working of Inhibitive/Passive Mechanism

Even while the coatings provide resistance to the transfuse of moisture, some moisture does still reach the substrate and is inevitably going to react with the steel substrate in the absence of any primer at the interface. This is the corrosion process' anodic reaction. If the primer has an inhibitive tendency at this point, the surface becomes passive and inhibits corrosion. Numerous known inhibitive pigments exist, including calcium chromate, zinc phosphate, red lead, and lead silico-chromate. They dissolve or leach in moisture to create a passive coating that covers the steel surface, preventing corrosion from occurring. The other pigments are being phased out due to environmental concerns, leaving just zinc chromate and zinc phosphate in use today.

13.3.3 Sacrificial Mechanism

Metallic pigments are used for sacrificial-type cathodic protection in primers, which is how the sacrificial mechanism operates [11,12].

13.3.3.1 Working of Sacrificial Mechanism

The zinc present at the interface, which is anodic to steel, becomes the anode and dissolves itself preferentially and protects the steel if the primer is of a sacrificial nature and contains fine zinc powder as pigment in an organic or inorganic medium (referred to as organic or inorganic zinc rich). Cathodic protection offered in sacrifice for this. Zinc ethyl silicate primer, an organic zinc-rich primer based on epoxies, polyurethanes, and chlorinated rubbers, and zinc ethyl silicate primer, an inorganic zinc-rich primer, are both common primers for these uses. Suppression of anodic reaction can be referred to as the two operations of inhibitive and sacrificial primers at the coating-metal interface.

13.3.4 Multi-coat Protective Coating System

For extremely corrosive situations, multi-coat layers made up of primer, undercoat, and topcoat are devised since it is not possible to satisfactorily realise both a barrier and a passivation/sacrificial mechanism inside the same coating layer.

13.3.4.1 A Multi-coat Protective Coating System Consists of:

I. An anticorrosive primer containing either an inhibitive or a sacrificial pigment.
II. A thick undercoat of a cross-linked polymer system optionally with a pigment of leafy structure.
III. An UV-resistant topcoat.

The methods mentioned above are used in combination by the multi-coat system to function. Moisture tries to pass through the coating of the coated metal when it is exposed to corrosive air, but the undercoat is more resistant to this because of its greater thickness and the cross-linked polymers' lower MVTR. This resilience is increased by the undercoat's inclusion of leafy pigment.

The undercoat inhibits the flow of corrosive ions through it to the substrate because of its high ionic resistance. When moisture eventually makes its way through the undercoat and primer to the substrate, the inhibitive or sacrificial pigment in the primer protects the metal by either a sacrificial or inhibitive method. The topcoat shields the bottom coats from sun-induced UV deterioration. The total protection provided by a multi-coat system is a cumulative effect of all the theories and mechanisms explained above.

13.4 NANOPARTICLES AND NANOCOMPOSITES FOR COATINGS

The majority of the time, 100 nm-sized clusters of atoms or molecules are referred to as nanoparticles. There are roughly 25 atoms in a cluster of 1-nanometre radius, but the majority of them are on the cluster's surface [13]. As a result of the majority of atoms being in a distinct environment, nanomaterials' intrinsic properties differ from those of ordinary materials as shown in Figure 13.2.

The number of surface molecules or atoms is quite great compared to the molecules or atoms in the bulk of the materials, making nanomaterials practically the ultimate in increasing surface area and chemically very active. High surface area materials offer improved mechanical, optical, chemical, and magnetic properties, which can be used for a variety of structural and non-structural applications. Nanoparticles/fillers are used in corrosion-, wear-, and erosion-resistant products. [14].

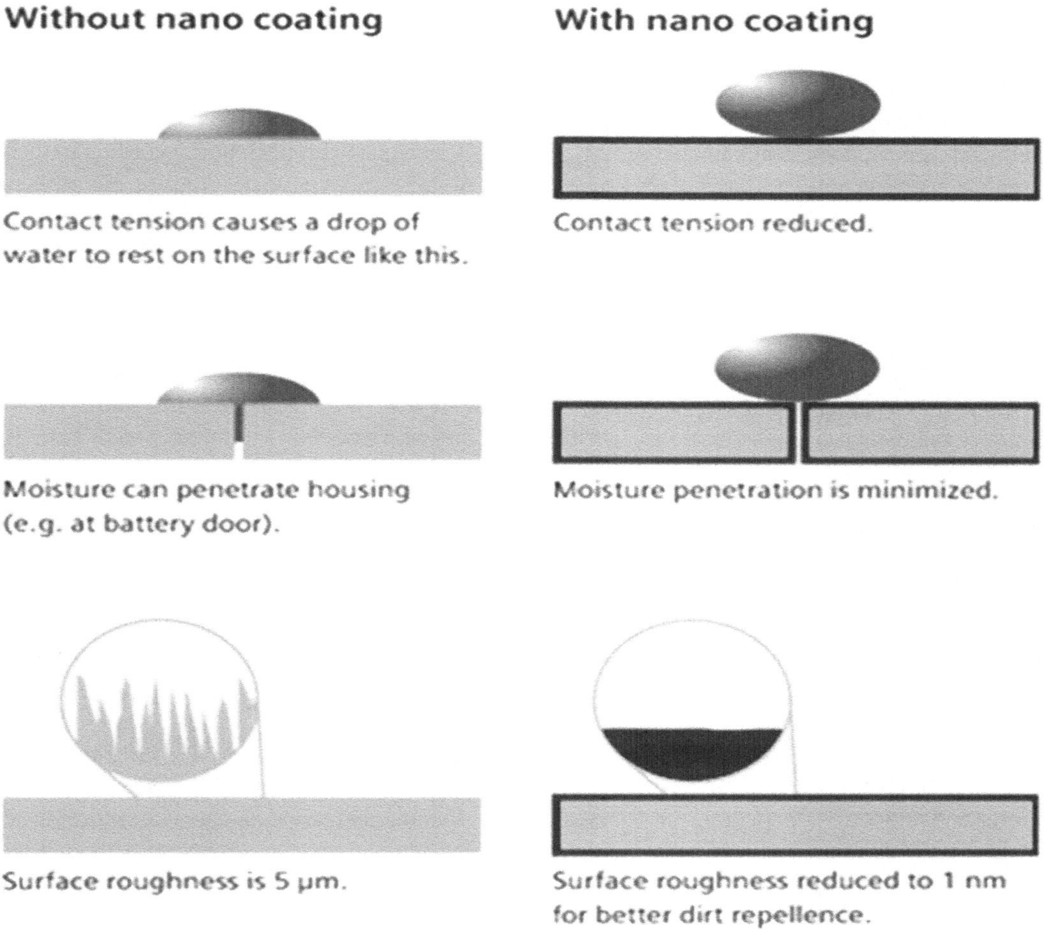

FIGURE 13.2 Nano coating vs. conventional coating [1,13].

Car tyres use common nanoparticles like carbon black to extend tyre life and produce the black colour. Nano coatings are coatings created by utilising a component at the nanoscale to achieve desired qualities [15]. The empty space between the pigment and the resin is reduced because nanoparticles may absorb more resins than regular pigments due to their higher surface activity. Nanoparticles reduce the route taken by corrosive species during transit while improving the effectiveness and qualities of protection. An emerging area of research in the various branches of polymer materials science and technology is nanocomposites. It has been demonstrated that adding nanoparticles under specific circumstances can improve the characteristics of resins generally. When a material slid across a smooth steel counterpart, Min Zhi Rong et al. showed the significance of TiO_2-nanoparticles dispersed in an epoxy resin matrix on the tribological properties of the material [16]. It was discovered that the dispersion states of the nanoparticles had a substantial impact on the friction and wear behaviour of polymeric nanocomposites in sliding environments, and their wear resistance could be significantly boosted by improving the nanocomposites' microstructural homogeneity [16]. The performance of sliding wear of epoxy composites containing nano-sized Al_2O_3 particles was studied. The obtained results showed that epoxy can have its frictional coefficient and wear rate lowered at relatively low nano-Al_2O_3 concentrations, and the pretreatment of the particles further enhanced this beneficial impact. The composites with 0.24 vol.% nano-Al_2O_3 grafted by polyacrylamide show the lowest specific wear rate, 1.6106 mm3/Nm, which is reduced by 97% when compared to the value of plain epoxy [17,18].

There have previously been several papers published on coating systems with nanofillers and nanocomposites such as on SiO_2/polyurethane nanocoatings [19], anatase-TiO_2 nanocomposite coatings [20], silicone/phosphorus/epoxy [21], silica/polysiloxane [22], epoxy/graphene oxide nanocomposite [23], TiO_2 nanoparticles surface modified with gallic acid esters [24], ZnO/epoxy-polyamide nanocomposites [25], silica nanoparticles-covered graphene oxide nanohybrids (SiO_2-GO) [26], Silica/epoxy nanocomposites [27], TiO_2-SiO_2 nanoparticles [28], and according to these research, using nanofillers has benefits over typical organic coatings that comprise microfillers, including increased corrosion mitigation performance, film adhesion, film stiffness, and reduced water permeability. It is crucial to spread the binder for an effective "nanoorganic coating." Agglomerates will form and the desired qualities won't be obtained if this isn't accomplished. When uniform integration is challenging, as it is with lamellar silicates, it may be essential to chemically modify the nanofiller surface [29]. Because of the extensive "inter-phase" that is produced surrounding the nanopigment particles, the presence of nanopigments in binder produces novel and intriguing macroscopic features. Nanotechnology, which deals with adding nanoparticles that enhance particular traits, has the possibility of unleashing enormous potential in the coatings industry. The addition of nanofillers can enhance the protective qualities of a coating by altering the structure of the coating, which has a significant impact on surface protection. Studies on the creation of epoxy-clay nanocomposite coatings [30] demonstrated that adding nanoclay particles, such as those shown in Figure 13.3, to organic coatings enhances the coating's anticorrosive capabilities. In their work, the appropriate volume of nanoclay and resin was combined. The blending procedure was carried out in an oil bath (50–70 degree). The combination was then put through an 8–12-hour sonication process. The stoichiometric amount of the hardener was added to the mixture for the creation of nanocoating after adding various additives to the epoxy-clay mixture.

The stoichiometric amount of the hardener was added to the mixture for the creation of nanocoating after adding various additives to the epoxy-clay mixture. It was discovered that the barrier and anticorrosive qualities increase as the clay loading increases. At clay concentrations of 3 and 5 wt%, the required corrosion mitigation efficacy of coatings was achieved. Although utilising 3 wt% is more practical economically, the EIS results with samples of clay concentrations of 3 and 5 wt% were comparatively similar. Using epoxy resin and nano-$CaCO_3$ composites produced by in situ and inclusion polymerisation, Yu et al. developed nano-$CaCO_3$ modified powder coatings [31]. They compared modified powder coatings to unmodified powder coatings and

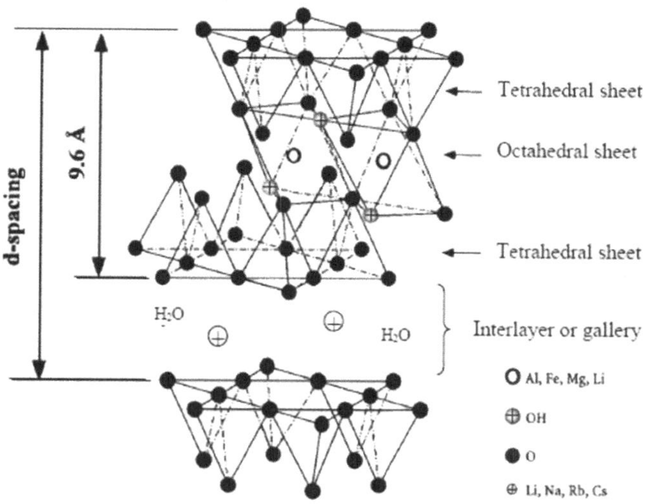

FIGURE 13.3 Structure of nanoclay [30].

discovered that the modified coatings had significantly better tensile characteristics and neutral salt spray corrosion resistance. They came to the conclusion that the properties of the resulting powder coatings are significantly impacted by the dispersion of nanoparticles in the resulting films, and that the in situ and inclusion polymerisation method is clearly an effective way to disperse nano-$CaCO_3$ in the powder coatings effectively. Additionally, different types of nanoparticle-modified powder coatings can be made using this technique as a guide [31].

By incorporating a second phase that is miscible with the polymer, reducing porosity, and zigzagging the diffusion path for harmful species, polymer coatings' physical barrier performance can be enhanced [32], as shown in Figure 13.4. Conventional polymer coatings' anticorrosive performance on steel substrates was greatly improved by the addition of nanoparticles (polyaniline/ferrite, Fe2O3, halloysite clay, etc.), as shown in Table 13.1. This table lists various polymer nanocomposite barrier coatings along with their anticorrosion capabilities for different metallic substrates [33–55]. In order to calculate the corrosion inhibition efficiency (*IE%*), the corrosion rate (*CR*) or the current density (i_{corr}) was typically derived from potentiodynamic polarisation curves or from charge transfer resistance (R_{ct}), electrochemical impedance spectroscopy was accomplished.

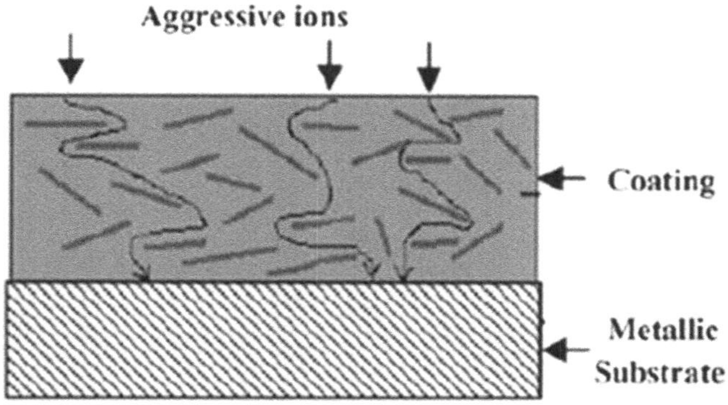

FIGURE 13.4 Nanoplatelets' inclusion causes the diffusion path of corrosive media to zigzag [32].

TABLE 13.1
A Comparison of Several Recently Created Polymer-Based Nanocomposite Coatings' Characteristics with Those of Their Non-Nanomodified Counterparts [33–55]

Metal surface	Polymer	Nanoparticles (NPs)/ Nanocomposites (NC)	Calculated Corrosion Inhibition Efficiency for Nanocomposite Coating Compared to Polymer Coating Without NM	Properties Other Than Corrosion Resistance and Mechanisms of Protection in Addition to Physical Barrier
Mild steel	Alkyd resin	Polyaniline/ferrite	48.6%, 28.1%, 50% for 1.5% NM in 5% HCl, 5% NaOH, 3.5% NaCl solution, respectively, after 960 h immersion	The nanostructured coating was dense and nonporous with continuous network-like structure
Mild steel	Alkyd-based waterborne polymer	Fe_2O_3, ZnO	N.B	UV resistance, scratch resistance, and abrasion resistance were improved
Steel	Epoxy resin	SiO_2, Zn, Fe_2O_3, and halloysite clay	96.9%, 90.9%, 99.9%, 99.8% for SiO_2, Zn, Fe_2O_3, halloysite clay (at 1% NM), respectively, in 3% NaCl solution after 28 days immersion	SiO_2 and Zn enhanced the Young's modulus while halloysite clay and Fe_2O_3 decreased it
Iron	Polyaniline	Clinoptilolite (Clino)	74.9%, 95.8%, –38.7% for 3% NM in 1 M H_2SO_4, 1 M HCl, 3.5% NaCl solution, respectively	The alignment of polyaniline chains was improved
Carbon steel	Polyurethane	Organoclay	88.2%, 98.7%, 99.7% for 1%, 2%, 3% NM, respectively, in 5% NaCl solution after 30 days immersion	Adhesion of coating decreased with increasing weight percent of NM
Galvanized steel	Epoxy-polyamide	ZnO	N.B.	The nanocomposite coating had less decrease in cross-linking density, indentation hardness, and adhesion after immersion than control
Carbon steel	Epoxy resin	Waterborne polyaniline	95.5% for 1% NM in 5% NaCl solution after 720 h immersion	Better physico-mechanical properties. Formation of passive oxide layer on metal
Carbon steel	Epoxy resin	Al	67.6% for 40% NM in 0.1 M NaCl solution after 16 days immersion	Nanoscale complex oxides containing Al and Fe increased the corrosion resistance
Carbon steel	Epoxy resin	Si	76.2% for 40% NM in 0.1 M NaCl solution after 8 h immersion	Si particles have sacrificial effect against the substrate corrosion
Steel	Polyurethane	Silica (SiO_2)	N.B.	The water absorption was reduced; the curing temperature and abrasion resistance were increased
Carbon steel	Epoxy resin	Urea-modified montmorillonite clay	30.3%, 34.5%, 77.7% for 1%, 2%, 3% NM, respectively, in 3.5% NaCl solution	The protectiveness of coating decreased with increasing temperature

Carbon steel	Epoxy resin	Hybrid nanozinc dust and halloysite	N.B.	Thermal stability of epoxy coating was improved. Nanozinc dust served as sacrificial anode
Aluminium grade 2024-T3	Polyurea/siloxanes/epoxy ester	Montmorillonite clay	N.B.	The presence of clay nanoparticles decreased diffusivity
Magnesium grade AZ31	Polyetherimide/diethylenetriamine	Hydroxyapatite	N.B.	Biocompatibility, adhesion, and uniformity were enhanced
Steel	Polystyrene	Organo-functionalized graphene oxide	91.4%, 96.4%, 99.2% for 1%, 1.5%, 2% NM, respectively, in 3.5% NaCl solution after 30 days immersion	The exfoliated dispersion of the graphene oxide reduced gas permeability and improved mechanical strength
Steel	Epoxy resin	Clay	N.B.	The microstructure has been modified to a homogeneous exfoliated structure. The mechanical properties have been improved. The thermal properties were improved
Steel	Waterborne epoxy resin	Polyaniline-mesoporous silica MCM-41	N.B.	The cross-linking density became higher. Polyaniline acts as an electronic and chemical diffusion barrier
Aluminium bipolar	Polyaniline	Carbon nanotube	71.5%, 87.7%, 92.9% for 0.4%, 0.6%, 0.8% NM, respectively, in 0.1 M H_2SO_4	The electrical conductivity of coating was increased. Increased the adhesion strength of coating and decreased the oxygen and water permeability
Mild steel	Polyaniline	Functionalized multi-walled carbon nanotube	91.9%, 96.9%, 99.2% for 1%, 2%, 3% NM, respectively, in 3.5% NaCl solution	The hardness was increased significantly and the surface became more hydrophobic
Mild steel	Epoxy resin	Polyaniline–camphorsulfonate	N.B.	Durable bond strength between coated rebar and concrete after 1 year. Presence of a protective passive layer on the steel surface
Mild steel	Epoxy resin	Polyesteramide-clay	N.B.	No significant improvement in thermal stability
Mild steel	Epoxy resin	Zirconia (ZrO_2)	N.B.	The tensile and hardness properties of the coatings were improved
Aluminium grade 1050	Polyaniline	TiO2, Ag, Zn	5.7%, 71.3%, 12.9% for TiO_2, Ag, Zn (1% NM), respectively, in 3.5% NaCl solution	Polyaniline/Ag film was the most homogeneous coating

13.5 SURFACE COATING NANOCOMPOSITES WITH SYNERGISTIC CORROSION-MITIGATING METHODS

Through their integration in polymeric coatings, a number of nanocomposite materials provided a synergistic anticorrosion performance. In order to reduce corrosion, multi-functional nanocomposites coatings were produced and tested on the surface of carbon steel. In order to modify various high build epoxy coating formulations, processed ilmenite/melamine formaldehyde composite (IMFC) in varied proportions that were then hardened by poly amine were created [56]. Accelerated corrosion experiments (salt spray chamber) were used to compare the anticorrosion performance characteristics of the proposed IMFC epoxy coating formulations to unaltered blank standard epoxy after 500 hours of exposure to 5% NaCl solution as shown in Figure 13.5. The obtained results showed that increasing the loading level (%) by weight of the nanocomposite modifier in the coating film significantly improved the rust grade, blistering size, colour tolerance, and adhesion strength of IMFC modified epoxy coated steel films.

As the rust grade grew, the inhibitive effectiveness of the IMFC epoxy coating films improved and was supported by an increase in the loading amount (%) by weight of the IMFC modifier in the following order: IMFC-E>IMFC-D>IMFC-C>IMFC-B>IMFC-A> Blank conventional epoxy coating, as shown in Figure 13.6.

Using the potentiodynamic polarisation technique, it was determined how the concentration of $FeTiO_3$/melamine formaldehyde composite (IMFC) particles in the coating layer affected the anodic and cathodic curves of carbon steel electrodes in aggressive formation water solution (severe circumstances). The secured results are illustrated in Figure 13.7. In comparison to the blank curve, the Tafel lines for the cathodic and anodic processes are shifted to more negative and positive potentials, respectively. This suggests that the IMFC modifier influences both the cathodic development of hydrogen and the anodic dissolution of the metal (i.e. mixed-type inhibitor) [56]. The E_{corr} value increased from -907 mV for a blank carbon steel electrode to -803 mV for a conventional coating before reaching its maximum value at -799 mV for an IMFC-E coated film. As a result, the corrosion mitigation behaviour of epoxy coating was improved by the positive

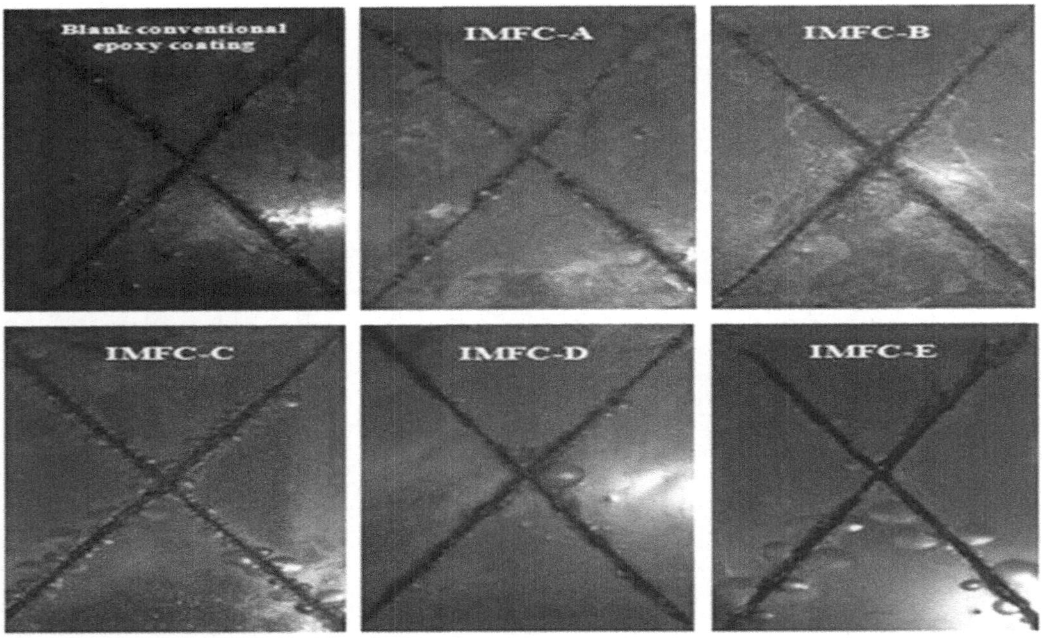

FIGURE 13.5 Test for accelerated corrosion under salt spray on cured epoxy coatings with and without IMFC modifications [56].

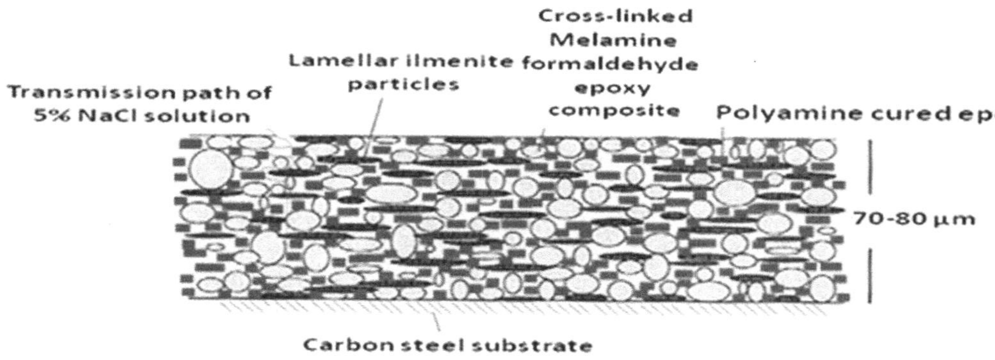

FIGURE 13.6 The proposed model for the IMFC polyamine-cured epoxy coating's defence mechanism [56].

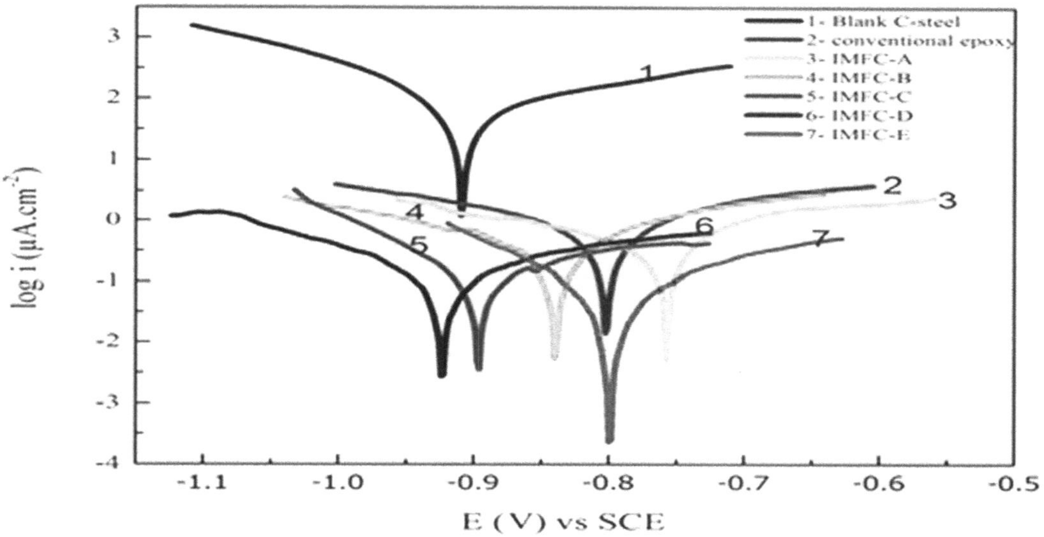

FIGURE 13.7 Various epoxy coated films were tested against blank carbon steel using tafels after being submerged in oil-well formation water [56].

change in the corrosion potential values brought about by the introduction of IMFC modifier particles. A robust barrier may be created during well-dispersion through the coated layer using IMFC modifier particles. In contrast to SCE, Tafel lines intercepts for coated panels were adjusted to values with lower corrosion current densities (anodic protection) as offered in Figure 13.7.

TiO_2 nanoparticles (TNPs) produced by sol-gel synthesis were combined with nonylphenol ethoxylate (NPhE) to create nonylphenol ethoxylate/TiO2 nanocomposite (NPhE/TNC), which served as a modifier for an epoxy coating that was applied to the surface of C-steel [57]. Different modified NPhE/TNC epoxy coating formulations were created, each with a different loading percentage (0.5, 1, 1.5, 2, and 2.5%) by weight [57]. By using a salt spray accelerated corrosion test, the anticorrosion performance of surface-modified NPhE/TNC coated films versus unmodified conventional epoxy was investigated. The acquired results, which were supported by an SEM morphological survey, showed a significant improvement in the rust grade, blistering size and frequency, and adhesion strength of modified NPhE/TNC epoxy nanocomposite coated films. The complete dispersion of NPhE/TNC particles through the epoxy vehicle matrix as a pigment that prevented the propagation of the aggressive salt spray fog through the coated layer, thereby enhancing the chemical bonding density and establishing an effective protective barrier, was credited with this improvement in the protective performance [57]. Additionally, the creation of Ti-O, Ti-O-C, and O-Ti-O linkages (inorganic phases) with epoxy (DGEBPA) and the formation of dense hybrid structures of polyamine-cured DGEBPA-TiO_2 and NPhE-TiO_2 nanocomposites were both influenced by the presence of well-dispersed nanocomposite molecules, as shown in Scheme 1. In order to minimise the transmission routes for the aggressive fog (NaCl solution vapours) to enter the coating layer and prevent steel corrosion, this performance caused TNPs to absorb more epoxy binder on their surfaces and increase the coating film density [58–60] as illustrated in the anticorrosion mechanism as shown in Figure 13.8. The amount of cross-linking between NPhE and epoxy terminals was increased, which improved the epoxy coating's

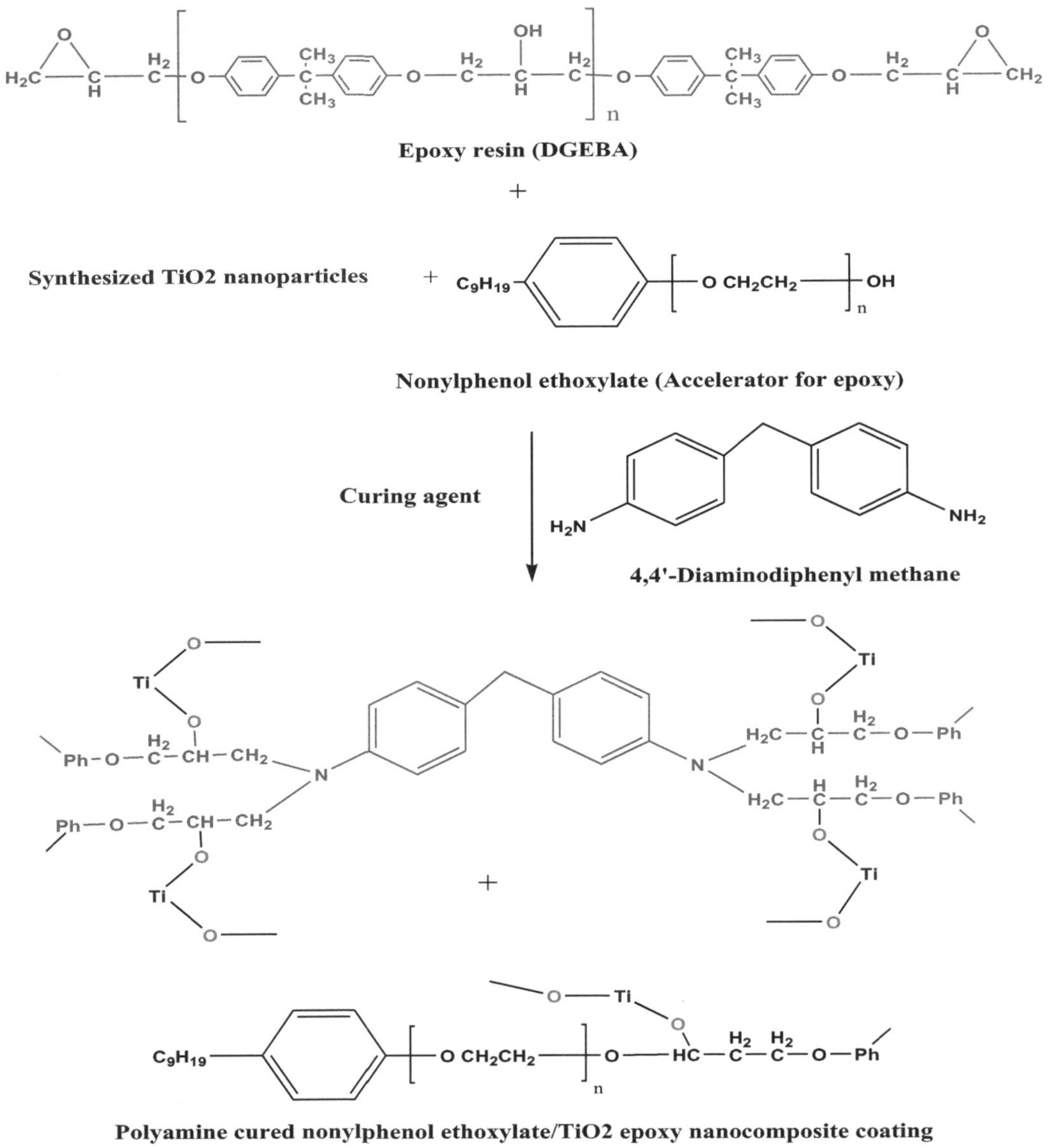

SCHEME 1 The creation of cross-linked three-dimensional networks in a nanocomposite made of nonylphenol ethoxylate and titanium dioxide [57].

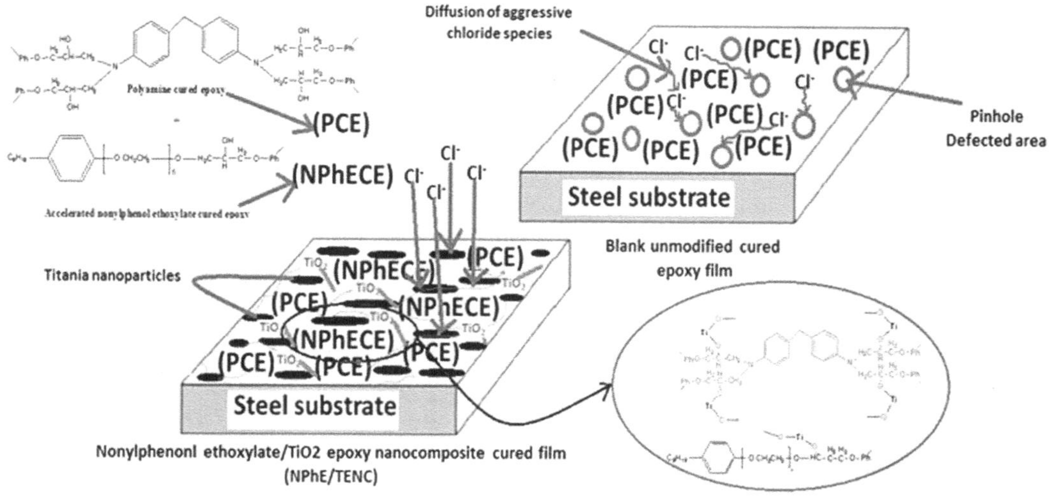

FIGURE 13.8 The suggested corrosion mitigation mechanism of the surface-modified NPhE/TNC epoxy coating applied on C-steel substrate against unmodified conventional epoxy [57].

hydrophobicity and reduced the wettability of the corrosive environment. The surface-treated NPhE/TNC epoxy coating's improved adhesion property can be attributed to the high compatibility of the fully-dispersed TNPs with the metal complex molecules that fill pinholes and voids in the coated film and serve as bridges (Ti-O, Ti-O-C, and O-Ti-O) in the epoxy matrix's interconnected structure. This resulted in a decrease in total free volumes and an increase in the cross-linking density of the cured film. After that, the layer has been cured, which has reduced segmental chain motions and increased rigidity [61].

Another study examined the use of epoxy coating modified with various metal (Co(II), Zr(IV), Ni(II), Cu(II), and Cr(III)) complexes of Schiff base versatile ligand known as

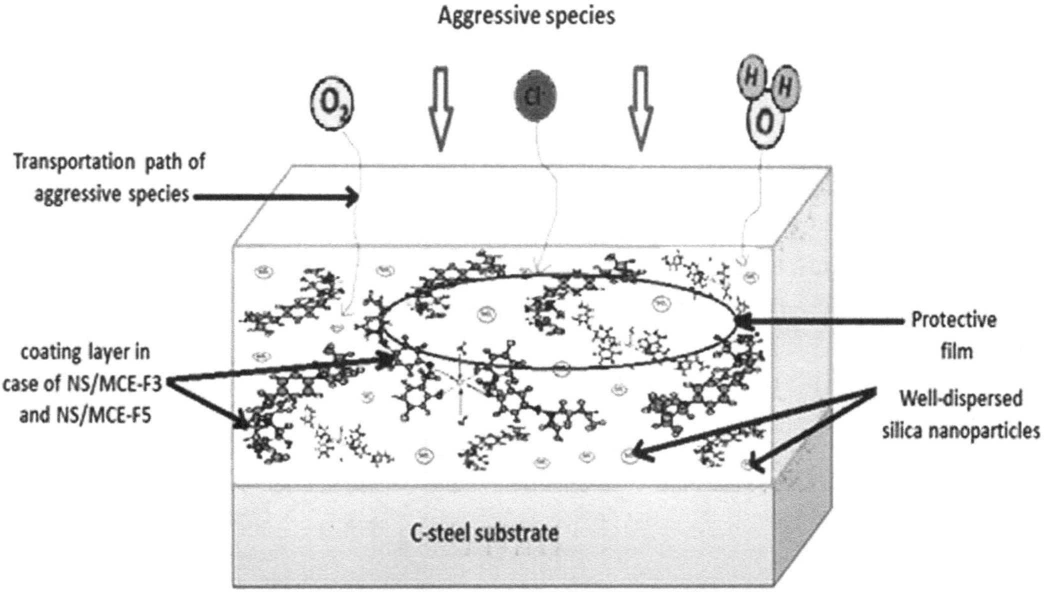

FIGURE 13.9 The anticipated defence system used by NS/MCE-F3 and NS/MCE-F5 coatings [62].

p-Phenylamine-N(4-chloro salicylaldenemine) (HL), composited with SiO_2 nanoparticles, to enhance its protective behaviour and mechanical resistance on the C-steel surface [62]. The modified cured Cr(III) and Ni(II) complex epoxy/SiO_2 nanocomposite coated films (NS/MCE-F3 and NS/MCE-F5) demonstrated the lowest corrosion rate (CR) at 0.00063 mm/y and a great protection efficiency (PE) recorded at 99.80%. The inhibitive mechanism provided the synergistic protective behaviour of NS/MCE-F3, NS/MCE-F5, and other modified metal complex epoxy nanocomposite coatings (NS/MCE-F1, NS/MCE-F2, and NS/MCE-F4) against NS/LE-F coated film in preventing the corrosion of C-steel as offered in Figure 13.9 depending on the molecular structure data as depicted in Table 13.2.

TABLE 13.2
Molecular Structures of the Investigated Molecules and Its 3D Models [62]

Molecular structure	3D model
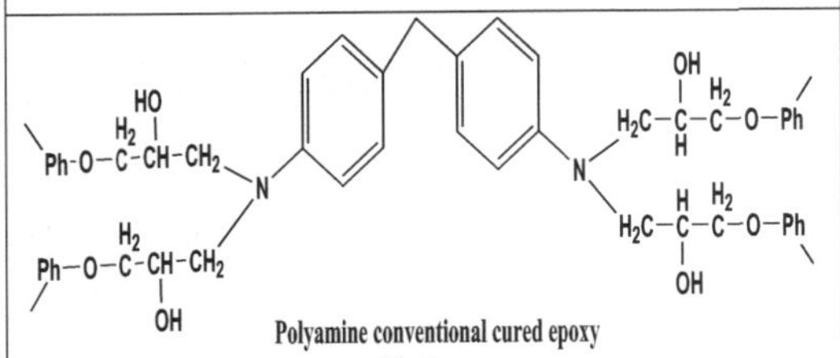 Polyamine conventional cured epoxy Blank epoxy	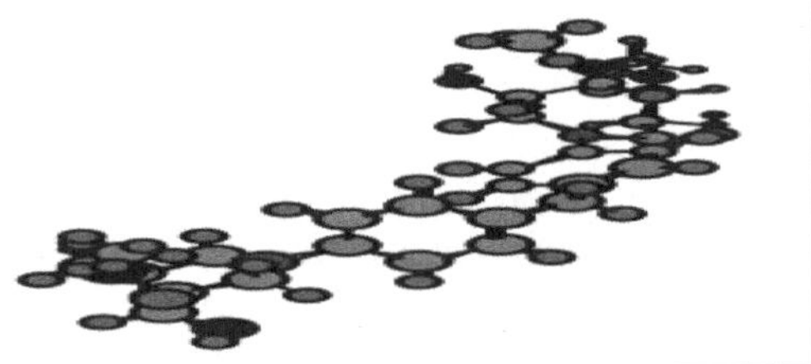
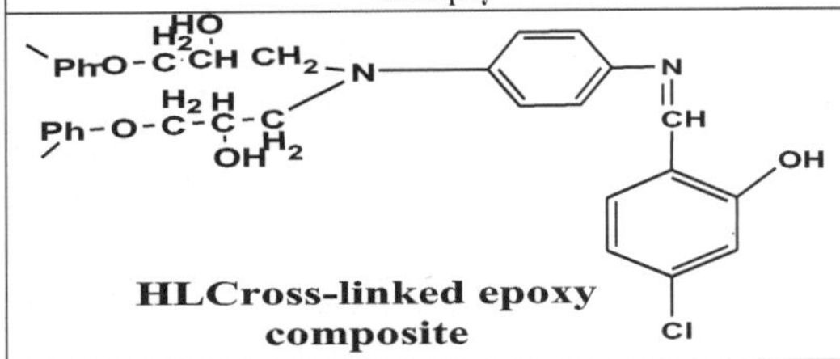 HLCross-linked epoxy composite	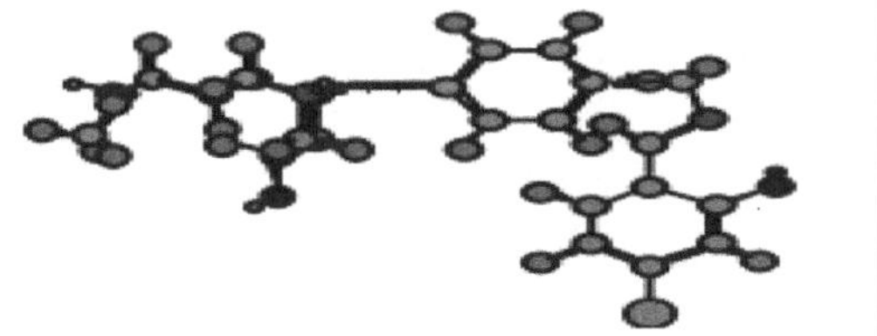
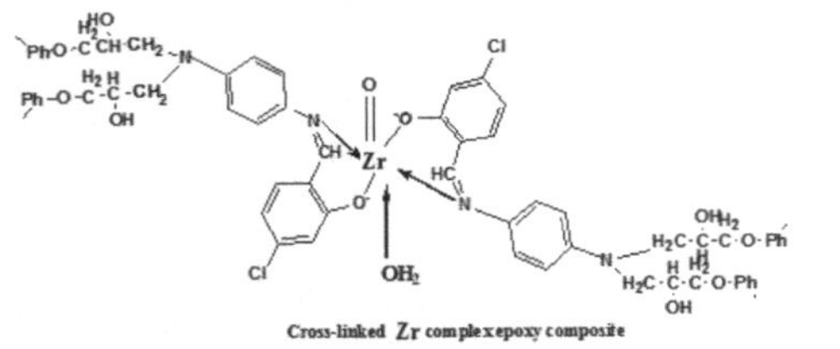 Cross-linked Zr complexepoxy composite	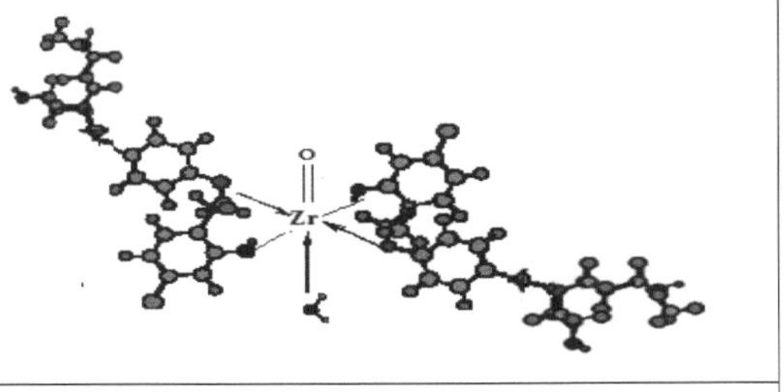
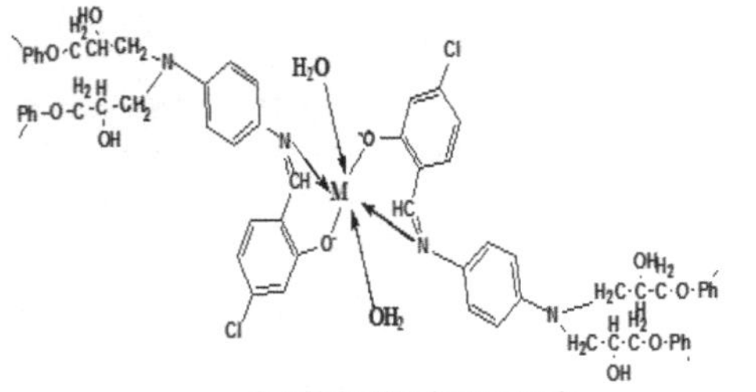 Cross-linked metal complexepoxy composite	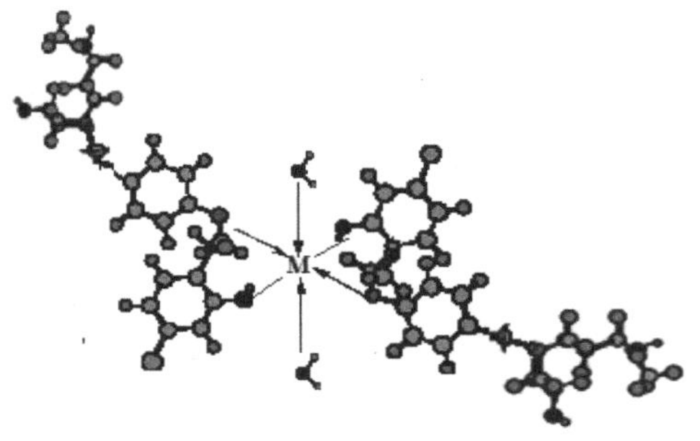

TABLE 13.3
Before and after 500 Hours of Exposure in a Salt Spray Cabinet Utilising 5% NaCl Solution, Weight Loss Data of Traditional Unmodified Epoxy and Modified HL and Its Metal Complexes Epoxy/SiO_2 Hybrid Nanocomposite Coated Films Were Recorded [62]

Coating Code	Initial Mass before Exposure (mg)	Final Mass after Exposure (mg)	Mass Loss (mg)	CR (mm/y)	PE (%)
Blank epoxy coating	85842.2	85178.8	663.4	0.29610	–
NS/LE-F	84176.4	84055.3	121.1	0.05400	81.70
NS/MCE-F1	81904.4	81847.6	56.8	0.02530	91.40
NS/MCE-F2	84395.8	84351.8	44	0.01960	93.30
NS/MCE-F3	86449.0	86447.6	1.40	0.00063	99.80
NS/MCE-F4	84441.9	84440.0	1.90	0.00080	99.71
NS/MCE-F5	83993.0	83991.6	1.40	0.00063	99.80

CR= Corrosion Rate.
PE= Protection Efficiency.

The scattered SiO_2 nanoparticles' small size allows them to absorb extra resin (epoxy binder) on their surfaces, strengthening the density of the epoxy film and reducing the transfer of corrosive salt spray fog species to the paint layer, which prevents corrosion of the undercoating [63,64]. By reducing the wettability of corrosive environments and boosting the cross-linking density between Ni(II) and Cr(III) complexes and epoxy terminals, the epoxy coating's hydrophobicity will be improved. In this study, the weight loss method was used to assess the rate of corrosion and level of protection of epoxy coated films modified with HL and its metal complexes composited with silica nanoparticles after being exposed to a severe salt spray fog for 500 hours (21 days) at a time. The data shown in Table 13.3 illustrated that the blank conventional epoxy-coated film had a corrosion rate (CR) of 0.29610 mm/y and an upper weight loss estimate of 663.4 mg. The observed results were attributed to the higher magnetic moments of Ni(II) and Cr(III) complexes compared to Co(II), Zr(IV), and Cu(II) complexes, which increased the electronegativity density before strengthening the cross-linking density of the coating layer and strengthened the adhesion between the C-steel and epoxy interface.

An increase in the cross-linking density of the cured film is supported by the fully dispersed SiO_2 nanoparticles with metal complex molecules through the coating vehicle occupying holidays like pinholes and voids in the thin-film coating. This reduces the total free volumes and enhances the film stiffness on the steel surface [65].

REFERENCES

1. Fadl, A.M. 2017. Preparation of Bi-system Synergistic Paints for Cold Galvanizing Technology to Prevent corrosion of Petroleum Steel Structures Using Egyptian Nano Ilmenite and Amorphous Silica. PhD. Thesis, Ain Shams University, Faculty of Science, Chemistry department, pp. 6–10.
2. Olajire, A.A. 2017. Corrosion inhibition of offshore oil and gas production facilities using organic compound inhibitors – A review. *J Mol Liq* 248:775–808.
3. https://www.pcimag.com/articles/105888-advanced-anti-corrosion-coating-utilized-on-two-north-sea-offshore-platforms, (2019).
4. Schremp, F.W. 1984. Corrosion prevention for Offshore Platforms. *J Pet Technol* 36: 605–612.
5. Ramezanzadeh, B., Karimi, B., Ramezanzadeh, M., Rostami, M. 2019. Synthesis and characterization of polyaniline tailored graphene oxide quantum dot as an advance and highly crystalline carbon-based luminescent nanomaterial for fabrication of an effective anti-corrosion epoxy system on mild steel. *J Taiwan Inst Chem Eng* 95:369–382.

6. Gibson, G. 2017. Epoxy Resins, in: Brydson's Plastics Materials. 8th Ed., pp. 773–797.
7. Su, Y., Qiu, S.H., Yang, D., Liu, S., Zhao, H., Wang, L., Xue, Q. 2020. Active anti-corrosion of epoxy coating by nitrite ions intercalated MgAl LDH. *J Hazard Mater* 391:122215.
8. Mohammadkhani, R., Ramezanzadeh, M., Saadatmandi, S., Ramezanzadeh, B. 2020. Designing a dual-functional epoxy composite system with self-healing/barrier anti-corrosion performance using graphene oxide nano-scale platforms decorated with zinc doped-conductive polypyrrole nanoparticles with great environmental stability and non-toxicity. *Chem Eng J* 382:122819.
9. Yang, F., Liu, T., Li, J., Qiu, S., Zhao, H. 2018. Anticorrosive behavior of a zinc-rich epoxy coating containing sulfonated polyaniline in 3.5% NaCl solution. *RSC Adv* 8:13237–13247.
10. Hare, C.H. 1998. Mechanisms of corrosion protection with surface treated wollastonite pigments. *J Protect Coat* 14:47–82.
11. Hammouda, N., Chadli, H., Guillemot, G., Belmokre, K. 2011. The corrosion protection behaviour of zinc rich epoxy paint in 3% NaCl solution. *Adv Chem Eng Sci* 1:51–60.
12. Abreu, C.M., Izquierdo, M., Keddam, M., Novoa, X.R., Takenouti, H. 1996. Electrochemical behaviour of zinc-rich epoxy paints in 3% NaCl solution. *Electrochimica Acta* 41:2405–2415.
13. Poole, C.P., Jr., Owens, F.J. 2003. Introduction to Nanotechnology, Jhon Wiley & Sons, INC Publications, Hoboken, New Jersey.
14. Wilson, M., Kannangara, K., Smith, G., Simmons, M., Raguse, B. 2002. Nanotechnology Basic Science and Emerging Technologies, A book, 1st Edition, Chapman and Hall/CRC.
15. Gleiter, H. 2000. Nanostructured materials: Basic concepts and microstructure. *Acta Mater* 48:1–29.
16. Rong, M.Z., Zhang, M.Q., Liu, H., Zeng, H., Wetzel, B., Friedrich, K. 2001. Microstructure and tribological behavior of polymeric nanocomposites. *Ind Lub Tribol* 53:72–77.
17. Shi, G., Zhang, M.Q., Rong, M.Z., Wetzel, B., Friedrich, K. 2004. Sliding wear behavior of epoxy containing nano-Al_2O_3 particles with different pretreatments. *Wear* 256:1072–1081.
18. Gao, Z., Zhao, L. 2015. Effect of nano-fillers on the thermal conductivity of epoxy composites with micro-Al_2O_3 particles. *Mater Des* 66:176–182.
19. Dolatzadeh, F., Moradian, S., Jalili, M.M. 2011. Influence of various surface treated silica nanoparticles on the electrochemical properties of SiO_2/polyurethane nanocoatings. *Corros Sci* 53:4248–4257.
20. Fu, G., Vary, P.S., Lin, C.T. 2005. Anatase TiO_2 nanocomposites for antimicrobial coatings. *J Phys Chem B* 109:8889–8898.
21. Kumar, S.A., Balakrishnan, T., Alagar, M., Denchev, Z. 2006. Development and characterization of silicone/phosphorus modified epoxy materials and their application as anticorrosion and antifouling coatings. *Prog Org Coat* 55:207–217.
22. Matin, E., Attar, M., Ramezanzadeh, B. 2015. Investigation of corrosion protection properties of an epoxy nanocomposite loaded with polysiloxane surface modified nanosilica particles on the steel substrate. *Prog Org Coat* 78:395–403.
23. Pourhashem, S., Vaezi, M.R., Rashidi, A., Bagherzadeh, M.R. 2017. Exploring corrosion protection properties of solvent based epoxy-graphene oxide nanocomposite coatings on mild steel. *Corros Sci* 115:78–92.
24. Radoman, T.S., DŽunuzović, J.V., Jeremić, K.B., Grgur, B.N., Miličević, D.S., Popović I.G., et al. 2014. Improvement of epoxy resin properties by incorporation of TiO_2 nanoparticles surface modified with gallic acid esters. *Mater Des* 62:158–167.
25. Ramezanzadeh, B., Attar, M., Farzam, M. 2011. A study on the anticorrosion performance of the epoxy–polyamide nanocomposites containing ZnO nanoparticles. *Prog Org Coat* 72:410–422.
26. Ramezanzadeh, B., Haeri, Z., Ramezanzadeh, M. 2016. A facile route of making silica nanoparticles-covered graphene oxide nanohybrids (SiO_2-GO); fabrication of SiO_2-GO/epoxy composite coating with superior barrier and corrosion protection performance. *Chem Eng J* 303:511–528.
27. Shi, X., Nguyen, T.A., Suo, Z., Liu, Y., Avci, R. 2009. Effect of nanoparticles on the anticorrosion and mechanical properties of epoxy coating. *Surf Coat Technol* 204:237–245.
28. Vázquez-Velázquez, A., Velasco-Soto, M., Pérez-García, S., Licea-Jiménez, L. 2018. Functionalization effect on polymer nanocomposite coatings based on TiO_2-SiO_2 nanoparticles with super-hydrophilic properties. *Nanomaterials* 8:369.
29. Giannelis, E.P., Krishnamoorti, R., Manias, E. 1999. Polymer-Silicate Nanocomposites: Model Systems for Confined Polymers and Polymer Brushes, Part of the Advances in Polymer Science book series (POLYMER, volume 138).

30. Bagherzadeh, M.R., Mahdavi F. 2007. Preparation of epoxy–clay nanocomposite and investigation on its anti-corrosive behavior in epoxy coating. *Prog Org Coat* 60:117–120.
31. Yu, H.J., Wang, L., Shi, Q., Jiang, G.H., Zhao, Z.R., Dong, X.C. 2006. Study on nano-$CaCO_3$ modified epoxy powder coatings. *Prog Org Coat* 55:296–300.
32. Nazari, M.H., Shi, X. 2016. Polymer-Based Nanocomposite Coatings for Anticorrosion Applications. In: Hosseini M., Makhlouf A. (eds.) Industrial Applications for Intelligent Polymers and Coatings. Springer, Cham.
33. Alam, J., Riaz, U., Ashraf, S.M., Ahmad, S. 2007. Corrosion-protective performance of nano polyaniline/ferrite dispersed alkyd coatings. *J Coat Technol Res* 5:123–128.
34. Dhoke, S.K., Khanna, A.S., Sinha, T.J.M. 2009. Effect of nano-ZnO particles on the corrosion behavior of alkyd-based waterborne coatings. *Prog Org Coat* 64:371–382.
35. Dhoke, S.K., Khanna, A.S. 2009. Effect of nano-Fe2O3 particles on the corrosion behavior of alkyd based waterborne coatings. *Corros Sci* 51:6–20.
36. Olad, A., Naseri, B. 2010. Preparation, characterization and anticorrosive properties of a novel polyaniline/clinoptilolite nanocomposite. *Prog Org Coat* 67:233–238.
37. Heidarian, M., Shishesaz, M.R., Kassiriha, S.M., Nematollahi, M. 2010. Characterization of structure and corrosion resistivity of polyurethane/organoclay nanocomposite coatings prepared through an ultrasonication assisted process. *Prog Org Coat* 68:180–188.
38. Ramezanzadeh, B., Attar, M.M. 2011. Studying the effects of micro and nano sized ZnO particles on the corrosion resistance and deterioration behavior of an epoxy-polyamide coating on hot-dip galvanized steel. *Prog Org Coat* 71:314–328.
39. Jadhav, R.S., Patil, K.J., Hundiwale, D.G., Mahulikar, P.P. 2011. Synthesis of waterborne nanopolyanilne latexes and application of nanopolyaniline particles in epoxy paint formulation for smart corrosion resistivity of carbon steel. *Polym Adv Technol* 22:1620–1627.
40. Madhankumar, A., Nagarajan, S., Rajendran, N., Nishimura, T. 2011. EIS evaluation of protective performance and surface characterization of epoxy coating with aluminum nanoparticles after wet and dry corrosion test. *J Solid State. Electrochem* 16:2085–2093.
41. Madhankumar, A., Rajendran, N., Nishimura, T. 2012. Influence of Si nanoparticles on the electrochemical behavior of organic coatings on carbon steel in chloride environment. *J Coat Technol Res* 9:609–620.
42. Mills, D.J., Jamali, S.S., Paprocka, K. 2012. Investigation into the effect of nano-silica on the protective properties of polyurethane coatings. *Surf Coat Technol* 209:137–142.
43. Toloei, A., Atashin, S., Bahrololoom, M.E. 2013. A New Approach in Modifying Polymeric Coatings to Increase Corrosion Resistance Properties. *Int Sch Res Not* 213:e648246.
44. Moazeni, N., Mohamad, Z., Faisal, N.L.I., Tehrani, M.A., Dehbari, N. 2013. Anticorrosion epoxy coating enriched with hybrid nanozinc dust and halloysite nanotubes. *J Appl Polym Sci* 130:955–960.
45. Singh-Beemat, J., Iroh, J.O., Feng, L. 2013. Mechanism of corrosion protection of aluminum alloy substrate by hybrid polymer nanocomposite coatings. *Prog Org Coat* 76:1576–1580.
46. Zomorodian, A., Garcia, M.P., Silva, T., Moura, E., Fernandes, J.C.S., Fernandes, M.H., Montemor, M.F. 2013. Corrosion resistance of a composite polymeric coating applied on biodegradable AZ31 magnesium alloy Acta. *Biomater* 9:8660–8670.
47. Yu, Y-H, Lin, Y-Y, Lin, C-H, Chan, C-C, Huang, Y-C. 2013. High-performance polystyrene/graphene-based nanocomposites with excellent anti-corrosion properties, *Polym Chem* 5:535–550.
48. Tomić, M.D., Dunjić, B., Likić, V., Bajat, J., Rogan, J., Djonlagić, J. 2014. The use of nanoclay in preparation of epoxy anticorrosive coatings. *Prog Org Coat* 77:518–527.
49. Wang, N., Wu, Y.H., Cheng, K.Q., Zhang, J. 2014. Investigation on anticorrosion performance of polyaniline-mesoporous MCM-41 composites in new water-based epoxy coating. *Mater Corros* 65:968–976.
50. Deyab, M.A. 2014. Corrosion protection of aluminum bipolar plates with polyaniline coating containing carbon nanotubes in acidic medium inside the polymer electrolyte membrane fuel cell. *J Power Sources* 268:50–55.
51. Madhan, K.A., Gasem, Z.M. 2015. In situ electrochemical synthesis of polyaniline/f-MWCNT nanocomposite coatings on mild steel for corrosion protection in 3.5 % NaCl solution. *Prog Org Coat* 78:387–394.
52. Pour-Ali, S., Dehghanian, C., Kosari, A. 2015. Corrosion protection of the reinforcing steels in chloride-laden concrete environment through epoxy/polyaniline–camphorsulfonate nanocomposite coating. *Corros Sci* 90:239–247.

53. Ganjaee, S.M., Ramezanzadeh, B., Shahbazi, M., Pakdel, A.S. 2015. Influence of nanoclay particles modification by polyester-amide hyperbranched polymer on the corrosion protective performance of the epoxy nanocomposite. *Corros Sci* 92:162–172.
54. Haddadi, S.A., Mahdavian, M., Karimi, E. 2015. Evaluation of the corrosion protection properties of an epoxy coating containing sol–gel surface modified nano-zirconia on mild steel. *RSC Adv* 5:28769–28777.
55. Ates, M., Topkaya, E. 2015. Review study of electrochemical impedance spectroscopy and equivalent electrical circuits of conducting polymers on carbon surfaces. *Prog Org Coat* 82:33–40.
56. Abdou, M.I., Ayad, M.I., Diab, A.S.M., Hassan, I.A., Fadl, A.M. 2017. Influence of surface-modified ilmenite/melamine-formaldehyde composite on the anti-corrosion and mechanical properties of conventional polyamine cured epoxy for internal coating of gas and oil transmission pipelines. *Prog Org Coat* 113:1–14.
57. Fadl, A.M., Abdou, M.I., Abo Al-Elaa, S., Hamza, M.A., Sadeek, S.A. 2019. Evaluation the anti-corrosion behavior, impact resistance, acids and alkali immovability of nonylphenol ethoxylate/TiO2 hybrid epoxy nanocomposite coating applied on the carbon steel surface. *Prog Org Coat* 136:105263.
58. Yang, L.H., Liu, F.C., Han, E.H. 2005. Effects of P/B on the properties of anticorrosive coatings with different particle size. *Prog Org Coat* 53:91–98.
59. Kalendova, A. 2003. Effects of particle sizes and shapes of zinc metal on the properties of anticorrosive coatings. *Prog Org Coat* 46:324–332.
60. Shi, H., Liu, F., Han, E., Wei, Y. 2007. Effect of nano pigments on the corrosion resistance of alkyd coatings. *J Mater Sci Technol* 23:551–558.
61. Ghosal, A., Ahmad, S. 2017. High-performance anti-corrosive epoxy–titania hybrid nanocomposite coatings. *New J Chem* 41:4599–4610.
62. Fadl, A.M., Abdou, M.I., Laila, D., Sadeek, S.A. 2019. Application insights of Schiffbase metal complex/SiO_2 hybrid epoxy nanocomposite for steel surface coating: Correlation the protective behavior and mechanical properties with material loading. *Prog Org Coat* 136:105226.
63. Rahman, O.U., Ahmad, S. 2014. Physico-mechanical and electrochemical corrosion behavior of soy alkyd/Fe_3O_4 nanocomposite coatings. *RSC Adv* 4:14936–14947.
64. Shi, H., Liu, F., Han, E., Wei, Y. 2007. Effect of nanopigments on the corrosion resistance of alkyd coatings. *J Mater Sci Technol* 23:551–558.
65. Ramezanzadeh, B., Ahmadi, A., Mahdavian, M. 2016. Enhancement of the corrosion protection performance and cathodic delamination resistance of epoxy coating through the treatment of steel substrate by a novel nanometric sol-gel based silane composite film filled with functionalized graphene oxides nanosheets. *Corros Sci* 109:182–205.

Index

Note: Locators in *italics* represent figures and **bold** indicate tables in the text.